T0189613

Advances in Intelligent Systems and Computing

Volume 547

Series editor

Janusz Kacprzyk, Polish Academy of Sciences, Warsaw, Poland
e-mail: kacprzyk@ibspan.waw.pl

About this Series

The series "Advances in Intelligent Systems and Computing" contains publications on theory, applications, and design methods of Intelligent Systems and Intelligent Computing. Virtually all disciplines such as engineering, natural sciences, computer and information science, ICT, economics, business, e-commerce, environment, healthcare, life science are covered. The list of topics spans all the areas of modern intelligent systems and computing.

The publications within "Advances in Intelligent Systems and Computing" are primarily textbooks and proceedings of important conferences, symposia and congresses. They cover significant recent developments in the field, both of a foundational and applicable character. An important characteristic feature of the series is the short publication time and world-wide distribution. This permits a rapid and broad dissemination of research results.

More information about this series at http://www.springer.com/series/11156

Kusum Deep · Jagdish Chand Bansal
Kedar Nath Das · Arvind Kumar Lal
Harish Garg · Atulya K. Nagar
Millie Pant
Editors

Proceedings of Sixth International Conference on Soft Computing for Problem Solving

SocProS 2016, Volume 2

 Springer

Editors
Kusum Deep
Department of Mathematics
Indian Institute of Technology Roorkee
Roorkee
India

Jagdish Chand Bansal
Department of Mathematics
South Asian University
New Delhi
India

Kedar Nath Das
Department of Mathematics
National Institute of Technology, Silchar
Silchar, Assam
India

Arvind Kumar Lal
School of Mathematics
Thapar Institute of Engineering
  and Technology University
Patiala, Punjab
India

Harish Garg
School of Mathematics
Thapar Institute of Engineering
  and Technology University
Patiala, Punjab
India

Atulya K. Nagar
Department of Mathematics
  and Computer Science
Liverpool Hope University
Liverpool
UK

Millie Pant
Department of Paper Technology
Indian Institute of Technology Roorkee
Roorkee
India

ISSN 2194-5357   ISSN 2194-5365   (electronic)
Advances in Intelligent Systems and Computing
ISBN 978-981-10-3324-7   ISBN 978-981-10-3325-4   (eBook)
DOI 10.1007/978-981-10-3325-4

Library of Congress Control Number: 2017931564

Printed on acid-free paper

This Springer imprint is published by Springer Nature
The registered company is Springer Nature Singapore Pte Ltd.
The registered company address is: 152 Beach Road, #21-01/04 Gateway East, Singapore 189721, Singapore

Contents

About the Editors

Prof. Kusum Deep is Professor at the Department of Mathematics, Indian Institute of Technology Roorkee, India. Over the past 25 years, her research has made her a central international figure in the area of nature inspired optimization techniques, genetic algorithms, and particle swarm optimization.

Dr. Jagdish Chand Bansal is Assistant Professor with South Asian University, New Delhi, India. Holding an excellent academic record, he is an outstanding researcher in the field of swarm intelligence at the national and international level, having written several research papers in journals of national and international repute.

Dr. Kedar Nath Das is Assistant Professor at the Department of Mathematics, National Institute of Technology, Silchar, Assam, India. Over the past 10 years, he has made substantial contributions to research on 'soft computing'. He has published several research papers in prominent national and international journals. His chief area of interest is in evolutionary and bio-inspired algorithms for optimization.

Dr. Arvind Kumar Lal is currently associated with the School of Mathematics and Computer Applications at Thapar University, Patiala. He received his BSc. Honors (mathematics) and MSc. (mathematics) from Bihar University, Muzaffarpur in 1984 and 1987, respectively. He completed his PhD (mathematics) at the University of Roorkee (now the IIT, Roorkee) in 1995. Dr. Lal has over 130 publications in journals and conference proceedings to his credit. His research areas include applied mathematics (modeling of stellar structure and pulsations), reliability analysis, and numerical analysis.

Dr. Harish Garg is Assistant Professor at the School of Mathematics at Thapar University, Patiala, Punjab, India. He received his BSc. (computer applications) and MSc. (Mathematics) from Punjabi University, Patiala, before completing his PhD (applied mathematics) at the Indian Institute of Technology Roorkee. He is currently teaching undergraduate and postgraduate students and is pursuing innovative and insightful research in the area of reliability theory using evolutionary algorithms and fuzzy set theory with their application in numerous industrial engineering areas. Dr. Garg has produced 62 publications, which include six book chapters, 50 journal papers, and six conference papers.

Prof. Atulya K. Nagar holds the Foundation Chair as Professor of Mathematical Sciences and is Dean of the Faculty of Science at Liverpool Hope University, UK. Professor Nagar is an internationally respected scholar working at the cutting edge of theoretical computer science, applied mathematical analysis, operations research, and systems engineering.

Dr. Millie Pant is Associate Professor at the Department of Paper Technology, Indian Institute of Technology Roorkee, India. She has published several research papers in national and international journals and is a prominent figure in the field of swarm intelligence and evolutionary algorithms.

Spammer Classification Using Ensemble Methods over Content-Based Features

Aaisha Makkar[✉] and Shivani Goel

Computer Science and Engineering Department,
Thapar University, Patiala, India
{aaisha.makkar,shivani}@thapar.edu

Abstract. As the web documents are raising at high scale, it is very difficult to access useful information. Search engines play a major role in retrieval of relevant information and knowledge. They deal with managing large amount of information with efficient page ranking algorithms. Still web spammers try to intrude the search engine results by various web spamming techniques for their personal benefit. According to the recent report from Internetlivestats in March (2016), an Internet survey company, states that there are currently 3.4 billion Internet users in the world. From this survey it can be judged that the search engines play a vital role in retrieval of information. In this research, we have investigated fifteen different machine learning classification algorithms over content based features to classify the spam and non spam web pages. Ensemble approach is done by using three algorithms which are computed as best on the basis of various parameters. Ten Fold Cross-validation approach is also used.

Keywords: Web spamming · Machine learning · Boosting · Ensemble

1 Introduction

As the Internet grew, retrieving the relevant and appropriate information became more difficult. Due to non-standard structure of web, complex styles of different Web data, the exponential growth, dynamic nature of the Web, searching for relevant information is a challenging task. The search engines are used to download, index and rank Web pages according to keywords, and presented to the user as Search Engine Result Pages (SERPs). Web spam can intrude in any kind of information system whether it is an e-mail, advertisement or online business websites [1]. Web spamming is an activity on the Web by spammers to gain a better ranking in the Search Engine result pages, try to harm the search engine ranking algorithms [2]. There are two categories of Web spam techniques: Boosting techniques and Hiding techniques and these are further explained in Fig. 1.

© Springer Nature Singapore Pte Ltd. 2017
K. Deep et al. (eds.), *Proceedings of Sixth International Conference on Soft Computing for Problem Solving*, Advances in Intelligent Systems and Computing 547,
DOI 10.1007/978-981-10-3325-4_1

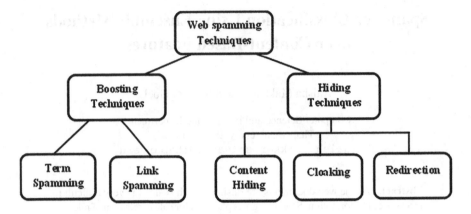

Fig. 1. Web spamming techniques

Boosting Techniques: It is the process of increasing the score for the web page. Boosting techniques can be further classified into Link spamming and Term spamming. In link spamming, link farms try to alter the web sites link structure [3]. To harm the ranking algorithm used by the search engine, link farm is created with the interconnected web pages. Spam mass is the concept to measure the effect of link spamming on the rank of the web page [4, 5]. In Term Spamming, the content of the Web pages is changed by keyword insertion at numerous places on the web known as Keyword Stuffing [6]. There are two types of Term Spamming namely Body spam which refers to excessive keywords used in the body of the page and Title spam which refers to excessive keywords used in the title of the page.

Hiding Techniques: Hiding techniques do not bother about the rank of web pages but make use of boosting techniques. Manipulating the anchor text by changing its color is one of the example. Cloaking is one such technique of presenting different contents to the user irrespective of search engine results [7]. While loading URL in the SERP's, another URL is opened without the knowledge of user and this is known as redirection. In content hiding, some content is hidden from the user in the webpage.

2   Related Work

This section gives a brief overview of different web spam detection algorithms and techniques which exists in the literature.

Nikita Spirin *et al.* stated that the Web Spamming Detection approaches started with Content-Based such as cloaking and later Link Based methods were introduced like trust or badness methods along with the behavior of the user. Link based web spamming detection techniques has given tremendous results; still web is growing at the rapid rate, so it is the open research issue. 90% Spam is being detected by using the present algorithms [1].

Hector Garcia *et al.* addressed boosting and hiding techniques of web spamming. Firstly, detect the web pages containing spam, stop crawling and even don't allow it to enter into the indexer module. Secondly prevent spamming by specific spamming techniques and finally taking counter measures. This procedure is still followed by the anti-web spamming teams [2].

Becchetti *et al.* proposed techniques based on link for web spam detection by computing the probability of the web page against the graph of web and estimating the rank propagation of the web page directing to it. To improve the accuracy of classification, algorithms of graph clustering and predicted labels propagation are used by the authors. The methodology used has resulted in 80% spam pages [3].

Gyongyi *et al.* introduced techniques to evaluate the effect of link spamming on the Ranking algorithm mainly PageRank known as mass spam. The pages that gets the home from link spamming can be measured with the help of spam mass can identify pages that benefit from link spamming. Two scores are taken into consideration, first is the normal pagerank score and second is the score of good nodes which are visited randomly. With the help of these two scores, mass spam is calculated. TrustRank resulted is not as good as this technique which is able to handle irregular estimation of link structure for web spam demotion [4].

Bin Zhou and Jian Pei proposed two page farm spam measures that are SpamRank and SpamMass which can easily detect that whether the page is targeted by the spam or not, by experimenting over WEBSPAM-2006 dataset. With the help of such measures the Precision and Recall gives the better classification results. The dataset being used here, has given tremendous results and also proved that this farm can be easily included in our calculations [5].

Wu *et al.* introduced a technique to identify link farm spam pages, consisting of mainly three steps: generating step, expansion step and ranking step. At first, this algorithm generates a spam seed set by its common incoming and outgoing links. Then Parent Penalty is used to expand the seed set. The assumption made is that the page is bad if it points to the large number of bad pages. Lastly, authors rank the web graph by down weighting elements in the adjacency matrix [6].

Basavaraju *et al.* performed the experiment using the approach of Text Based Clustering which is the one of the Web Spam Detection developed in 1999. Vector space model is used in this technique and able to detect spam effectively. He named that algorithm as BIRCH, by comparing it with the K-NNC method, it gave more accuracy. The accuracy was high because of the working while processing the dataset, which is scanned once [7].

3   Problem Formulation

Web spamming is an emerging influence in online websites that degrades the performance of search engines. Due to the development of various web spamming techniques, quality of search results by the search engine are getting inaccurate and leading to the high cost incurred by searching time and bandwidth wastage. To detect web spam occurring frequently in the web data, it is difficult to confine to the already developed web detection techniques [8] as the spammers always try to intrude new spamming techniques.

Table 1. Dataset features Description

Dataset Feature	Description	Dataset Feature	Description
Anchor text	Amount of anchor text in the web page	Average length	Average length of words
Meta character	Number of meta characters in meta element	Title word	Number of characters in the title element
Meta word	Number of meta words within the webpage	Comp_ratio	Counted complexity factor
Unique word	Number of unique words in webpage	Img	Number of images
Total word	Number of total words in webpage	Spam	0 for non-spam and -1 for spam
Max length	Maximum length of words		

4 Proposed Methodology

In the proposed work various approaches have been used for detection of web spam in online websites data [10]. In the proposed work UK-2011 web spam dataset has been used for detection of spam websites. This dataset contains different Arabic websites information about various web pages from April-2011 to August-2011. Various features have been used for detection of web spam in datasets. These features are available in the dataset and have different importance factor for detection of web spam available in dataset. Table 1 defines various features that are available in the dataset. On the basis of these features dataset has been preprocessed [13] and evaluated for detection of spam data.

- **Pre-processing:** In the pre-processing of dataset, all the attributes have been analyzed and used for processing by removing redundant information and null values with dummy substitution technique in the dataset. Dataset character values have been converted by applying filter of nominal to numeric conversion that convert all character value to numeric segment.
- **Feature Selection:** Feature selection has been done by using random forest machine learning approach. This approach utilizes decision tree based approach for selection of best features available in the dataset. Random forest approach takes data as an input and provides all tree based decision making steps on all the attributes for computation of accuracy and assign calculated Gini to each feature. On the basis of Gini, feature selection has been done so that optimal features can be utilized for detection of web spam information available in the dataset.
- **Machine Learning Models for classification:** In the process of classification various models have been developed in recent researches for detection of web spam data available in the network. Some of these models are CART, KNN, MARS, Linear model, Parallel Random Forest, Partial DSA and Tree Based Models. To detect spam

more accurately we proposed various new machine learning models described in Table 2. These different models have been utilized by using various prediction strategies for detection of spam information in the dataset.

Table 2. Machine learning models

Model	Method argument value	Packages	Tuning parameters
Bagged CART	Treebag	ipred, plyr, e1071	None
Bagged Model	Bag	Caret	Vars
Bayesian Generalized Linear Model	bayesglm	Arm	None
Boosted Linear Model	BstLm	bst, plyr	mstop, nu
Conditional Inference Tree	Ctree	Party	Mincriterion
Bagged MARS	bagEarth	Earth	nprune, degree
eXtreme Gradient Boosting	xgbLinear	Xgboost	nrounds, lambda, alpha,
Generalized Linear Model with Stepwise Feature Selection	glmStepAIC	MASS	None
k-Nearest Neighbors	Kknn	Kknn	kmax, distance, kernel
Least Angle Regression	Lars	Lars	Fraction
Parallel Random Forest	parRF	e1071, randomForest, foreach	Mtry
partDSA	partDSA	partDSA	cut.off.growth, MPD
Partial Least Squares	kernelpls	Pls	Ncomp
Partial Least Squares Generalized Linear Models	plsRglm	plsRglm	nt, alpha.pvals.expli
Tree-Based Ensembles	nodeHarvest	nodeHarvest	maxinter, mode

During the training process, three best models named Bagged CART, eXtreme Boosting Technique, and Parallel Random Forest out of 15 different classification models have been utilized for Ensemble approach which are computed as best among all after the evaluation of different parameters. Prediction strategies of these entire models have been combined for evaluation of average stronger prediction strategy so that malicious or spam information can be detected in efficient manner.

5 Results and Discussions

In the proposed work various models have been utilized for the detection of spam information available in the dataset so that search engines can be optimized to provide any user query results [12]. After training all the models, the evaluated results are given in Table 3.

Table 3. Evaluation results

Models	H	Gini	AUC	Precision	Recall	Accuracy	Total time
Bagged_CART.R	0.601	0.851	0.926	0.526	1	84.93	14.64
Bagged_Model	0.312	0.598	0.799	0.515	1	71.44	386.594
Bayesglm	0.096	0.207	0.603	0.53	1	56.85	4.886
Boosted_Linear_Model	0.075	0.179	0.589	0.529	1	54.83	42.786
Conditional_inference_tree	0.512	0.776	0.888	0.517	1	81.33	10.821
Earth	0.27	0.539	0.769	0.513	1	68.95	300.626
eXtreme_Gradient_Boosting	0.735	0.921	0.96	0.53	1	89.86	162.355
Generalised_linear_model_with_stepwise_feature_selection	0.091	0.192	0.596	0.53	0.999	56	12.364
kknnModel	0.57	0.828	0.914	0.532	1	83.6	24.879
Lars	0.102	0.203	0.601	0.541	1	58.33	7.098
Parallel_Random_Forest	0.73	0.93	0.965	0.534	1	88.91	233.327
partDSA	0.122	0.333	0.666	0.524	1	65.39	79.58
Partial_Least_Squares	0.057	0.15	0.575	0.523	1	55.1	5.348
Partial_Least_Squares_Generalized_Linear_Models	0.058	0.118	0.559	0.534	1	54.78	192.532
Tree-Based_Ensembles	0.365	0.654	0.827	0.535	1	73.78	1142.816

On the basis of accuracy, the Bagged CART, eXtreme Boosting Technique, and Parallel Random Forest are the best three models examined. These three models are thus ensemble and the simulation results are given below in Table 4:

Table 4. Ensemble results

H	Gini	AUC	Precision	Recall	Accuracy	Total time
0.682	0.799	0.899	0.528	1	89.17	356.53

- Cross validation: In this methodology validation approach of cross validation is used, in which data get shuffled on random basis. The goal of the cross validation is to define a test dataset which is used for testing the system and it also reduces the problem of over fitting. The dataset is shuffled ten times and the results are cross validated. Cross validation results with respect to various evaluation parameters are shown in the Table 5 and in Fig. 2.

Table 5. Cross validation

	H	Gini	AUC	Precision	Recall	Accuracy	Total time
Round 1	0.687	0.802	0.901	0.54	1	89.86	363.907
Round 2	0.672	0.791	0.896	0.518	1	89.54	363.298
Round 3	0.655	0.781	0.891	0.531	1	89.17	363.914
Round 4	0.632	0.765	0.882	0.53	1	87.58	350.834
Round 5	0.675	0.794	0.897	0.524	1	89.54	355.089
Round 6	0.676	0.794	0.897	0.539	1	90.13	364.363
Round 7	0.636	0.767	0.884	0.529	1	88.11	357.27
Round 8	0.657	0.783	0.891	0.537	1	88.91	375.386
Round 9	0.634	0.766	0.883	0.52	1	88	350.354
Round 10	0.72	0.823	0.912	0.532	1	90.39	353.143

The above Table 5 contains the evaluated measures after the last round of Cross-Validation.

Figure 2 show the graphs plotted with respect to prediction.

(a) ROC curve: It predicts how well the model is performing. Area under this curve depicts the true positive values with respect to false positive values.

(b) H-measure: It is used to check the cost of the proposed model in the normalized form.

(c) AUC (w): The cost incurred by different features which is affected by the weights assigned to them with filter method.

(d) Smoothed score distribution: It depicts the score earned by both the classes with proportion to each other.

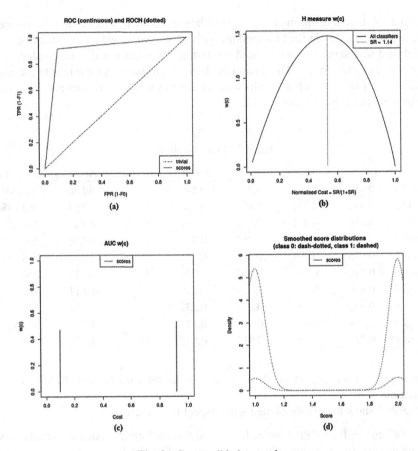

Fig. 2. Cross validation results

6 Conclusion

Web Spamming activities are increasing in this internet era [11]. Though it is not possible to eliminate the spam pages in the web completely but the detection of spam pages is also difficult task. In this paper, detection of spam pages is performed with the help of machine learning classification models over UK-2011 web spam dataset. Numerous machine learning approaches using content-based features have been used in literature to detect spammed web pages. Ensemble learning approaches like bagging and eXtreme-boosting that aims to improve the performance of individual classifiers exist in literature, but they have not been extensively evaluated for the spammer detection task. We have build fifteen Machine learning Models and among them three models with the highest accuracies (Bagged CART-84.93, eXtreme_Gradient_Boosting-89.86 and Parallel Random Forest-88.91) are ensemble. After applying this approach the accuracy is raised to 89.17. Cross-validation Validation set approach has resulted in 90.39% accuracy in

10 rounds. In future, the spam pages can be detected using much larger dataset over link-based features [9].

References

1. Spirin, N., Jiawei, H.: Survey on web spam detection: principles and algorithms. ACM SIGKDD Explor. Newsl. **13**(2), 50–64 (2012)
2. Gyongyi, Z., Garcia-Molina, H.: Web spam taxonomy. In: First International Workshop on Adversarial Information Retrieval on the Web, pp. 1–9 (2005)
3. Becchetti, L., Castillo, C., Donato, D., Baeza-Yates, R., Leonardi, S.: Link analysis for web spam detection. ACM Trans. Web (TWEB) **2**(1) (2008)
4. Gyongyi, Z., Berkhin, P., Garcia-Molina, H., Pedersen, J.: Link spam detection based on mass estimation. In: Proceedings of the 32nd International Conference on Very Large Data Bases, pp. 439–450 (2006)
5. Zhou, B., Pei, J.: Link spam target detection using page farms. ACM Trans. Knowl. Discovery Data (TKDD) **3**(3), 1–38 (2009). Article No. 13
6. Bhattarai, A., Rus, V., Dasgupta, D.: Characterizing comment spam in the blogosphere through content analysis. In: IEEE Symposium on Computational Intelligence in Cyber Security, CICS 2009, pp. 37–44 (2009)
7. Basavaraju, M., Prabhakar, D.R.: A novel method of spam mail detection using text based clustering approach. Int. J. Comput. Appl. **5**(4), 15–25 (2010)
8. Wu, B., Davison, B.D.: Identifying link farm spam pages. In: Special Interest Tracks and Posters of the 14th International Conference on World Wide Web, pp. 820–829. ACM (2005)
9. Xing, W., Ghorbani, A.: Weighted pagerank algorithm. In: Proceedings of Second Annual Conference on Communication Networks and Services Research, pp. 305–314. IEEE (2009)
10. Kleinberg, J.M., Kumar, R., Raghavan, P., Rajagopalan, S., Tomkins, A.S.: The web as a graph: measurements, models, and methods. In: Asano, T., Imai, H., Lee, D.T., Nakano, S., Tokuyama, T. (eds.) COCOON 1999. LNCS, vol. 1627, pp. 1–17. Springer, Heidelberg (1999). doi:10.1007/3-540-48686-0_1
11. Bidoki, A.M.Z., Yazdani, N.: DistanceRank: an intelligent ranking algorithm for web pages. Inf. Process. Manag. **44**(2), 877–892 (2008)
12. Lempel, R., Moran, S.: The stochastic approach for link-structure analysis (SALSA) and the TKC effect. Comput. Netw. **33**(1), 387–401 (2000)
13. Sangeetha, M., Joseph, K.S.: Page ranking algorithms used in Web Mining. In: International Conference on Information Communication and Embedded Systems (ICICES), pp. 1–7. IEEE (2014)

A Modified BPDHE Enhancement Algorithm for Low Resolution Images

Pooja Kaushik and Unnati Gupta[✉]

Department of Electronics and Communication, MMEC, Mullana, Ambala, India
guptaunnati25@yahoo.co.in

Abstract. Image enhancement is an important image processing task that effectively improves the visual quality of an image. This image processing technique performs the operations on the input image in order to get an enhanced image. Histogram technique is a simple enhancement technique that leads to high degree of enhancement and reduces unnatural artifacts. In this research article, a color image enhancement algorithm for low resolution images is proposed. Also, a comparison of performance of various existing histogram enhancement techniques with the proposed algorithm is assessed using three parameters i.e. Mean Square Error (MSE), Peak Signal to Noise Ratio (PSNR), and Bit Error Rate (BER). From these parameters, it has been observed that the proposed algorithm well enhances the low resolution images by preserving original hue.

Keywords: Histogram equalization · Image enhancement · Image processing

1 Introduction

Image enhancement processes consist of number of algorithms/techniques that are meant to enhance or improve the visual appearance of digital images and also these methods tend to reconstruct the image in a suitable form for human and machine analysis. This is basically the process of changing the intensity of pixels present in that image [1]. In the process of enhancement of a digital image, the various features of an image are varied or filtered in order to reduce the noise contents present in that image and to make it more pleasant for eyes. There are number of such methods for image contrast enhancement. Global Histogram Equalization is one of the basic methods for the image enhancement which makes the use of image histogram [2]. Histogram Equalization is a method for image enhancement in accordance with the distribution of intensity over the entire range of histogram.

A number of researchers and academicians have contributed their efforts in improving the quality of the degraded images using various histogram modification techniques [4–8]. In 1987, Adaptive Histogram Equalization (AHE) technique was proposed in which the image was made to partition in several blocks and then these blocks were enhanced independently to highlight the local region [4]. The Brightness preserving Bi-Histogram Equalization (BBHE) was another algorithm in this chain proposed by Kim in 1997. In this technique the mean value is calculated in order to divide the histogram into smaller histograms [5]. Histogram technique was further

© Springer Nature Singapore Pte Ltd. 2017
K. Deep et al. (eds.), *Proceedings of Sixth International Conference on Soft Computing for Problem Solving*, Advances in Intelligent Systems and Computing 547,
DOI 10.1007/978-981-10-3325-4_2

extended by Ibrahim. The idea behind splitting the histogram into multiple sub-histograms is based on local maxima [6]. In this method, the loop holes of BBHE are removed and it improves the mean brightness of the output image. The recursive mean separated histogram considers the mean of the input image for sub-dividing the histogram into parts [7]. The process of sub-division continues until each part gets equalized independently. The number of sub-divisions occurred are based on the parameter called minimum mean square error (calculated between the input and output image). The method has a disadvantage that it has very high computational cost that was resolved by Dynamic Histogram Equalization (DHE) [8].

The complete paper is arranged as follows: Sect. 2 deals with the detailed study of BPDHE algorithm. Section 3 briefs out the proposed algorithm. Section 4 shows the results of the proposed algorithm and also comparison of the proposed method with the various existing techniques. Finally, Sect. 5 deals with the conclusion based on PSNR, MSE, and BER.

2 Related Work

The base of the proposed algorithm works on the theory of Brightness Preserving Dynamic Histogram Equalization (BPDHE) and Discrete Wavelet Transformation (DWT).

2.1 Brightness Preserving Dynamic Histogram Equalization

The major steps of the BPDHE are as follows:
Step 1: The histogram is leveled up by using 1-D Gaussian filter. The Gaussian filter is specified by the following equation:

$$G(x) = exp\left(-x^2/2\sigma^2\right) \tag{1}$$

where x represents the coordinate relative to the centre of the kernel, and σ represents the standard deviation.
Step 2: The leveled histogram is used for the process of splitting original histogram into sub-histogram and this process is based on local maxima. Here local maxima are used for separating the pixel intensities instead of local minima.
Step 3: Now, assign each sub-divided histograms a new dynamic range using Eq. (2).

$$width_i = L_h - L_l \tag{2}$$

$$f_i = width_i \times log10\,M \tag{3}$$

where M represents the total number of pixels in a width.

$$Span_i = Span_i / \sum_{k=1}^{n+1} f_i \tag{4}$$

The range $Span_i$ starts from P and ends at Q.

$$P_i = \sum_{k=1}^{i-1} Span_{k+1} \tag{5}$$

$$Q_i = \sum_{k=1}^{i-1} Span_k \tag{6}$$

Step 4: The next processing step is to equalize each sub-divided histograms separately. For sub-histogram with the range $[P_i, Q_i]$, the equalization is followed by the transformation (Fig. 1).

a) Original image   b) Enhanced using BPDHE

Fig. 1. Enhancement of an image using BPDHE.

Step 5: Finally, the reconstructed image is obtained using the Eq. (7), where M_i is the mean brightness of the input image, M_o is the mean brightness of the output image and $f(x,y)$ is the original input image

$$g(x, y) = (M_i/M_o)f(x, y) \tag{7}$$

This technique of enhancement normalizes the intensity of the output image by carrying out the mean intensity of the output image near to the mean intensity of the input image. But the histogram in this technique is divided into too many histograms that lead to the insignificant amount of improvement.

2.2 Discrete Wavelet Transformation

Output from the DWT has two components:

a) Approximate coefficients (from low pass filter).
b) Detailed coefficients (from high pass filter).

The pictorial representation is shown in the Fig. 2 where $f(x, y)$ is an input image. The input is processed using two different filters i.e. low pass filter (LPF) and high pass filter (HPF). The LPF produces the approximate coefficient while HPF produces the detailed coefficients. The low-frequency component contains approximation information of the image. The high-frequency component contains most noises and all the detail information of image [9]. The output of the LPF and HPF are down sampled by 2 (Fig. 3).

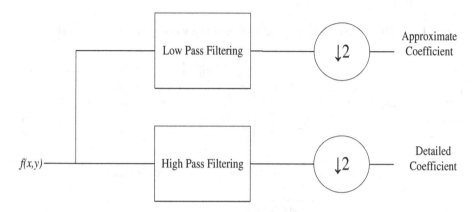

Fig. 2. Illustration of DWT [10].

Fig. 3. Image obtained using DWT.

3 Work Methodology of Proposed Work

This section provides the complete steps of the proposed work for the image enhancement (Fig. 4).

The basic steps are given below:

Step I: Decompose the acquired image into low and high components in frequency domain by single level DWT using Haar wavelet transformation. This wavelet makes the hardware implementation easier as it is simple.

Step II: Enhance the lower coefficients using Brightness Preserved Dynamic Histogram Equalization.

Step III: The high-frequency coefficients are filtered using Gaussian filter so as to remove the noise.

Step IV: Reproduce the improved image by taking inverse DWT using modified coefficients.

Here, LF and HF represent the low and high frequency components respectively.

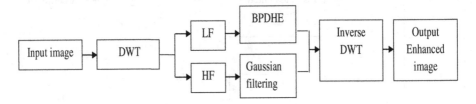

Fig. 4. A typical block diagram of the proposed algorithm.

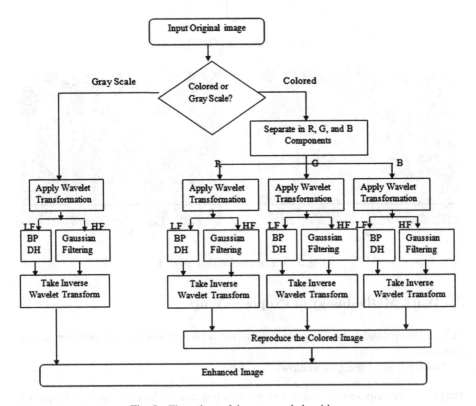

Fig. 5. Flow chart of the proposed algorithm.

Detailed Steps of the proposed algorithm described in Fig. 5 are given under as:

1. Acquire the input image.
2. Check whether it is a colored or gray scale image.

3. If gray scale image then follow steps 3(a)–(e)
    a) Divide the image into low and high frequency components using wavelet transformation.
    b) Apply Enhancement technique BPDHE on the lower frequency component to enhance it.
    c) Higher frequency components are then filtered using Gaussian filter.
    d) Take inverse wavelet transform using modified low and high components.
    e) Go to step 5.
4. If image is colored then follow steps 4(a)–(d)
    a) Separate the image in Red, Green and Blue components.
    b) Follow steps 3(a)–(d) for all the three components separately.
    c) Reproduce the single colored image from three components.
    d) Go to step 5
5. Output the final image as an enhanced image.

4   Results and Discussions

This research proposes a modified algorithm for color improvement of digital images. The performance of proposed algorithm is compared with the four existing color enhancement algorithms i.e. Global Histogram Equalization (GHE) (or simply Histogram Equalization) [1], Contrast Limited Adaptive Histogram Equalization (CLAHE) [3], Brightness Preserved Dynamic Histogram Equalization (BPDHE) [6] and

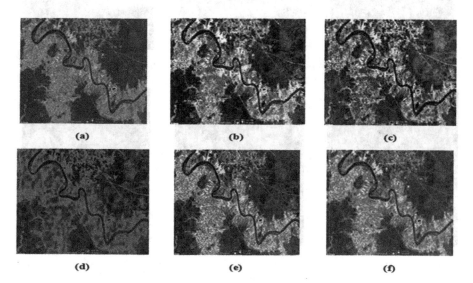

Fig. 6. (a) Original input image 1, Enhanced using (b) Global Histogram Equalization (GHE), (c) Contrast Limited Adaptive Histogram Equalization (CLAHE), (d) Combination of Contrast Limited Adaptive Histogram Equalization and Discrete Wavelet Transformation (CLAHE-DWT), (e) Brightness Preserved Dynamic Histogram Equalization (BPDHE), (f) Proposed algorithm.

Combination of Contrast Limited Adaptive Histogram Equalization and Discrete Wavelet Transformation (CLAHE-DWT) [9]. Figures 6, 7, 8 and 9 are colored images of size 256 x 256. The evaluation of different techniques has been done on the basis of MSE, PSNR and BER.

Figures 6, 7, 8 and 9 show the original and the image enhanced by the different algorithms. It is clear from the figure that colors of the image are changed in case of CLAHE and CLAHE-DWT techniques but the color has been preserved by BPDHE and proposed method and at the same time image is better enhanced by the proposed algorithm with the better PSNR, MSE and BER. The different parameters are tabulated in Table 1. It shows that the improvement in terms of PSNR is best in case of proposed algorithm. The CLAHE has least PSNR and CLAHE-DWT has more PSNR value than CLAHE but less than BPDHE and BPDHE has a less PSNR value than proposed technique. The comparison of different parameters for the different images due to various algorithms is shown in the Figs. 10, 11 and 12.

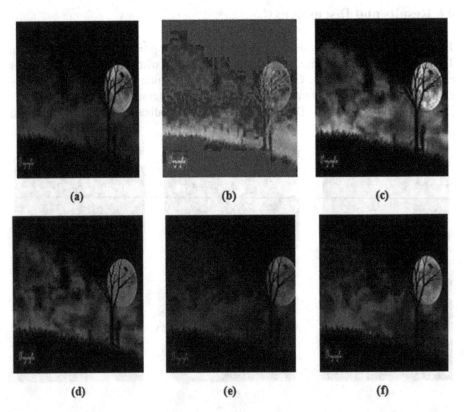

Fig. 7. (a) Original input image 2, Enhanced using (b) Global Histogram Equalization (GHE), (c) Contrast Limited Adaptive Histogram Equalization (CLAHE), (d) Combination of Contrast Limited Adaptive Histogram Equalization and Discrete Wavelet Transformation (CLAHE-DWT), (e) Brightness Preserved Dynamic Histogram Equalization (BPDHE), (f) Proposed algorithm.

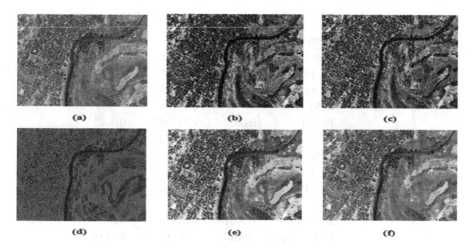

Fig. 8. (a) Original input image 3, Enhanced using (b) Global Histogram Equalization (GHE), (c) Contrast Limited Adaptive Histogram Equalization (CLAHE), (d) Combination of Contrast Limited Adaptive Histogram Equalization and Discrete Wavelet Transformation (CLAHE-DWT), (e) Brightness Preserved Dynamic Histogram Equalization (BPDHE), (f) Proposed algorithm.

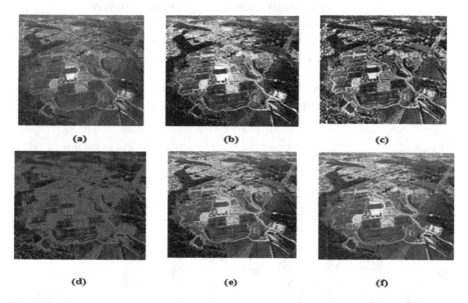

Fig. 9. (a) Original input image 4, Enhanced using (b) Global Histogram Equalization (GHE), (c) Contrast Limited Adaptive Histogram Equalization (CLAHE), (d) Combination of Contrast Limited Adaptive Histogram Equalization and Discrete Wavelet Transformation (CLAHE-DWT), (e) Brightness Preserved Dynamic Histogram Equalization (BPDHE), (f) Proposed algorithm.

Table 1. Comparison of different algorithms with the proposed algorithm for different images.

Image	Method	MSE	PSNR	BER
Image 1	**GHE**	0.0313	62.4165	0.0160
	CLAHE	0.0198	65.1676	0.0153
	CLAHE-DWT	0.0177	65.6532	0.0152
	BPDHE	0.0057	70.5497	0.0142
	PROPOSED	**0.0028**	**73.6144**	**0.0136**
Image 2	**GHE**	0.1691	55.8498	0.0179
	CLAHE	0.0159	66.1212	0.0151
	CLAHE-DWT	0.0059	70.4041	0.0142
	BPDHE	0.0027	73.8727	0.0135
	PROPOSED	**0.0019**	**75.2985**	**0.0133**
Image 3	**GHE**	0.0523	60.9499	0.0164
	CLAHE	0.0401	62.1001	0.0161
	CLAHE-DWT	0.0369	62.4657	0.0160
	BPDHE	0.0128	67.0734	0.0149
	PROPOSED	**0.0042**	**71.8707**	**0.0139**
Image 4	**GHE**	0.0325	63.0153	0.0159
	CLAHE	0.0259	64.0028	0.0156
	CLAHE-DWT	0.0232	64.4772	0.0155
	BPDHE	0.0056	70.6865	0.0141
	PROPOSED	**0.0039**	**72.2671**	**0.0138**

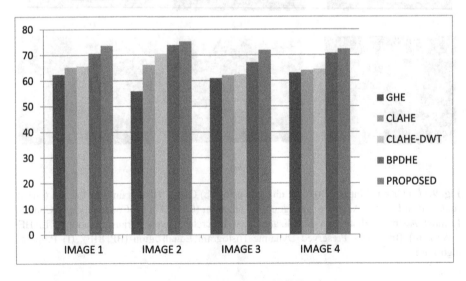

Fig. 10. Comparison of PSNR

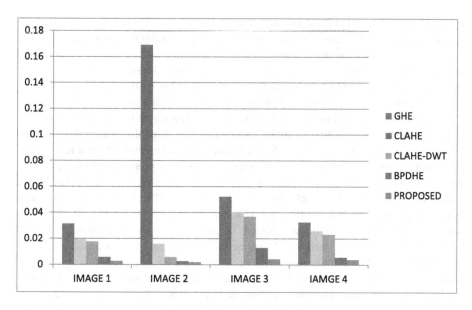

Fig. 11. Comparison of MSE

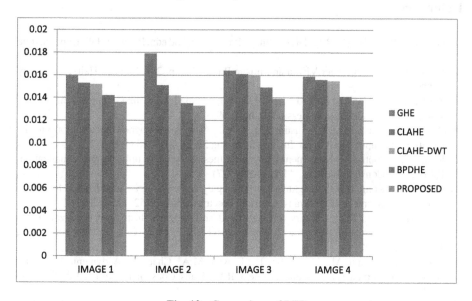

Fig. 12. Comparison of BER

5 Conclusion

This research article proposes a modified algorithm for the color enhancement of low resolution digital images. As discussed in Sect. 2, the high-frequency component contains most noises and all the detail information of image. Therefore, to limit the noise enhancement and protect the detail information from over-enhancement, only the low-frequency component is enhanced and the high-frequency component is kept unchanged. The enhancement of the lower components is only done and the higher components which contain detailed components are only filtered using Gaussian filter. The division into the lower and detailed components is done using DWT [9]. The enhancement of the lower component is performed using the BPDHE algorithm for image enhancement. The method is thus called as the modified BPDHE technique. In this, the comparison of the proposed algorithm with the existing techniques is made on the basis of three parameters called as Peak- Signal to Noise Ratio, Mean Square Error and Bit Error Rate. From the results in Figs. 6, 7, 8 and 9, it has been observed that proposed method can very efficiently preserves the original color of the image and capable of having higher PSNR values than other histogram techniques as CLAHE, CLAHE-DWT and BPDHE. The graphs of parameters PSNR, MSE and BER are shown in Figs. 10, 11 and 12 respectively. As shown in Fig. 10, higher PSNR is obtained for proposed technique compared to the other equalization methods. It has been observed that the proposed algorithm yields maximum PSNR, minimum MSE and BER. The comparison from visual appearance of Figs. 6, 7, 8 and 9, it has been observed that the proposed algorithm improves the information contents present in that image while preserving the original hue.

References

1. Gonzalez, R.C., Woods, R.E.: Digital Image Processing, 2nd edn. Prentice Hall, Upper Saddle (2002)
2. Umbaugh, S.E.: Computer Vision and Image Processing, p. 209. Prentice Hall, New Jersey (1998)
3. Zuiderveld, K.: Contrast limited adaptive histogram equalization. In: Graphics Gems IV, pp. 474–485. Academic Press Professional, Inc. (1994)
4. Pizer, S.M., Amburn, E.P., Austin, J.D., et al.: Adaptive histogram equalization and its variations. Comput. Vis. Graph. Image Processing $39(3)$, 355–368 (1987)
5. Kim, Y.T.: Contrast enhancement using brightness preserving bi-histogram equalization. IEEE Trans. Consum. Electron. $43(1)$, 1–8 (1997)
6. Ibrahim, H., Kong, N.S.P.: Brightness preserving dynamic histogram equalization for image contrast enhancement. IEEE Trans. Consum. Electron. $53(4)$, 1752–1758 (2007)
7. Kim, M., Chung, M.G.: Recursively separated and weighted histogram equalization for brightness preservation and contrast enhancement. IEEE Trans. Consum. Electron. $54(3)$, 1389–1397 (2008)
8. Abdullah-Al-Wadud, M., Md. Kabir, H., Dewan, M.A., Chae, O.: A dynamic histogram equalization for image contrast enhancement. IEEE Trans. Consum. Electron. $53(2)$, 593–600 (2007)
9. Lidong, H., Wei, Z., Jun, W., Zebin, S.: Combination of contrast limited adaptive histogram equalization and discrete wavelet transform for image enhancement. IET Image Process. $9(10)$, 908–915 (2015)
10. Kociołek, M., Materka, A., Strzelecki, M., Szczypiński, P.: Discrete wavelet transform – derived features for digital image texture analysis. In: Proceedings of International Conference on Signals and Electronic Systems, Lodz, Poland, pp. 163–168 (2001)

FP-Growth Implementation Using Tries
for Association Rule Mining

Manu Goel^(⊠) and Kanu Goel

CSED, Thapar University, Patiala, Punjab, India
manugoel467@gmail.com, kanugoel10@gmail.com

Abstract. With the advent of technology the past few years have seen the rise in the field of data mining. Data mining generally considered as the process of extracting the useful information by finding the hidden and the non-trivial information out of large chunks of dataset. Now here comes the role of association rule mining which forms a crucial component of data mining. Many recent applications like market basket analysis, text mining etc. are done using this approach. In this paper we have discussed the novel approach to implement the FP Growth method using the trie data structure. Tries provide a special feature of merging the shared sets of data with the number of occurrences that were already registered as count. So this paper widely gives an idea about how the interesting patterns are generated from the large databases using association rule mining methodologies.

Keywords: FP Growth · Trie · Association rule mining · Data mining

1 Introduction

Data mining refers to the extraction of predictive information that is in the hidden form from huge databases. Association rule mining is intended for the purpose of identifying the important rules that will be discovered inside the databases by performing some techniques of data mining and after using measures of interestingness [1]. In this paper we have designed an algorithm for FP Growth technique by taking into the consideration of the repeating patterns (items) in the frequent item sets. Here, by using trie data structure we can reduce the space and time complexity by efficiently grouping the overlapping items and thus perform the techniques on the data in this form.

Section 2 gives an overview of the work that has been done so far by the researchers in the field of data mining and association rule mining. Section 3 briefly describes the basic concepts related to the topic. Section 4 represents the fundamentals of FP-Growth discussing the advantages of using TRIES for implementing the same. Section 5 gives the actual implementation of the proposed approach. Then we discuss the results in the following Sect. 6. Finally giving the conclusion and future scope in Sect. 7.

© Springer Nature Singapore Pte Ltd. 2017
K. Deep et al. (eds.), *Proceedings of Sixth International Conference on Soft Computing for Problem Solving*, Advances in Intelligent Systems and Computing 547, DOI 10.1007/978-981-10-3325-4_3

2 Related Work

In literature, FP growth algorithm has been implemented using the tries data structure proposing new fundamentals for generating the association rules with the help of frequent item sets. Taktak and Slimani [2] have proposed a multi-support variation in the versions of FP Growth by handling two cases of increasing and decreasing the minsup values. They have generated rare rules carrying out their work [2]. Hipp et al. [3] have described the fundamentals of the association rule mining by deriving a general framework for the same. They have identified the concerns for various strategies to find out the support values frequents item sets [3]. Borgelt [4] has compared the performance of the FP-Growth algorithm with the other three primary approaches of association rule mining namely Eclat, Apriori and Relim [4]. Kumar and Rukmani [5] have explained the research work of implementing all the phases of web mining and study the associated pattern of usage [5]. Then Racz [6] has dealt with the issues related to the implementation of the FP Growth approach along with the memory layout and data structures for fast traversal. Liet et al. [7] in their work have proposed a model to perform the parallelization of the FP Growth algorithm in a way that it can work on the distributed machines.

3 What Is Data Mining?

With the help of data mining we can discover patterns based upon the computational methods present in large datasets that involve the intersection of techniques pertaining to subjects like machine learning, artificial intelligence, database systems and even the statistics for mathematical and probabilistic calculations. Now the overall process of applying data mining to any of the datasets apart from applying the basic raw analysis involves the pre-processing of the data, later considering it on basis of model and inference, calculating the interestingness of metrics and finally the post-processing of the discovered visualization, structures, and also the online updating. Therefore the Data mining is also known as the step of analysis in the "knowledge discovery in databases" process, or shortly called as the KDD.

Also the prospective and automated analyses that is offered by data mining helps in moving beyond analyses of the past events given by retrospective tools that were typically belonging to the decision support systems. It predicts the information that also the experts may miss by scrutinizing the huge databases because sometimes it may lie beyond the human expectations [6].

3.1 Association Rule Mining

Association rule mining is the process meant to discover the interesting relations that exist between the variables that form a part of large databases [7]. This type of information is primarily used for the process of decision making in the marketing sectors for the promotional activities and placement of the products. These decisions also help to increase the sales of the consumer products in supermarkets etc.

Now to note here the association rule mining does not take into account the order of the items either within the transaction or across the transaction unlike the sequence mining.

3.2 Parameters of Association Rule Mining

Association rule learning finds out whether there exists any relation between the varied products or any frequent patterns. Popularly-known two of the constraints are minimum value of thresholds on two parameters are support and confidence to select interesting rules form dataset. Support and confidence are the parameters that are essentially used to conduct the association rule mining. Now here the term support refers to how frequent a particular item has occurred in the database while the confidence represents the count of the times the rule appeared to be true. Now for instance to let there be an item W and number of transactions denoted by T. So here the support of W will be the frequency of occurrence of item W in T as given in (1). It will be the proportion of the transactions present in our database that contain the item-set W.

$$\text{Support (W)} = \text{Frequency of item set (W)} / \text{Total number of transactions T} \quad (1)$$

And support for two items V and W occurring together is given by (2).

$$\text{Support (VW)} = \text{Frequency of item sets (V) and (W) occurring together}$$
$$/ \text{Total no. of Transactions} \quad (2)$$

The parameter confidence is used as an indication of how often the rule was found to be true in our transaction database. Now here the confidence value associated with a rule, $V \Rightarrow W$, with respect to the set of transactions T will be the proportion of the transactions in our dataset that include V and W both simultaneously. Confidence is given by (3).

$$\text{Confidence } (V \Rightarrow W) = \text{Support (VW)} / \text{Support (V)} \quad (3)$$

4 FP-Growth

Algorithm of FP-Growth that was initially proposed by Han [8] has been an efficient and scalable method to mine the complete set of frequent patterns with the process of pattern fragment growth. Now major difference between the frequent pattern-growth and other algorithms is found out to be that FP-growth does not essentially generate the candidates, rather it just tests them. Whereas the Apriori algorithm not only just tests the candidate item sets but beforehand generates them also [9].

The motivation behind FP-tree methods has been the following:

- Only frequent items are simply required to locate the association rules, hence it is best for finding the frequent items while ignoring the others.

- If in case of multiple transactions which share a particular set of frequent items, it may be a possible case to allow merging of the shared sets along with number of occurrences that are registered as count.

  Advantages of FP Growth:

- FP Growth clearly avoids the scanning of database more than two times for finding out the count of support.
- Eliminates the costly candidate generation that can prove to be an expensive operation in reference to the Apriori algorithm for the candidate sets [10].
- Also proves to be better when the support count is low and transaction database is huge enough, it would satisfy count of the support thus size of candidate sets for the case of Apriori would be many and large.

4.1 TRIE Data Structure

In computer science as we know there exists plenty of data structures to represent various forms of data depending upon the purpose of storage and retrieval of the data. One of these largely used is Trie. Trie is basically an ordered tree data structure which is primarily used for storing a dynamic form of associative array or set whereby the keys form usually the strings of data. Now unlike a normal binary search tree where no node in tree data structure stores any of the key associated with that node, here its position in that tree will be defining the key with which it is being associated. Also to note all the descendants of a particular node will have common prefix of data string associated to that specific node, consequently the root will be associated with an empty string. However the values may not necessarily be associated with every single node. Instead of this values tend to be associated with just the leaves, and also with some of the inner nodes corresponding to the keys of our interest.

4.2 Advantages of Tries Over Other Data Structures

Now here we shall discuss as to how the Tries are advantageous over the other tree data structures. Due to their unique features they can even be used while replacing the standard hash table. Now let us say k is the length of the string that needs to be searched in our application. So the worst case complexity for doing so will be of the order of $O(k)$. As the imperfect hash table may have the key collisions (hash function mapping of dissimilar keys to the one same position in the hash table). But instead there would be no collisions for the case of different keys in a trie.

5 Implementation of the Proposed Approach

Here we describe the steps of the procedure that is used while implementing the proposed approach of FP-Growth using TRIES.

- Enter the number of transactions, number of Items, and support percentage.
- Calculation of Support = No. of transaction * Support Percentage.
- Input the transaction database in the form of Boolean 2-D array.
- All the Transaction items are assigned number ranging from 0 to number of items-1.
- For each transaction enter-
  - 1: Item is present
  - 0: Item is not present.
- Initialize the frequent item set array to 0.
- As the transaction database is stored in the horizontal representation, so by adding the number of 1's in each column find the frequency of each item.
- Initialize the Boolean bit vector array to 0.
- Now set the bit vector corresponding to each item as 1 if the frequency of that item is greater than or even equal to the minimum support specified.
- Now modify the transaction table by putting 0 in each transaction for all non-frequent items if it was originally 1.
- From the bit vector find the number of frequent items by adding the number of 1's.
- Make header table showing the ranking of items w.r.t. the frequency of frequent items.
- Now insert the items for each transaction from the modified transaction table in the Trie in accordance to their ranking.
- Finally apply the data mining/association rule mining function on the Trie structure formed after insertion.
- Print the frequent item pair set satisfying the minimum support condition.

5.1   Code for Insertion into the Trie

```
void insert(trieNode* iter, bool transactionArray[], int numOfFrequentItems, int rankArray[])
{
    int digit;
    int index;

    for(digit=0;digit<numOfFrequentItems;digit++)
        if(transactionArray[rankArray[digit]] == 1)
        {
            index = rankArray[digit];
            if( (iter->children[index]) == NULL )
                iter->children[index] = createNode();

        iter = iter->children[index];
        (iter->count)++;
        }
}
```

5.2   Code for Finding Frequent Item Sets Pairs

```
int miningFunc(trieNode* iter, int miningArray[], int itemNum,int child)
{
    if(iter == NULL) return 0;
    if(child == itemNum)
    {
        miningArray[child] = miningArray[child] + iter->count;
        return iter->count;
    }
    int val = 0;
    for(int childNum=0; childNum<MAX _ITEMS; childNum++)
        if(iter->children[childNum])
            val = val + miningFunc(iter->children[childNum], miningArray, itemNum, childNum);
    if(val > 0)
        miningArray[child] = miningArray[child] + val;
    return val;
}
```

Here, we have used an array (miningArray[ ]) to store the frequency of items that occur together with the item that is to be associated. Each time the miningFunc() is called, we are inserting the value of frequency (i.e. count) of items that are present in the path to the final item (i.e. itemNum: the item that is selected to form association rules.). After each call the miningArray[ ] is checked for the values where the frequency of items satisfies the given support. And then all the supported items are selected with corresponding itemNum. And thus the association rule for frequent items is generated consisting of 2 item pairs.

6   Results and Discussions

Now let us take an example on which we have applied our algorithm. The transaction database is as follows (Table 1):

Table 1. Transaction database

Transaction ID	Items
1	Bread, Cheese, Eggs, Juice
2	Bread, Cheese, Juice
3	Bread, Milk, Yoghurt
4	Bread, Juice, Milk
5	Cheese, Juice, Milk

Table 2. The frequency of items

Item	Frequency
Bread	4
Cheese	3
Eggs	1
Juice	4
Milk	3
Yoghurt	1

Table 3. Frequent items

Item	Frequency
Bread	4
Juice	4
Cheese	3
Milk	3

Now, the first step is to find the frequency of each item set (Table 2). And then remove the item sets that are less frequent than the minimum support required (Table 3).

Then we have modified the transaction table by removing the non-frequent item sets and ordering the items in non-increasing order of frequency (Table 4).

Now, we apply the algorithm to insert transaction items to form the trie structure (Fig. 1).

Now, after applying the data mining function on the trie formed in above step, we get three conditional trees as our frequent item count = 4. This has been depicted by (Figs. 2, 3 and 4).

Figure 5 shows the input and output of the program. The output here shows that there are 2 frequent item pairs i.e. (3,1) – (Juice, Cheese) and (0,3) – (Bread, Juice).

Table 4. Modified transaction table after removal of non-frequent items & ordered according to frequency.

Transaction ID	Items
1	Bread, Juice, Cheese
2	Bread, Juice, Cheese
3	Bread, Milk
4	Bread, Juice, Milk
5	Juice, Cheese, Milk

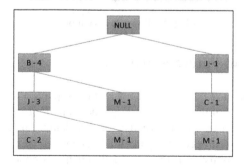

Fig. 1. FP-Tree implementation using Trie

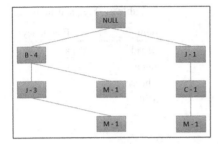

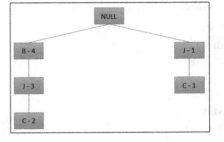

Fig. 2. Conditional tree for M   **Fig. 3.** Conditional tree for C

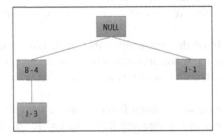

Fig. 4. Conditional tree for J

```
Enter the number of transactions : 5
Enter the number of items : 6
Enter the support percentage : 50

Enter transaction details
Transaction 1 : 1 1 1 1 0 0
Transaction 2 : 1 1 0 1 0 0
Transaction 3 : 1 0 0 0 1 1
Transaction 4 : 1 0 0 1 1 0
Transaction 5 : 0 1 0 1 1 0

The Frequent Itemset : 3 1

The Frequent Itemset : 0 3
```

Fig. 5. Program execution

7 Conclusion and Future Scope

In the field of data mining the association rule mining plays a crucial role to analyze
and predict the behavior of the customers. Implementing FP growth using the tries data
structure has thus increased the performance of association rule mining undoubtedly.
For now the given approach generates only the frequent two item pair sets. This can be
further improved by modifying it to generate the item sets containing more than two
item pair sets.

References

1. Agarwal, R., Srikant, R.: Mining sequential patterns. In: ICDE 1995, Taipei, Taiwan, pp. 3–14, March 1995
2. Taktak, W., Slimani, Y.: MS-FP-Growth: a multi-support version of FP-Growth agorithm. Int. J. Hybrid Inf. Technol. **7**(3), 155–166 (2014)
3. Hipp, J., Güntzer, U., Nakhaeizadeh, G.: Algorithms for Association Rule Mining-A General Survey and Comparison. University of Tubingen, Tubingen (2000)
4. Borgelt, C.: An implementation of the FP-Growth algorithm. In: Proceedings of the 1st International Workshop on Open Source Data Mining: Frequent Pattern Mining Implementations, pp. 1–5. ACM (2005)
5. Kumar, B.S., Rukmani, K.V.: Implementation of web usage mining using APRIORI and FP growth algorithms. Int. J. Adv. Netw. Appl. **1**(6), 400–404 (2010)
6. Rácz, B.: nonordfp: an FP-Growth variation without rebuilding the FP-Tree. In: 2nd International Workshop on Frequent Itemset Mining Implementations, FIMI (2004)
7. Li, H., Wang, Y., Zhang, D., Zhang, M., Chang, E.Y.: Pfp: parallel FP-Growth for query recommendation In: Proceedings of the ACM Conference on Recommender Systems, pp. 107–114. ACM (2008)
8. Han, J., Pei, J.: Mining frequent patterns by pattern-growth: methodology and implications. ACM SIGKDD Explor. Newsl. **2**(2), 14–20 (2000)
9. Tjioe, H.C., Taniar, D.: A framework for mining association rules in data warehouses. In: Yang, Z.R., Yin, H., Everson, R.M. (eds.) IDEAL 2004. LNCS, vol. 3177, pp. 159–165. Springer, Heidelberg (2004). doi:10.1007/978-3-540-28651-6_23
10. Becuzzi, P., Coppola, M., Vanneschi, M.: Mining of association rules in very large databases: a structured parallel approach. In: Amestoy, P., Berger, P., Daydé, M., Ruiz, D., Duff, I., Fraysse, V., Giraud, L. (eds.) Euro-Par 1999. LNCS, vol. 1685, pp. 1441–1450. Springer, Heidelberg (1999). doi:10.1007/3-540-48311-X_204

Cost Optimization of 2-Way Ribbed Slab Using Hybrid Self Organizing Migrating Algorithm

Piyush Vidyarthi[1], Dipti Singh[2(✉)], Shilpa Pal[2], and Seema Agrawal[3]

[1] Department of Civil Engineering, Gautam Buddha University, Greater Noida,
Uttar Pradesh, India
vid.piyush@gmail.com
[2] Gautam Buddha University, Greater Noida, Uttar Pradesh, India
diptipma@rediffmail.com, shilpa@gbu.ac.in
[3] Department of Mathematics, S.S.V.P.G. College, Hapur, Uttar Pradesh, India
Seemagrwl7@gmail.com

Abstract. In this paper, an optimization problem of 2-way ribbed slab or waffle square slab has been solved using a hybrid variant of self organizing migrating algorithm, C-SOMAQI. The main objective of this problem is to design a 2-way ribbed waffled slab of dimension 10 m * 10 m with optimum cost of steel as well as concrete. The design variables of 2-way ribbed waffled slab are taken as the effective depth of the slab (j), ribs width (p), the spacing between ribs (b), effective depth of ribs (d), and the area of reinforcement. Various population based techniques are available in literature to solve optimization problems. In this paper, a low population based hybrid technique C-SOMAQI has been taken to solve this problem that provides the solution at an optimum cost. To validate the claim, this problem has also been solved theoretically and a comparative analysis between these two solutions has been made. The study concludes that C-SOMAQI provides better results as compare to theoretical method.

Keywords: Ribbed slab · Waffle slab · Optimization · Self organizing migrating algorithm

1 Introduction

Waffle slab is typical concrete slab system in which the concrete joists/beams are laid in both directions. These concrete joists are cast monolithically with slab hence, there is a decrease in the dead load of the structure as the thickness of the structure also decreases which enhances the load bearing capacity. In case of normal slabs, due to its high weight there is limitation of span which is the main obstacle in the concrete construction. Therefore, to overcome this problem development in the reinforced structure have focused on enhancing the span which is done either by overcoming the concrete's natural weakness in tension or by reducing the weight. By utilizing the approach for reducing the weight of normal slab, waffled slab is developed. It has drawn the attention of various researchers. The need for optimum design of waffle slab is

© Springer Nature Singapore Pte Ltd. 2017
K. Deep et al. (eds.), *Proceedings of Sixth International Conference on Soft Computing for Problem Solving*, Advances in Intelligent Systems and Computing 547,
DOI 10.1007/978-981-10-3325-4_4

because of its various properties such as light weight & long spans, economical (when reusable formwork pans are used), vertical penetrations between ribs is easy, profile may be expressed architecturally, or used for heat transfer, it allows a considerable reduction in dead load as compared to conventional flat slab construction since the slab thickness can be minimized due to the short span between the joists.

Cost optimum design of reinforced concrete structures is receiving more and more attention from the researchers. Pretzer et al. [8] investigated unusual applications of pre-stressed waffle slabs and composite beams and concluded that waffle slabs are generally employed in long-span constructions. Practical spans vary between (5 m and 15 m) for R.C.C. under normal office or industrial loading. With the use of pre-stressing, longer spans or higher loads are possible. Reiss et al. [9] designed a rectangular ribbed flat slab and analyzed it as orthotropic plates with zero twisting rigidity. As a result of the missing twisting rigidity, relatively large moment concentrations were found in the column strips, which tend to act as supporting beams. Galeb et al. [4] carried out a research on the optimization process using MATLAB of waffle slab, The result indicated that the optimum value of the total cost is enhanced with the increase in population size. M. Khalaf et al. [6] used a 3-D finite element method to carry out a study on influencing parameters such as slab aspect ratio, slab boundary conditions, steel sheet depth, corrugation cell aspect ratio and its orientation, to investigate the overall behavior of two-way composite waffle slab under different conditions. The result showed that waffle composite slabs is superior and is strongly recommended to manufacture new steel sheet. Mohd. Shahezad et al. [5] presented the comparison of R.C.C Beam Slab, One-Way Continuous Slab and Grid Slab and results revealed that a R.C.C. One-Way Continuous slab is more economical than other two types of slabs for the considered span. Indrajit et al. [7] proposed a semi analytical method for the analysis of waffle slab with any arbitrary boundary conditions; fixed, free and simply supported and stated that it can be adapted for analysis of waffle slab for generalized boundary conditions. Hatindera Singh et al. [3] presented a paper on the cost optimization of one way slab and the result implies that it is not always true that higher grade will always results in minimum cost.

Minimization of construction cost of waffle slab is one of the most important factors for civil engineers without making any change in the value of its load bearing capacity. Design optimization [2] of various structures is a vast topic of research in the field of structure design. The motive of this work is to develop optimal solution for the structure in terms of cost and for the better utilization of construction materials which is directly related to the increase of the load bearing capacity of the structure. Optimization directly lowers the cost of the structure without compromising in its functional capacity to serve the various loads. The theoretical results are obtained by calculating serviceability loads as per IS: 456-2000 (Plain and reinforced concrete design) and dimensioning has been done as per BS: 8110 (design and construction of reinforced and pre-stressed concrete structures).

In this paper, optimization of a square (10 m * 10 m) span, 2 way ribbed slab has been done using SOMAQI which is a hybrid variant of Self Organizing Migrating Algorithm (SOMA). The main objective of this paper is optimum design of reinforcement & concrete waffled (2-way ribbed) square slab of (10 * 10) m using M20 grade of concrete and Fe415 grade of steel. The problem has also been solved theoretically and comparative analysis between these two solutions obtained theoretically and analytically has been made. The rest of the paper has been organized as follows: In Sect. 1, Introduction of the problem with a brief summary of Literature has been given; in Sect. 2, 2-Way Ribbed Slab or Waffle Slab Problem has been modeled; in Sect. 3, the methodology of C-SOMAQI has been discussed; Sect. 4 discusses the results and analysis and conclusion is presented in Sect. 5.

2   2-Way Ribbed Slab or Waffle Slab Problem

Design [1] a 2-way ribbed slab (waffle slab) for interior span of (10×10) m as shown in Fig. 1 considering that there is no shear reinforcement. Dimensioning of the components of waffle slab is done by using BS: 8110 and analysis is done by using IS: 456. By taking grade of steel is Fe415 and for concrete is M20.

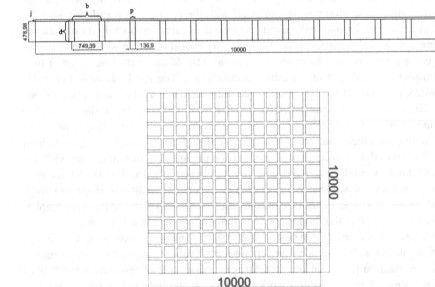

Fig. 1. 2 way ribbed slab (all dimensions are in mm).

2.1   Objective Function

2.1.1   Reinforcement
Optimization of reinforcement (m^3) present in waffle slab.

For steel grade Fe415 cost (Rs 3, 53,250/m^3).

$$Ast\,(m^3) = ((30 * j * C_1) + ((f1/f2) * ((b * d)/2) * (1 - sqrt(1 - 4.6 * m)/(f1 * b * d * d)))$$
$$* 2 * ((C_1/b) - 1) * C_1) + ((f1/f2) * ((b * d)/2) * (1 - sqrt(1 - 4.6 * n)/(f1 * b * d * d)))$$
$$* 2 * ((C_1/b) - 1) * C_1))/(10^9) + (p * 0)$$

Where,

$C_1 = 10000$ mm

j = Depth of the slab (mm)

f1 = characteristic strength of concrete (M20 = 20 N/mm2)

f2 = yield strength of steel (Fe415 = 415 N/mm2)

b = the spacing between ribs (mm)

d = Projection of rib downward from topping slab (mm)

m & n = moment for which area of steel is calculated (Nmm)

p = ribs width (mm)

2.1.2 Concrete

Optimization of concrete in (m^3)

For M20 cost (Rs 4800/m^3)

Concrete volume $(m)^3 = ((C_1^2 * (j+d) - ((C_1/b)^2 * (b-p)^2 * d))/(10^9)$; Where,

$C_1 = 10000$ mm

j = Depth of the slab (mm)

d = Projection of rib downward from topping slab (mm)

b = the spacing between ribs (mm)

p = ribs width (mm)

2.1.3 Total Cost

$$\text{MIN(Cost)} = (((((c*j) + (x*y*z*2*u*s) + (x*y*v*2*u*s))/(w)) + (p*0)) * 353250)$$
$$+ ((((s*s*r) - ((s/b)*(s/b)*(b-p)*(b-p)*d))/(w)) * 4800)$$

Where,

c = 300000

x = 0.0482

y = (b*d)/2

z = $(1 - \text{sqrt}(1 - (4.6*m)/(f1*b*d*d)))$

u = $((10000/b)-1)$

v = $(1 - \text{sqrt}(1 - (4.6*n)/(f1*b*d*d)))$

w = 1000000000

r = (j + d)

s = 10000

2.1.4 Constraints

To control the geometric configuration as well as to maintain the load bearing capacity of slab constraints have been formulated as per the basic rules of dimensioning of 2-way ribbed or waffle slab given in BS:8110 and IS:456.

(1) depends upon the spacing between ribs (b) & width of ribs (p)

$$b - p - 750 =< 0$$

(2) depends upon the depth of ribs (d) & width of ribs (p)

$$d - (3.5 * p) =< 0$$

(3) depends upon width of ribs (p)

$$-p + 100 =< 0$$

(4) depends upon moment according to which reinforcement being designed, depth of topping slab (j), spacing of ribs (b), depth of rib from topping (d)

$$jbd - 0.416bj^2 - 15.5 * 10^6 =< 0$$

2.1.5   Upper Bound and Lower Bounds
There are 6 variables and hence, the ranges of each variable is given below

$$45 =< j =< 75$$
$$240 =< d =< 700$$
$$60 * 10^6 =< m =< 110 * 10^6$$
$$46 * 10^6 =< n =< 110 * 10^6$$
$$100 =< p =< 200$$
$$350 =< d =< 700$$

3   Methodology of C-SOMAQI

A number of population based evolutionary algorithms (EAs) are available in literature to solve engineering optimization problems. In this paper, a hybrid variant of SOMA that is C-SOMAQI has been used to solve 2-way ribbed slab or waffle slab optimization problem. C-SOMAQI combines the feature of SOMA and quadratic interpolation cross-over operator [10]. The computational step of this algorithm is as follows:

Step 1:   generate the initial feasible population;
Step 2:   evaluate all individuals in the population;
Step 3:   generate PRT vector for all individuals;
Step 4:   sort all of them;
Step 5:   select the best fitness individual as leader and worst as active;
Step 6:   for active individual a new population of size n is created. This population is nothing but the new positions of the active individual towards the leader in n steps of defined length. The movement of this individual is given as:

$$x_{i,j}^{MLnew} = x_{i,j,start}^{ML} + \left(x_{L,j}^{ML} - x_{i,j,start}^{ML} \right).t.\ PRTVector_j$$

Where $t \in < 0$, by step to, PathLength > and ML is actual migration loop

$x_{i,j}^{MLnew}$ is the new positions of an individual.

$x_{i,j,start}^{ML}$ is the positions of active individual.

$x_{L,j}^{ML}$ is the positions of leader.

Step 7:   sort new population with respect to fitness;

Step 8:   for each individual in the sorted population, check feasibility criterion;

Step 9:   if feasibility criterion is satisfied replace the active individual with the new position, else move to next position in the sort order and go to step 8;

Step 10: create new point by quadratic interpolation crossover operator as follows:
Select three distinct points R_1 (with best fitness value), R_2 and R_3
Then A new trial point of minima $x' = (x_1', x_2', \ldots \ldots, x_n')$ is given as

$$x' = \frac{1}{2} \frac{\left[\left(R_2^2 - R_3^2\right) * f(R_1) + \left(R_3^2 - R_1^2\right) * f(R_2) + \left(R_1^2 - R_2^2\right) * f(R_3)\right]}{\left[(R_2 - R_3 * f(R_1) + (R_3 - R_1) * f(R_2) + (R_1 - R_2) * f(R_3)\right]}$$

Where $f(R_1), f(R_2)$ and $f(R_3)$ are the objective function values at R_1, R_2 and R_3 respectively.

Step 11: if feasibility criterion is satisfied replace the active individual with the new position, else go to step 9 until prescribed number of iterations are exhausted;

Step 12: if new point is better than active replace active with the new one;

Step 13: if termination criterion is satisfied stop else go to step 5;

Step 14: report the best individual as the final optimal solution;
The parameters used in C-SOMAQI are Population size: 20, PRT: 1, Step Size: .85, Path Length: 3.

4   Results and Analysis

This section discusses the comparative study of the values obtained by C-SOMAQI and theoretical study. The design variables obtained by theoretical calculations have been used as pre-set values for comparative study. C-SOMAQI optimized the cost of steel and also determined the optimum values of design variables. Comparison of the values obtained theoretically and by C-SOMAQI has been shown in Table 1. From Table 1, it is observed that the pre-set value of depth of slab which was taken as 60 mm has been reduced to 45 mm by the optimization technique with projection of rib as 353 mm and ribs width as 101 mm. With the optimized values of depth of slab, moment resisting capacity is increasing by 20% as shown in Table 1. The most important parameter i.e. cost of concrete and steel used is also reduced by 18% which is an appreciable amount. The same results have been shown in graphically form in Figs. 2 and 3 .

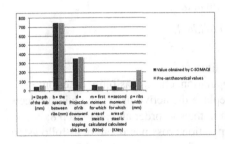
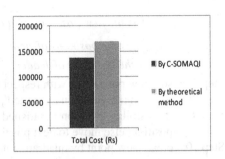

Fig. 2. Comparison of cost obtained by C-SOMAQI and theoretical method.

Fig. 3. Comparison of different parameter values obtained by C-SOMAQI and pre-set theoretical values.

Table 1. Comparison between C-SOMAQI and theoretical study

Different variable with their descriptions	Value obtained by C-SOMAQI (obtained values)	Theoretical calculation (pre-set values)	% increase or decrease
j = Depth of the slab (mm)	45	60	25% decrease
b = the spacing between ribs (mm)	750	750	No change
d = Projection of rib downward from topping slab (mm)	353	367	–
m = first moment for which area of steel is calculated (Nmm)	60000000	47520000	20% increase
n = second moment for which area of steel is calculated (Nmm)	46000000	35000000	23.9% increase
p = ribs width (mm)	101	220	–
Total cost of concrete & steel	Rs 139683	Rs 171765	18.6% decrease

5   Conclusion

The following conclusions are drawn from the present study:

1. The parameters obtained from the C-SOMAQI are much more optimized then the pre-set theoretical values. Hence, the results obtained from C-SOMAQI are more appropriate in terms of cost of concrete and steel approximately by 18%.

2. The value of depth of topping slab obtained by C-SOMAQI is 25% lesser than the pre-set theoretical values which eventually enhance the load bearing capacity of 2-way ribbed waffled slab by 20–23%. Thus, construction cost is minimized without compromising with the load bearing capacity.

3. Theoretical analysis is much more time consuming than the optimization technique. The optimization technique directly gives the better results on entering the desired objective function.

4. In optimization technique, the results can directly be changed according to the requirement. This technique is not only limited to cost optimization but can also be used to obtain optimum quantity of concrete and steel reinforcement.

5. This study is limited only for a square waffle slab of 10 m * 10 m span.

References

1. Gambhir, M.L.: Design of Reinforcement Concrete Structure. PHI Learning Pvt. Ltd., New Delhi (2010)
2. Deb, K.: Optimization for Engineering Design (Algorithm and Examples). PHI Learning Pvt. Ltd., New Delhi (2004)
3. Singh, H., Rai, H.S., Singh, J.: Discrete optimisation of one way slab using genetic algorithm. Int. J. Eng. Bus. Enterp. Appl. (IJEBEA) 9(2), 116–121 (2014)
4. Galeb, A.C., Atiyah, Z.F.: Optimum design of reinforced concrete waffle slabs. Int. J. Civ. Struct. Eng. 1(4), 862–880 (2011)
5. Shahezad, M., Shaikh, U.A.: Study on economical aspect of R.C.C beam slab construction and grid slab construction. Int. J. Eng. Sci. Eng. Technol., 3(8), 284–287 (2014)
6. Khalaf, M., El-Shihy, A., Shehab, H., Mustafa, S.: Structural behavior analysis of two-ways (Waffle) composite slabs. Int. J. Eng. Innovative Technol. (IJEIT) 2, 47–54 (2013)
7. Chawdhury, I., Singh, J.P.: Analysis and design of waffle slab with different boundary conditions. Indian Concr. J., 86(6), 43–52 (2012)
8. Pretzer, C.A.: Unusual application of prestressed waffle slabs and composite beams. J. Am. Concr. Inst. 69, 765–769 (1972)
9. Reiss, M., Sokal, J.: Design of ribbed flat slabs. J. Inst. Struct. Eng. 50, 303–307 (1972)
10. Singh, D., Agrawal, S., Deep, K.: C-SOMAQI: self organizing migrating algorithm with quadratic interpolation crossover operator for constrained global optimization. In: Davendra, D., Zelinka, I. (eds.) Self-Organizing Migrating Algorithm. SCI, vol. 626, pp. 147–165. Springer, Cham (2016). doi:10.1007/978-3-319-28161-2_7

A Complete Ontological Survey of Cloud Forensic in the Area of Cloud Computing

Shaik Khaja Mohiddin[1(✉)], Suresh Babu Yalavarthi[2],
and Shaik Sharmila[3]

[1] Acharya Nagarjuna University, Guntur, AP, India
mail2mohiddin@gmail.com
[2] Jagarlamudi Kuppuswamy Choudary College, Guntur, India
yalavarthi_s@yahoo.com
[3] Montessori Mahila Kalasala, Vijayawada, India
sharmilamca2011@gmail.com

Abstract. In order to arm yourself and assess risk correctly in the cloud, carrying out investigations are necessary. Though Cloud Computing is the buzzing technology now a days, with the advent with which cloud computing is developing and providing solutions to the upcoming technologies where on the other side of it there also lies a down fall with respect to theft of data in the cloud, loss of data in the cloud which are related to forensic, these things create an distrustful relation between cloud user and cloud providers, during the past half decade cloud forensic has emerged as a challenging point that has to be solved in cloud computing, the prevailing existing solutions which were suggested at the initial stage of Cloud Forensics were satisfied at that time but those solutions cannot rule now also, due to which they are not up to the mark to give much satisfaction which there by strengthen the faith of users in cloud, and makes Cloud Service Providers to provide the services and make a path to be laid for the coming technologies to utilize the flavors of cloud computing. This paper inculcates the contents related to cloud forensic which are traced out which are being known, to be known. In this paper we have traced out the major causes of Cloud Forensics, and thrown a light on the causes and origins which are helpful for laying a success ladder for Cloud Forensic concepts, and provided relevant content which would be helpful to trace out the solutions which are causing obstacles in the field of Cloud Forensics, utilizing which one can get an overview of the challenges and if considered can definitely lead to a good solution methods which if implemented in a proper way can lead to recovery of lost data as well as to know the real cause of the effect and also in order to reduce their effect.

Keywords: Cloud computing · Cloud forensic · Cloud Service Providers · Digital Forensic Readiness · Forensic Tool Kit (FTK)

1 Introduction

Cloud Forensic is a product of Cloud Computing and digital forensics. Many government organizations are implementing cloud solutions, Digital Forensic and analyze huge collection of data, and converts them to meaningful information, NIST defines

© Springer Nature Singapore Pte Ltd. 2017
K. Deep et al. (eds.), *Proceedings of Sixth International Conference on Soft Computing for Problem Solving*, Advances in Intelligent Systems and Computing 547,
DOI 10.1007/978-981-10-3325-4_5

Digital Forensic as a branch of applied science which is used for identification, collection examination and analysis of data, cloud forensic can also be considered as a subset of network forensics, about 95% of entire information of world is generated and stored in digital format, but only one third of the evidence taken in the form of print outs. Cloud forensic also deals with registry history, file systems, process, and cash. For the past decades with an advent revolutions in cloud computing various methods has been revolutionized which are being used for the transmission, processing and storing of digital data in the cloud. Forensic is one of the applications of branch of science used in order to solve a problem legally; one can solve Computer Crime as a profitable thing which grows continuously. For the past recent years the big challenge which has risen in front of cloud is Cloud Forensics. Digital forensics or Computers forensics are the synonyms for Cloud forensic, Cloud forensics is a fusion of cloud computing and digital forensics. For the support in order to organize Cloud Forensic successfully these challenges should be co related to some of the characteristics of the cloud computing. Our everyday digital devices such as gaming consoles, cell phones and many more consist of a treasure of evidence, but none of them are given importance until they are being traced out during the crime scene. Some of the important issues in cloud forensics are

- Difficulties in Acquisition of data.
- Co operation from the cloud providers is must.
- Lagging of forensic attributes in cloud data.
- More complex chain of custody.
- Present available Tools are not up to the mark.

2 Digital Forensic Process

The process in which digital forensic is carried out was described in step by step by Ken Zatkyto in 2007 in Forensic Magazine, he traced out the eight phases which are involved in this, deploying digital forensic services which are cloud dependent system. They are

- **Authority:** It is the first step in any forensic process, evidence which is collected without a proper searching authority is likely to be changed.
- **Chain of Custody:** This is carried out to maintain the evidence integrity for a well documentation.
- **Imaging/Hashing Function:** A unique value is produced, by a mathematical process this is unique for any file, such as finger print of DNA, and they are then compared with the original evidence of the forensic image.
- **Validated Tools:** Tools may be either hardware or software related they should be tested before using them for testing their accuracy of the results.
- **Analysis:** Experienced examiners use the available tools along with their skills to trace out the artifacts in the forensics, sometimes these analysis depends on the investigation circumstances and the available facts which are being traced out. This analysis includes some of the following things

- a specific users account is linked with an activity
- Events are arranged according to the timeline.
- Tracking out whether an USB is being connected to a storage device or not.
- break the encryption
- Identifying the websites which have been visited.

- **Repeatability:** It is related to quality assurance it's a pain taking task to generate a correct and accurate report, it may also be considered as the hallmark of the forensic process. When all the steps are being repeated by a separate examiner with the same evidences, with same tools he should also get the same results, this shows the quality assurance to get the same report even when the process is carried out by another person.
- **Reporting:** For every concept in digital forensic reports has to be generated some of the reports are too long and some of them are too short so it depends on the client and the organization in which they want there report to get generated.
- **Presentation by and Expert:** Explaining a technological concept of forensic to a non technical person such as Judge is a challenge job, if it is not done properly then all the hard work which has been carried out to achieve the result would be wasted, a failure at this stage will result in high cost.

3   Categorization of Cloud Forensic Challenges

- *Architecture:* They deal with the variations in the architectures of cloud and their providers; they also include proliferations of systems, endpoints in the locations which are used to store the data, secure and accurate provenance for the preservation of chain of custody, capturing of the cloud resources with disturbing other tenants. Existing technologies does not highlight problems related to data segregations (Fig. 1).

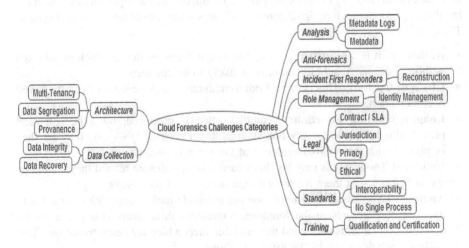

Fig. 1. Mind map of classification of categories and subcategories of cloud forensic challenges

- *Data collection:* They include the successful location of artifacts of forensics, dynamic and distributed systems, for the maintenances of integrity of data in a multi-tenant environment where the data is being shared between computers which are located in different locations for multiple parties, to access the data of one tenant so that no disturbance occurs with respect to the confidentiality of other tenants, recovering of deleted and damaged data in a distributed and shared virtual environment.
- *Analysis:* They deal with the forensic co relation artifacts within and around the cloud providers. They also deal with the events reconstructions and storage of virtual images.
- *Anti-forensics:* These are the set of techniques which are used for the prevention of forensic analysis; they also deal with data hiding, malware, and obfuscation.
- *Incident First responders:* They deal with the challenges related to competence, confidence and trustworthiness within the cloud providers; they also act as first responders and carry out the collection of data.
- *Role management:* They deal with identifying cloud account owner, showing the separation of physical users and user credential, tracing out the secrecy of creating variations in online identities, determining the exact data owners, they also helpful in tracing out the access control authentication.
- *Legal:* They include legal challenges in order to address and identify the issues which are related to jurisdictions of legal access of the data. They also deal with the issuing of subpoenas respect of the location of the data.
- *Standards:* Due to lack of standard operating procedures, neither standard tool exists which are being helpful in order to carry on test, validations and inter operability.
- *Training:* These challenges include misusing of digital forensic training materials, lack of experts in the area of cloud forensics in both the categories of instructors and investigators so trainers should be well qualified and well certified.

4 Why the Clouds Does not Exhibit Friendly Forensic Nature

As cloud are associated with many characteristics this makes a little trouble for cloud forensics, the data which the users store in the cloud are not local, a suspects computers cannot be impound even though the digital evidence is accessed by the agents of law enforcement, a cloud maintains several servers, separation of these servers from the data center is not possible because it may scratch the remaining data even during the cases where suspect has been identified. After a broad survey it has found that the following reasons are considerable as cloud forensic challenging.

- *Cloud Cartography:* In traditional computing the system is owned by users, where as cloud computing resembles a multitenant system, here a malicious user launches his VM beside regular VM of a cloud users and he steals away the secret information stored in regular users VM this is called as Cross VM side channel attacks.
- *Providing false information:* Investigators are completely dependent on CSP, during the collection of the evidence they have to depend on the employee of CSP

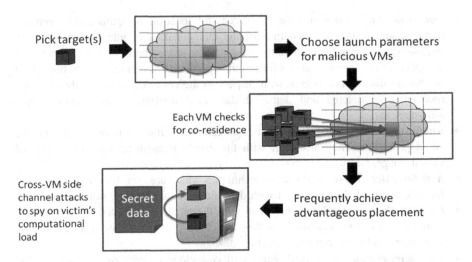

Fig. 2. Cloud Cartography

to get the logged information who are not a licensed investigator due to which there are chances that the information provided by the employee is a false information, even some time CSP may also give false information, even there are cases where if the employee of CSP, and even when CSP provides the information there is a chance that investigator may provide false information report these are the three major ways where provided information is false.

- *Control over the resources:* In case of traditional forensics the investigators have full control on the resources of the systems, where are sin cloud forensics as the resources are not located at same place so it creates some difficulty for the investigators to have control over the resources of the cloud.
- *Volatile nature of data:* As the data stored in VM is volatile in nature, i.e. data is lost when the power is off, a malicious user after completing his task can delete the data in VM due to which it sometimes becomes difficult to trace out exact data, due to this valuable data loss occurs and this makes impossible for the investigators to move forward.

5 Digital Evidences Are Stored in Storage Networks

There are two main technologies which are being used for the storage of Digital Evidence; they are Network Area Storage (NAS) and Storage Area Network (SAN). Digital Investigators in UK follow the guidelines given by "Association of Chief Police Officers (ACPO), Practice Guide for Electronic Evidence, Vacca defines that CF is composed of reconstruction activities which lead to determination and event answer to "what did they do".

- *Network Area Storage*: Here enormous quantities of data are stored in a centralized position, it was an extension of file server and made fashionable by Microsoft's Windows NT and Netware server. The negligence of the Comprehensive operating led a path for the creation of NAS storage devices; with the help of standard protocols these networks are being accessed.
- *Storage Area Network:* The above Fig. 2 shows SAN Architecture, here fiber switches are attached to RAID arrays of disk in the SAN, these are again connected to fiber channel cards to the computers presents in the network. as the computers and RAID arrays are connected to more fiber switches this makes the disk arrays and SAN for generating redundant paths in order to accessing the data during the failure of hardware.
- *Combining NAS and SAN Technologies:* The combined technologies give better performance then when considered individually. With the help of a server a Logical Unit Number (LUN) is shared by user workstations. In this combination only a part of SAN is allocated to the server which then provides services to the clients? Here data is transferred easily and quickly to many servers by reassigning LUNs. The efficiency of SAN and NAS solutions was verified by FBI's, (NT-RCFL).

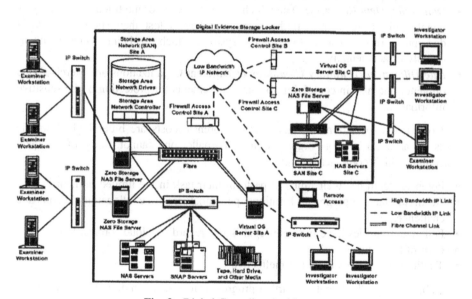

Fig. 3. Digital Custodian Architecture

- *Digital Evidence Custodian:* Architecture of DEC is dominant, it is designed to support investigations in electronic crimes, digital evidences which are being collected from multiple locations are streamlined and collaborative analysis is also made using DEC, the architecture is helpful for the physical and logical separations with respect to the forensic investigators. From the distribution of stored data in CC, the evidence can be collected (Fig. 3).

6 Uses of Cloud Forensics

Cloud Forensic has number of uses, Security as a service is emerged in cloud computing, there are advantages of anti-virus software which are cloud based, and forensic computing platforms. In general most of the computer systems are being owned by a trusting hosted company where the end users are leased these computer systems. During a Subpoena the CSP locates the correct computer which is being requested during the investigation.

- *Investigations:* Reconstruction of events in the cloud, to perform operations on suspect transactions, provision and acquisition of evidence in the court, in resource confiscation in collaborating law enforcement.
- *Troubleshooting:* It deals with security handling incidents in the cloud, monitoring and tracing events in order to access the current state of an event, to resolve operational and functional issued which are arising in the clouds, finding out the root cause of both isolated incidents and trends.
- *Log monitoring:* Collecting, analyzing and correlating of log entries are gathered across multiple systems which are being hosted in the cloud.
- *Data and system Recovery:* Data which is either deleted or modified accidentally or intentionally is recovered, if the encryption key is lost then encrypted data is decrypted., recovery and repair of the systems which are being damaged intentionally or accidentally.
- *Due Diligence/Regulatory Compliance:* Different Organizations are exercised due diligence and also in complying for the requirements for protection of the sensitive information.

Compared to traditional forensics cloud forensic costs are bitter high, Forensic Costs are afforded by large organizations but smaller one cannot. Some of the main constraints which come across are:

- Forensics team has their limits to perform their task in the limited domain.
- At dissimilar geographical locations personnel resource allocation becomes a challenge.
- Different countries have different legal issues regarding cloud forensics.
- Cost of forensics is high.
- Traditional forensics tools are ineffective

Table 1 shows the literature, further progress can be made in digital forensic by the development of standard DF systems, a huge amount of work has to be done further to standardize DF process which requires more attention by the researchers.

Table 1. Showing Cloud Forensic System Evaluation [✔ = satisfied, ✗ = not satisfied, - = nature to be detected]

References	Implements a standardized digital forensic process			Live cloud forensic investigation	Cloud based			Aid and Lead investigator through a standard digital forensic process	Aimed at investigating cloud environments	Digital Forensic Readiness
	Standardized processes	Has defined forensic process and procedures	Incorporates standard process in implementation	Retrieve data from live system	Hardware resources can scale on demand	Investigators run similar service versions at all time	Collaborative investigators have access to the same service sequence	System renders tasks to investigator based on their standard sequence	Incident scene capability which is designed with remote access	Proactive forensic approach
[1]	✗	✗	✗	✔	✔	✔	✔	✗	✔	✗
[2]	✗	✗	✗	✗	-	-	-	✔	✔	✔
[3]	✗	✗	✗	-	✔	✔	✔	✗	-	✗
[4]	✔	✔	✗	✗	✔	✗	✔	-	✔	✔
[5]	✗	✔	✗	✗	✔	✗	✔	✗	✗	✗
[6]	✗	✔	✔	✗	✔	✔	✔	-	✔	✗
[7]	-	✗	✗	✗	✔	✗	✔	✗	✗	✗
[8]	-	✔	-	-	✔	✔	✔	✗	✔	✗
[9]	-	-	-	✗	✔	✔	✔	✗	✗	✔
[10]	✗	✗	✗	✗	✔	✗	✔	✗	✗	✗
[11]	✗	✗	✗	✗	✔	✔	✔	✗	✔	✔
[12]	✗	✗	✔	✔	✔	✔	✔	✔	✗	✔
[13]	✗	✗	✗	✔	✔	✔	✗	✗	✔	✔
[14]	✗	✗	✗	✗	✗	✗	✔	✔	✗	✗
[15]	✗	✔	✗	✗	✗	✗	✗	✗	✗	✗
[16]	✗	✗	✗	✗	✗	✗	✗	✗	✗	✔
[17]	✗	✗	✗	✔	✔	-	-	✗	✗	✔
[18]	✗	✗	✗	✗	-	-	-	✗	✔	✔
[19]	✗	✗	✗	✔	-	-	-	✗	✔	-
[20]	✗	✗	✗	-	✗	✗	✗	✗	✗	✔

7 Conclusion

Volatility and fragility nature of evidences are one of the classical problems in forensics. The three main factors which are effecting digital Forensic investigations are time, evidence and volume. As cloud is having a flexible nature this becomes a major hurdle for the present forensic tool and methodology. The importance of Digital Evidence is now recognized widely, these evidences which are being collected should be authenticated, precise and admissible and accurate such that court also accepts them without any litigation. The frontiers of Cloud Forensics are pushed by Cloud Computing. The fragile nature of the digital evidences creates an importance for proper and careful handling them, though there exists many such models in carrying out digital forensics, but not all are supporting to the existing cloud computing architectures and also they do not satisfy the elastic nature of the cloud these thing creates the importance of a proper structure frame work this leads to rapid development of Cloud Forensic in the area of Cloud Computing, with the present trend as the developments are carried out in cloud computing in the same parallel way the cloud forensics should also be carried out, multi-tenancy challenges and multi-jurisdictional challenges are observed by professionals in traditional digital forensics, with the creation of any type of imbalance may make the culprit to escape and create a space to carry the irreparable damaging nature of works to the society, this imbalances should be overcome for an unexpectedly created error in Cloud Forensic its solutions should also be traced out maintaining high standards at the same time rather than waiting for another error creation for the same existing faults. Beside many challenges which are faced by cloud forensics at the same time for the advance forensic investigations opportunities are leveraged, such as Cost Effectiveness, Data Abundance, Flexibility and Scalability, Overall Robustness.

References

1. Ting, Y.H., et al.: Design and implementation of a cloud digital forensic laboratory (2013)
2. Zawoad, S., et al.: I have the proof: providing proofs of past data possession in cloud forensics. In: 2012 International Conference on Cyber Security, pp. 75–82. IEEE, December 2012
3. Yan, C.: Cybercrime forensic system in cloud computing. In: 2011 International Conference on Image Analysis and Signal Processing, pp. 612–615. IEEE, October 2011
4. Thorpe, S., Grandison, T., Campbell, A., Williams, J., Burrell, K., Ray, I.: Towards a forensic-based service oriented architecture framework for auditing of cloud logs. In: 2013 IEEE Ninth World Congress on Services, pp. 75–83. IEEE, June 2013
5. Lee, T., Lee, H., Rhee, K.-H., Shin, U.: The efficient implementation of distributed indexing with Hadoop for digital investigations on Big Data. Comput. Sci. Inf. Syst. 11(3), 1037–1054 (2014)
6. van Baar, R., van Beek, H., van Eijk, E.: Digital forensics as a service: a game changer. Digital Invest. 11, S54–S62 (2014)

7. Lee, J., Un, S.: Digital forensics as a service: a case study of forensic indexed search. In: 2012 International Conference on ICT Convergence (ICTC 2012), pp. 499–503. IEEE, October 2012

8. Zeng, G.: Research on digital forensics based on private cloud computing, vol. 2, no. 9, pp. 24–29 (2014)

9. Shende, J.R.G.: Cloud forensics (2014)

10. Wen, Y., Man, X., Le, K., Shi, W.: Forensics-as-a-Service (FaaS): computer forensic workflow management and processing using cloud, pp. 208–214 (2013)

11. Sang, T.: A log based approach to make digital forensics easier on cloud computing. In: 2013 Third International Conference on Intelligent System Design and Engineering Applications, pp. 91–94. IEEE, January 2013

12. Lee, J.Y., et al.: Remote forensics system based on network (2011)

13. Dykstra, J., Sherman, A.T.: Design and implementation of FROST: digital forensic tools for the Open Stack cloud computing platform. Digital Invest. **10**, S87–S95 (2013)

14. Deepak, N., et al.: Digital forensics service platform for internet videos, vol. 2, pp. 456–464 (2013)

15. Li, J., Wang, Q., et al.: Performance overhead among three hypervisors: an experimental study using hadoop benchmarks. In: 2013 IEEE International Congress on Big Data, pp. 9–16, June 2013

16. Reddy, K., et al.: The architecture of a digital forensic readiness management system. Comput. Secur. **32**, 73–89 (2013)

17. Shirkhedkar, D., et al.: Design of digital forensic technique for cloud computing, vol. 2, no. 6, pp. 192–194 (2014)

18. Reichert, Z., Richards, K., Yoshigoe, K.: Automated forensic data acquisition in the cloud. In: 2014 IEEE 11th International Conference on Mobile Ad Hoc and Sensor Systems, pp. 725–730. IEEE, October 2014

19. Belorkar, A., Geethakumari, G.: Regeneration of events using system snapshots for cloud forensic analysis. In: 2011 Annual IEEE India Conference, pp. 1–4. IEEE, December 2011

20. Shields, C., et al.: A system for the proactive, continuous, and efficient collection of digital forensic evidence. Digital Invest. **8**, S3–S13 (2011)

Optimal Path Determination in a Survivable Virtual Topology of an Optical Network Using Ant Colony Optimization

Kaushlya Kumari$^{(\boxtimes)}$, Satyasai Jagannath Nanda, and Ravi Kumar Maddila

Department of Electronics and Communication Engineering, Malaviya National Institute of Technology Jaipur, Jaipur 302017, Rajasthan, India
meelkaush@gmail.com, nanda.satyasai@gmail.com, rkmaddila.ece@mnit.ac.in

Abstract. In optical networks the terminals are connected by higher bandwidth fibers. For effective use of bandwidth the capacity of these optical fibers is divided into several channels. Any damage to an optical fiber will affect the data communication on the associated channels ultimately results in data loss between the terminals. Therefore a survivable virtual mapping is required over the physical topology which can still connect the terminals in case of a link failure. Here popular ant colony optimization is used to determine the optimal paths in a survivable virtual topology. Simulation studies on a five node and a ten node network reveal the determination of virtual optimal path between all the nodes considering link-cost and hop-count. The computational time required to establish communication between the nodes of these networks using the real physical link and proposed virtual link are reported. It is observed that though the virtual connection take either equal or more time than the physical link, it can able to establish the connection even in case of any physical link failure.

Keywords: Survivable virtual topology · Optical networks · Ant colony optimization

1 Introduction

In today's market, the most favorable technology is optical networking to fulfill the requirement of high bandwidth. So different algorithms are proposed so far to make faster and secure communication over a long distance. An evolutionary approach is used to design a survivable network in order to minimize the resource usage of the network in [1]. In optical domain the routing of lightpaths in a survivable topology for a given traffic matrix is reported in [2]. This problem is known as multilayer resilient layer-1 VPN design problem. This is a part of ILP formulation.

In [3] new ant based algorithm is reported under the capacity constraint for routing and wavelength assignment (RWA) problems in optical networks. Foraging task was used for route selection by the cooperation of ants. The connection

© Springer Nature Singapore Pte Ltd. 2017
K. Deep et al. (eds.), *Proceedings of Sixth International Conference on Soft Computing for Problem Solving*, Advances in Intelligent Systems and Computing 547,
DOI 10.1007/978-981-10-3325-4_6

route is obtained by the constant update of routing table. It was reported that the algorithm has better adaptability of traffic variations and performs for fixed as well as for dynamic routing algorithms in case of blocking probabilities also. A static RWA problem is solved in [4] by considering single objective optimization problem in optical networks using evolutionary algorithm. Dynamic RWA problem is described in [5] and tried to reduce the blocking probability of a WDM optical network.

Fault management techniques are proposed in [6] for single link failure in the IP-over-WDM network to make the virtual topology survivable. As normal communication happens in electrical domain, any failure will interrupt the whole network. So to resolve this problem a modified integer linear programing (ILP) formulation is reported in [7] by routing the lightpaths of given virtual topology over the physical topology to avoid any disturbance of data in any optical network. It results in lesser computational time and concludes it as a NP-hard problem. ILP works effectively on small sized network.

To solve the problem of ILP a new methodology is reported in [8] by using a local search and a Fastsurv algorithm which provides a feasible communication in case of single or multiple nodes or links failure for large networks too. A survivable mapping of virtual topology over a mesh of fibers in IP-over-WDM is obtained in [9] for double link failure case.

In this manuscript an ACO technique is used along with survivability constraint. Thus the proposed virtual link remains connected in the case of single or multiple link failure. Extensive simulation studies on 5 node and 10 node networks establish the claim. The rest portion of paper is organized as follows. Section 2 deals with the fundamental problem of designing a survivable mapped network. The design using proposed ACO is dealt in Sect. 3. The simulation studies using 5 node and 10 node optical networks are shown in Sect. 4. Analysis of the obtained results is also given in this section. Section 5 concludes the significant outcome of the experiment.

2   Basic Design of a Survivable Mapped Optical Network

The virtual topology becomes unsurvivable (Fig. 1(d)) if routing of the lightpath a is routed in such a way that node four will become isolated (means the node connected to a and e lightpaths will not find any way to connect to the remaining nodes).

The objective here is to design a survivable virtual topology, which do not disconnect even in case of a link failure. The physical topology consists of N nodes and E number of edges, where (l, m) lies in E if any optical connection is established between these two nodes l and m. The capacity of each link is fixed i.e. W (no. of wavelengths). The virtual topology of the given network consists of N_L nodes where $N_L \in \boldsymbol{N}$ and a set of wavelength links E_L, where $(j, k) \in \boldsymbol{E_L}$ if there is an virtual optical connection exists.

The survivable mapping of a virtual topology over a given physical network by ILP formulation is proposed in [3,4]. First objective of this formulation is to

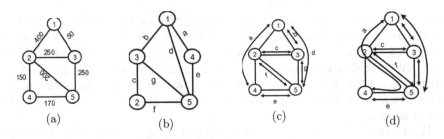

Fig. 1. Representation of optical network: (a) physical topology (b) virtual topology (c) survivable mapping (d) unsurvivable mapping

achieve a minimum number of physical links in the whole physical network and another objective is to achieve the least possible number of lightpaths used in the virtual topology. The illustration of this information of the network (which is used in this paper) is 7 physical links and 9 lightpaths.

3 Proposed Optimal Path Determination of a Survivable Virtual Topology Using ACO

The optimum routing of lightpaths are considered as a virtual topology mapping problem. In which initially n number of possible shortest paths are evaluated. The amount of pheromone deposited by an ant is

$$\delta_{lm}^k = \begin{cases} \frac{1}{D_k} & \text{if edge(l,m)} \in T_k \\ 0 & \text{otherwise} \end{cases} \tag{1}$$

where D_k is the cost function and T_k built by the k-th ant.

The proposed design consists of two main stages:

Stage 1. Construction of graph:

(a) By considering capacity and survivability constraints.
(b) Pheromone trail and heuristic information i.e. lightpath pheromone trails and the shortest path pheromone trail.

Stage 2. Determining the solution: The step associated are as follows:

1. Choose any random start node and shortest path.
2. Update the tabu list of unvisited nodes.
3. Update the lightpath pheromone information and probability below to select the next node

$$\tau_{ij} \leftarrow (1 - \rho)\tau_{ij} \tag{2}$$

where τ_{ij}-pheromone level and ρ is evaporation constant.

$$\tau_{ij} \leftarrow \tau_{ij} + \sum_{k=1}^{m} \delta \tau_{ij}^{k} \qquad (3)$$

$$P_{ij}^{k} = \begin{cases} \dfrac{\tau_{ij}^{\alpha} \eta_{ij}^{\beta}}{\sum_{i \in N_{i}^{k}} \tau_{ij}^{\alpha} \eta_{ij}^{\beta}} & \text{if } j \in N_{i}^{k} \in T_{k} \\ 0 & \text{otherwise} \end{cases} \qquad (4)$$

where η_{ij} is heuristic information between components i and j, α, β are ACO parameters.

4. Repeat step 2–3 till the tabu list becomes empty.

The main objective of the proposed algorithm is to achieve minimal cost of resources used in the optical network. This can be calculated in two ways (1) using link-cost, and (2) using hop-count by taking the survivability and the capacity constraints into account. The survivability of any solution is tested in such a way that one link is deleted from the given optical network at a time and if all the nodes still able to communicate to the remaining network, then the solution is considered as survivable.

4   Results and Discussions

Here the survivable virtual topology is designed for a five node and ten node network using the proposed ACO algorithm. The design of both networks is outlined below:

Case 1: Optimal path determination in a five node survivable optical network

The design topology of a five node survivable optical network is given in Fig. 1. Considering the shortest distance of all the nodes in Fig. 1(a), the link cost of the physical topology is expressed in matrix form as follows:

$$M = \begin{bmatrix} 0 & 300 & 50 & 450 & 300 \\ 300 & 0 & 250 & 150 & 200 \\ 50 & 250 & 0 & 400 & 250 \\ 450 & 150 & 400 & 0 & 170 \\ 300 & 200 & 250 & 170 & 0 \end{bmatrix}$$

Using M the possible paths between all the nodes of Fig. 1(a) along with corresponding distance are presented in Table 1. The shortest path and it's corresponding minimal distance is highlighted in bold letters.

The results of optimal path and distance obtained using proposed ant colony optimization are presented in Table 2. In this table the comparative analysis is also carried out with the optimal result obtained from Table 1. From Table 2 it is observed that though the results of column 1 are optimal they do not cover all the nodes. These paths can be used in link failure condition. In all other

Table 1. All possible paths between all the nodes of the five node network in Fig. 1(a) along with corresponding distance

S-D pair	p1 and dist	p2 and dist	p3 and dist	p4 and dist	p5 and dist	p6 and dist	p7 and dist
1-2	1-3-5-4-2 620	1-3-5-2 500	**1-3-2 300**	1-2 400	-	-	-
1-3	1-2-4-5-3 870	1-2-5-3 750	1-2-3 550	**1-3 50**	-	-	-
1-4	1-3-5-2-4 650	1-3-2-5-4 670	1-2-3-5-4 970	**1-3-2-4 450**	1-2-5-4 670	**1-2-4 450**	1-3-5-4 470
1-5	1-3-2-4-5 620	1-3-2-5 500	1-2-3-5 800	1-2-4-5 620	1-2-5 500	**1-3-5 300**	-
2-1	2-4-5-3-1 620	2-5-3-1 500	**2-3-1 300**	**2-1 300**	-	-	-
2-3	2-1-3 350	2-4-5-3 570	2-5-3 450	**2-3 250**	-	-	-
2-4	2-1-3-5-4 770	2-3-5-4 670	2-5-4 370	**2-4 150**	-	-	-
2-5	2-1-3-5 600	2-3-5 500	2-4-5 320	**2-5 200**	-	-	-
3-1	3-5-4-2-1 870	3-5-2-1 750	3-2-1 550	**3-1 50**	-	-	-
3-2	3-1-2 350	3-5-4-2 570	3-5-2 450	**3-2 250**	-	-	-
3-4	3-1-2-5-4 720	3-1-2-4 500	3-5-2-4 600	3-2-5-4 620	**3-2-4 400**	3-5-4 420	-
3-5	3-1-2-4-5 670	3-1-2-5 550	3-2-4-5 570	3-2-5 450	**3-5 250**	-	-
4-1	4-5-3-2-1 970	4-5-2-3-1 670	4-2-5-3-1 650	**4-2-3-1 450**	4-5-2-1 670	4-2-1 450	4-5-3-1 470
4-2	4-5-3-1-2 770	4-5-3-2 670	4-5-2 370	**4-2 150**	-	-	-
4-3	4-5-2-1-3 720	4-2-1-3 500	4-5-2-3 620	4-2-5-3 600	**4-2-3 400**	4-5-3 420	-
4-5	4-2-1-3-5 750	4-2-3-5 650	4-2-5 350	**4-5 170**	-	-	-
5-1	5-4-2-3-1 620	5-3-2-1 800	5-2-3-1 500	5-4-2-1 620	5-2-1 500	**5-3-1 300**	-
5-2	5-3-1-2 600	5-3-2 500	5-4-2 320	**5-2 200**	-	-	-
5-3	5-4-2-1-3 670	5-2-1-3 550	5-4-2-3 570	5-2-3 450	**5-3 250**	-	-
5-4	5-3-1-2-4 750	5-3-2-4 650	5-2-4 350	**5-4 170**	-	-	-

cases the optimal results with ant colony are to be used as they cover the entire network. Among all the optimal paths derived between all the nodes by ACO the 1-3-5-4-2 and 1-3-2-4-5 both are most appropriate for communication having lowest link cost of 620.

The possible routs for each light path of virtual topology (shown in Fig. 1(b)) based on link-cost and hop count of network is given in Table 3. The link establishment time required for physical topology and survivable virtual topology are computed based on link cost and speed of light in fiber (fiber is made up of silica with refractive index 1.4) are presented in Table 4. It is observed that the link establishment time for virtual topology is equal or greater than that of the physical one. Even though the virtual topology takes slightly more time it is helpful to establish the link in case of physical link failure.

Case 2. Optimal path determination in a ten node survivable optical network:

The design topology of a ten node survivable optical network is given in Fig. 2. Considering the shortest distance of all the nodes in Fig. 2(a), the link cost of the physical topology is expressed in a matrix form as follows:

$$M = \begin{bmatrix} 0 & 200 & 230 & 270 & 120 & 420 & 370 & 290 & 140 & 70 \\ 200 & 0 & 50 & 150 & 200 & 500 & 450 & 370 & 220 & 150 \\ 230 & 50 & 0 & 100 & 210 & 510 & 460 & 380 & 230 & 160 \\ 270 & 150 & 100 & 0 & 150 & 450 & 400 & 420 & 270 & 200 \\ 120 & 200 & 210 & 150 & 0 & 300 & 250 & 270 & 120 & 50 \\ 420 & 500 & 510 & 450 & 300 & 0 & 450 & 550 & 420 & 350 \\ 370 & 450 & 460 & 400 & 250 & 450 & 0 & 100 & 250 & 300 \\ 290 & 370 & 380 & 420 & 270 & 550 & 100 & 0 & 150 & 220 \\ 140 & 220 & 230 & 270 & 120 & 420 & 250 & 150 & 0 & 70 \\ 70 & 150 & 160 & 200 & 50 & 350 & 300 & 220 & 70 & 0 \end{bmatrix}$$

Using M the possible paths between node 'one' to all the nodes of Fig. 2(a) along with corresponding distance are presented in Tables 5 and 6. The possible optimal paths between the rest nodes are calculated in similar manner. But due to lack of space only 'node one' results are presented. The best results from node 1 to any node are highlighted in dark letters in Tables 5 and 6.

Table 2. Comparative results of optimal path and distance from Table 1 and achieved with ant colony optimization (covering all the nodes)

S-D pair	Optimal path from Table 1	Optimal path using ant colony
1-2	**1-3-2 300**	**1-3-5-4-2 620**
1-3	**1-3 50**	1-2-4-5-3 870
1-4	**1-2-4 450**	1-3-5-2-4 650
1-5	**1-3-5 300**	**1-3-2-4-5 620**
2-1	**2-3-1 300**	2-4-5-3-1 620
2-3	**2-3 250**	2-4-5-1-3 670
2-4	**2-4 150**	2-1-3-5-4 770
2-5	**2-5 200**	2-4-3-1-5 900
3-1	**3-1 50**	3-5-4-2-1 870
3-2	**3-2 250**	3-1-5-4-2 670
3-4	**3-2-4 400**	3-1-2-5-4 720
3-5	**3-5 250**	3-1-2-4-5 670
4-1	**4-2-3-1 450**	4-2-5-3-1 650
4-2	**4-2 150**	4-5-3-1-2 770
4-3	**4-2-3 400**	4-5-2-1-3 720
4-5	**4-5 170**	4-2-1-3-5 750
5-1	**5-3-1 300**	5-4-2-3-1 620
5-2	**5-2 200**	5-4-3-1-2 920
5-3	**5-3 250**	5-4-2-1-3 670
5-4	**5-4 170**	5-3-1-2-4 750

Table 3. First three shortest paths calculated based on link-cost and hop count for a five node network.

Lightpaths	Paths based on link-cost			Paths based on hop-count		
lightpath	sp1	sp2	sp3	sp1	sp2	sp3
1-3 (b)	1-3	1-2-3	1-2-5-3	1-3	1-2-3	1-2-5-3
3-5 (g)	3-5	3-2-5	3-2-4-5	3-5	3-2-5	3-2-4-5
4-5 (e)	4-5	4-2-5	4-2-3-5	4-5	4-2-5	4-2-3-5
2-5 (m)	2-5	2-4-5	2-3-5	2-5	2-3-5	2-4-5
2-3 (c)	2-3	2-5-3	2-1-3	2-3	2-1-3	2-5-3
1-5 (d)	1-3-5	1-3-2-5	1-3-2-4-5	1-3-5	1-2-5	1-2-3-5
1-4 (a)	1-3-2-4	1-2-4	1-3-2-5-4	1-2-4	1-3-2-4	1-3-5-4

Table 4. Link establishment time for physical and survivable virtual topology five node network (refractive index of silica fiber is 1.4 at 1550 nm)

S-D pair of physical topology	Time of PT (ms)	Wavelength link of virtual topology	Time of VT (ms)
1-2	0.1904	1-3-5-2	0.238
1-3	0.238	1-3	0.238
2-3	0.119	2-3	0.119
2-4	0.0714	2-5-4	0.176
2-5	0.0952	2-5	0.0952
4-5	0.0808	4-5	0.0808
3-5	0.119	3-5	0.119

The best optimal link cost computed using ACO algorithm for the 10 node network is shown in Fig. 3. It is observed that the algorithm achieved convergence on 12 iterations with optimal link cost covering all the nodes of the network as 1740 and the optimal path found to be '9-10-5-4-3-1-2-8-7-6'.

The possible routes for each lightpath of a virtual topology (shown in Fig. 2(b)) using link cost are presented in Table 7. The optimal routes along with link cost are highlighted in bold letters. In Fig. 2(c) if the routing of light-path c (through 6-7-8) is changed to 6-7-5-10-9-8 (shown in Fig. 2(d)) it leads to un-survivability of the network.

The comparison of link establishment time of physical and virtual topology of ten node network is represented in Table 8. As observed in 5 node network here also it is observed that the time taken by the virtual topology is greater or equal to that of the physical network. Though virtual takes more link establishment time but provides survivable solution.

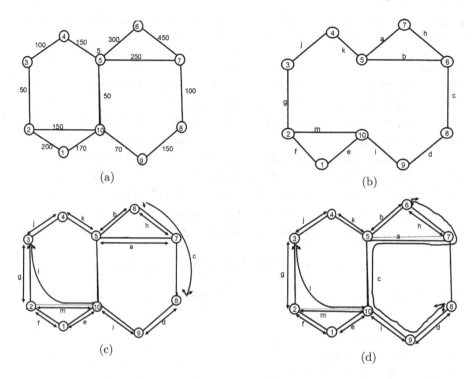

Fig. 2. Representation of 10 node optical network: (a) physical topology (b) virtual topology (c) survivable mapping (d) unsurvivable mapping

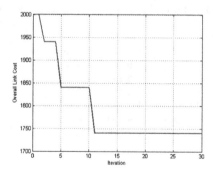

Fig. 3. Result of optimal link cost of entire network vs. number of iteration for ten node network using 'ACO'.

Table 5. All possible paths along with corresponding distance between node one to all rest nodes (sp1-sp5) of the network shown in Fig. 2(a)

S-D pair	sp1	sp2	sp3	sp4
1-2	**1-2 200**	1-10-2 320	1-10-5-4-3-2 520	1-10-9-8-7-5-4-3-2 1040
1-3	1-2-10-5-4-3 650	1-2-10-3 510	**1-2-3 250**	-
1-4	**1-2-3-4 350**	1-10-5-4 370	1-10-2-3-4 470	1-10-9-8-7-4 890
1-5	**1-10-5 220**	1-2-10-5 570	1-2-3-4-5 500	1-10-9-8-7-5 740
1-6	**1-10-5-6 520**	1-2-10-5-6 700	1-2-3-4-5-6 800	1-10-9-8-7-6 940
1-7	**1-10-5-7 470**	1-10-5-6-7 870	1-10-9-8-7 490	1-2-10-9-8-7 670
1-8	**1-10-9-8 390**	1-2-10-9-8 570	1-10-5-7-8 570	1-10-5-6-7-8 1070
1-9	**1-10-9 240**	1-2-10-9 420	1-2-3-4-5-10-9-620	1-10-5-7-8-9 720
1-10	1-2-3-4-5-10 550	1-2-3-10 410	**1-2-10 350**	-

Table 6. Rest part of Table 5 representing sp5-sp8.

Lightpath	sp5	sp6	sp7	sp8
1-2	-	-	-	-
1-3	-	-	-	-
1-4	1-10-9-8-7-6-5-4 1390	-	-	-
1-5	1-10-9-8-7-6-5 1240	-	-	-
1-6	1-10-9-8-7-5-6 1040	-	-	-
1-7	1-2-10-5-7 650	1-2-10-5-6-7 1150	1-2-3-4-5-7 750	1-2-3-4-5-6-7 1250
1-8	1-2-3-4-5-7-8 850	1-2-3-4-5-6-7-8 1350	-	-
1-9	1-10-5-6-7-8-9 1220	1-2-3-4-5-7-8-9 1000	1-2-3-4-5-6-7-9 1500	-
1-10	-	-	-	-

Table 7. All possible routes for each lightpath of a virtual topology shown in Fig. 2(b) using link cost

Lightpath	p1 and dist	p2 and dist	p3 and dist	p4 and dist	p5 and dist	p6 and dist
1-2 (f)	**1-2 200**	1-10-2 320	1-10-5-4-3-2 520	-	-	-
1-10 (e)	1-2-3-4-5-10 550	1-2-3-10 410	**1-2-10 350**	-	-	-
2-3 (g)	2-10-5-4-3 450	2-10-3 310	**2-3 50**	-	-	-
2-10 (m)	2-3-4-5-10 350	2-3-10 210	**2-10 150**	-	-	-
3-4 (j)	3-2-10-5-4 400	3-10-5-4 360	**3-4 100**	-	-	-
3-10 (l)	3-2-10 200	3-4-5-10 300	**3-10 160**	-	-	-
4-5 (k)	4-3-2-10-5 350	4-3-10-5 310	**4-5 150**	-	-	-
5-6 (b)	5-7-6 700	**5-6 300**	-	-	-	-
5-7 (a)	**5-7 250**	5-6-7 750	-	-	-	-
6-7 (h)	**6-7 450**	6-5-7 550	-	-	-	-
6-8(c)	6-7-5-4-3-2-10-9-8 1370	6-7-5-4-3-2-10-9-8 970	6-7-5-4-3-10-9-8 1330	6-5-4-3-10-9-8 930	6-7-5-10-9-8 970	**6-5-10-9-8 570**
8-9 (d)	8-9	**8-9 150**	8-7-5-10-9 470	8-7-6-5-10-9 970	-	-
9-10 (i)	9-10	**9-10 70**	9-8-7-5-10 550	9-8-7-6-5-10 1050	9-8-7-5-4-3-2-10 1020	-

Table 8. Link establishment time for physical and survivable virtual topology for ten node network (refractive index of fiber is 1.4 at 1550 nm)

S-D pair of physical topology	Time of PT (ms)	Wavelength link of virtual topology	Time of VT (ms)
1-2	0.0952	1-2	0.0952
1-10	0.0808	1-10	0.0808
2-3	0.0238	2-3	0.0238
3-4	0.0476	3-4	0.0476
4-5	0.0714	4-5	0.0714
5-6	0.1428	5-6	0.1428
5-7	0.119	5-7	0.119
6-7	2.142	6-7	2.142
7-8	0.0476	7-5-6-8	5.238
8-9	0.0714	8-9	0.0714
9-10	0.0332	9-10	0.0332
5-10	0.238	5-4-3-2-10	2.1428

5 Conclusion

In this manuscript a survivable virtual topology is designed which can able to connect the nodes of an optical network in case of physical link failure. The ant colony optimization is used to determine the optimal route in the network. Simulation studies on 5 node and 10 node networks reveals that though the virtual path take little more time for link establishment but it is effective in case of physical damage of fiber. The optimal paths obtained with shortest lightpath and optimal path obtained by covering all the nodes of the networks with ACO are compared. For the given example of five node the optimal paths covering all the nodes found with ACO are 1-3-5-4-2 and 1-3-2-4-5 with link cost 620. For the given example of 10 node network with ACO the optimal path is 9-10-5-4-3-1-2-8-7-6 with lowest link cost 1740.

References

1. Corut Ergin, F., Yayımlı, A., Uyar, Ş.: An evolutionary algorithm for survivable virtual topology mapping in optical WDM networks. In: Giacobini, M., Brabazon, A., Cagnoni, S., Caro, G.A., Ekárt, A., Esparcia-Alcázar, A.I., Farooq, M., Fink, A., Machado, P. (eds.) EvoWorkshops 2009. LNCS, vol. 5484, pp. 31–40. Springer, Heidelberg (2009). doi:10.1007/978-3-642-01129-0_4
2. Cavdar, C., Yayimli, A.G., Mukherjee, B.: Multilayer resilient design for layer-1 VPNs. In: Optical Fiber Communication Conference and Exposition (OFC), California, USA (2008)
3. Ngo, S., Jiang, X., Horiguchi, S.: An ant-based approach for dynamic RWA in optical WDM networks. Photon. Netw. Commun. **11**, 39–48 (2006)

4. Banerjee, N., Sharan, S.: An EA for solving the single objective static RWA problem in WDM networks. In: 2nd International Conference on Intelligent Sensing and Information Processing (2004)
5. Garlick, R.M., Barr, R.S.: Dynamic wavelength routing in WDM networks via ant colony optimization. In: Dorigo, M., Caro, G., Sampels, M. (eds.) ANTS 2002. LNCS, vol. 2463, pp. 250–255. Springer, Heidelberg (2002). doi:10.1007/3-540-45724-0_23
6. Sahasrabuddhe, L., Ramamurthy, S., Mukherjee, B.: Fault management in IP-over-WDM networks: WDM protection versus IP restoration. IEEE J. Sel. Areas Commun. **20**, 21–33 (2002)
7. Modiano, E., Narula-Tam, A.: Survivable lightpath routing: a new approach to the design of WDM-based networks. IEEE J. Sel. Areas Commun. **20**, 800–809 (2002)
8. Ducatelle, F., Gambardella, L.: A scalable algorithm for survivable routing in IP-over-WDM networks. In: Proceedings of the First International Conference on Broadband Networks, BroadNets, pp. 54–63 (2004)
9. Kurant, M., Thiran, P.: Survivable mapping algorithm by ring trimming (SMART) for large IP-over-WDM networks. In: BroadNets (2004)

Major Issues in Spectral Clustering Algorithm to Improve the Quality of Output Clusters

S.V. Suryanarayana[1(✉)], G. Venkateswara Rao[2], and G. Veereswara Swamy[3]

[1] Department of CSE, GITAM University, Visakhapatnam, AP, India
suryahcu@gmail.com
[2] Department of I.T, GITAM University, Visakhapatnam, AP, India
vrgurrala@yahoo.com
[3] Department of Physics, GITAM University, Visakhapatnam, AP, India
Drgv_swamy@yahoo.co.in

Abstract. Spectral clustering is playing a vital role in day-to-day technology in different forms since recent past. Spectral Clustering techniques are developed by using concepts of graph, algebra theories and conventional clustering methods. The theories and experiments conducted in various areas like, pattern recognition and image segmentation prove that it is having global reputation. This paper is to provide a framework for spectral clustering in different phases and its advantages when compared with traditional data clustering algorithms. In lieu of this paper, it is upgraded by intervening the inputs such as to making the bounds of clustering i.e., providing the issues through its relevance factors-produced by spectral clustering. On the other hand, some fundamental issues related to improving the quality of the output clusters are also covered. Constructing a sound similarity matrix is the most direct and efficient way to improve the quality of clusters. For building the similarity matrix sometimes we may choose the distance measures for measuring the similarity. This paper provides the issues related to improving the quality of the clusters produced by spectral clustering. We implemented spectral clustering in WEKA tool and produced the results for different similarity measures and observed how the quality of the clusters improved. For practical purpose we have used the datasets available in UCI machine library.

Keywords: Spectral clustering · Eigen values · Eigen vectors · Similarity matrix · Laplacian matrix · Scaling factor

1 Introduction

Data clustering is one of the main concepts in data mining, it is used to discover and extract patterns or rules hidden behind data. This concept is mainly based on statistical classification theories and its object is to divide datasets into clusters according to the principle that data points in the cluster should be similar while those in different clusters should be dissimilar. Since 1990s, this technology is already widely applied in business and scientific areas such as marketing, retailing and bioengineering, and gains great achievements. The algorithms based on classification methods rely on greedy algorithms [1]. Meanwhile,

© Springer Nature Singapore Pte Ltd. 2017
K. Deep et al. (eds.), *Proceedings of Sixth International Conference on Soft Computing for Problem Solving*, Advances in Intelligent Systems and Computing 547,
DOI 10.1007/978-981-10-3325-4_7

spectral clustering is superior in dealing with graph partition problems as the main idea of spectral clustering is to convert data clustering problems into graphic partitioning ones [2] and then to solve this problem by means of matrix theory. Nowadays, spectral clustering emerges as a matured and flexible data clustering method, through a series of discovery, experiments, refine and extension. It also gains solid theoretical supports after involving graph, algebra theories, and traditional clustering methods. Therefore, spectral clustering can generate varied solutions to clustering problems in various areas. Spectral clustering, normally, draws a similarity illustration from datasets when dealing with data clustering problems. Admittedly, there are a number of partition approaches which can be utilized in spectral clustering algorithms such as min-cut, ratio-cut and normalized-cut [13, 17]. In fact, the computational complexity is huge when applying them on similarity graph directly. Moreover, these partition approaches always fail to find a global and optimal solution as these algorithms lead to NP-hard problems which cannot be solved within polynomial time. Thus, spectral clustering needs to do a relaxation by converting the output clusters from discrete ones to continuous ones.

This paper is organized as follows: Working of the spectral clustering algorithm is introduced in next section. In Sect. 3, we briefly outline the framework for spectral clustering [16] in different phases. Section 4 formulates the spectral clustering problem to improve the quality of clusters. Experimental results with two well-known data sets, namely, *iris* and *birthday* databases are reported in Sect. 5. Finally, Sect. 6 presents some concluding remarks and suggestions for further work.

2   Spectral Clustering

Spectral clustering is a predominant branch of clustering. Tracing back to 1973, Donath and Hoffman [4] firstly proposed to cast graph partition problems into calculating Eigen structure of its corresponding adjacency matrix. In the same year, Fiedler [4] found a relationship between the second-smallest Eigen value (zero is the smallest Eigen value) of the Laplacian matrix and bi-partitions of that graph [6]. Nowadays, spectral clustering emerges as a matured and flexible data clustering method, through a series of discovery, experiments, refine and extension [5]. The Laplacian matrix is the method used to conduct this relaxation. In fact, a Laplacian matrix [7] refers to a type of matrix with a close relationship with graph partition approaches. According to the research [8], the Laplacian matrix normally has m non-negative real-valued eigen values. The structure of them is $0 = \lambda_1 < \lambda_2 < \ldots < \lambda_m$. It is a proper way to choose first $k(k \leq m)$ smallest eigenvectors as clusters.

2.1   Spectral Clustering Algorithm

Now we would like to state the most common spectral clustering algorithm. We assume that our data consists of n points $x_1, \ldots, x_n$ which can be arbitrary objects. We measure their pairwise similarities $s_{ij} = s(x_i, x_j)$ by some similarity function which is symmetric and non-negative, and we denote the corresponding similarity matrix by $S = (s_{ij})_{i,j=1,\ldots,n}$.

Input: Similarity matrix S belongs R^{n*n}, number k of clusters to construct.

1. Construct a similarity matrix using any one of the proximity measures and let W be its weighted adjacency matrix.
2. Compute the Unnormalized Laplacian L.
3. Compute the first k eigenvectors $u_1, \ldots, u_k$ of L.
4. Let U belongs R^{n*k} be the matrix containing the vectors $u_1, \ldots, u_k$ as columns.
5. For $i = 1, \ldots, n$, let y_i belongs R^k be the vector corresponding to the i-th row of U.
6. Cluster the points $(y_i)_{i=1,\ldots,n}$ in R^k with the k-means algorithm into clusters $C_1, \ldots, C_k$.

Output: Clusters $A_1, \ldots, A_k$ with $A_i = \{j \mid y_j \in C_i\}$.

Currently, the majority of spectral clustering algorithms fall into two categories: recursive spectral and multi-way spectral. In a recursive way, similarity graphs are separated into two subgraphs and this separating process is repeated on subgraphs until meeting your final conditions. On the other hand, the multi-way spectral, unlike the recursive spectral, conducts a separation on similarity graph directly into the specified number of clusters. Finally, spectral clustering has several notable benefits. In the first place, spectral clustering lessens the requirement to datasets and it can deal with complex, high-dimensional and non-convex data samples [3]. Meanwhile, it has a strong ability to process large datasets because the computational complexity of spectral clustering mainly concentrates on matrix calculation. Thus, spectral clustering can be deployed on distributed systems as to achieve parallel calculation [9]. Moreover, it improves the quality of clustering as it achieves globally optimal solution. In addition, spectral clustering is more like a clustering framework rather than a single and fixed algorithm. After combining with varied theories and algorithms, spectral clustering has a more broad application prospects.

3   Similarity Matrix Based on Distance Measures

Distance measures are flexible: Resemblance can be measured either as a distance (dissimilarity) or a similarity. Most distance measures can readily be converted into similarities and vice-versa. All of the distance measures described below can be applied to either binary (presence-absence) or quantitative data. There are many distance measures. A selection of the most commonly used and most effective measures are described below. It is important to know the domain of acceptable data values for each distance measure. Many distance measures are not compatible with negative numbers. Other distance measures assume that the data are proportions ranging between zero and one, inclusive.

Distance measures can be categorized as metric, semi-metric. or non-metric. A metric distance measure must satisfy the following rules:

1. The minimum value is zero when two items are identical.
2. When two items differ, the distance is positive (negative distances are not allowed).

3. Symmetry: The distance from objects A to object B is the same as the distance from B to A.
4. Triangle inequality axiom: With three objects the distance between two of these objects cannot be larger than the sum of the two other distances. Seim-metrics are extremely useful in community ecology but obey a non-Euclidean geometry Nonmetric violate one or more of the other rules and are seldom used in ecology.

Distance is another way to calculate the similarity based on nodes and this occasion, the attributes of vertices acquire a higher weight. The most common approach to calculating distances, in this context, is Euclidean distance, also called squared distance. It is normally used to calculate the similarity of quantitative valuable because data in this type is continuous real-valued-numbers. Certainly, expect Euclidean distance, there are some other distance functions [9] like Manhattan distance (absolute distance), Minkowski distance and Gaussian distance. In most cases, different attributes contribute differently to the distance. Here a_k is the adjusted value and the sum of these adjusted values is 1.

$$d(x_i, x_j) = \sum_{k=1}^{n} a_k d(x_{i-k}, y_{j-k}) \, (i = 1, 2, \dots, n, k = 1, 2, \dots, m) \text{ and } \sum_{k=1}^{n} a_k = 1 \tag{1}$$

The rationale for our choice is primarily empirical: we should select measures that have shown superior performance, based on the other criteria listed here One important theoretical difference between Euclidean and city-block distance is, however, apparent. Long Euclidean distances in species space are measured through an uninhabitable portion of species space. in other words, the straight-line segments tend to pass through areas of impossibly species-rich and overly full communities. In contrast, city-block distances are measured along the edges of species space exactly where sample units lie in the dust bunny distribution.

Minkowski Distance is a generalization of Euclidean Distance

$$dist = (\sum_{k=1}^{n} |pk - qk|^r)^{\frac{1}{r}} \tag{2}$$

where r is a parameter [11], n is the number of dimensions (attributes) and p_k and q_k are, respectively, the kth attributes (components) or data objects p and q. If $r = 1$, the distance measure is named as City block (Manhattan, taxicab, L_1 norm) distance. A common example of this is the Hamming distance, which is just the number of bits that are different between two binary vectors. If $r = 2$, the distance measure is named as Euclidean distance. Similarly, if $r \to \infty$, the distance measure is named as "supremum" (L_{max} norm, L_∞ norm) distance. This is the maximum difference between any components of the vectors. Do not confuse r with n, i.e., all these distances are defined for all numbers of dimensions. Similarities can be measured by the concept of distance. The most common similarity graphs are neighborhood graph and connected graph.

However, when calculating similarity matrices, it leads to several problems when using neighbor graphs or fully connected graph. Firstly, both ε-neighborhood and K-nearest neighbor fail to construct a similarity graph, if the points in a dataset cannot distribute evenly. Secondly, input parameter ε and K are always hard to choose. If they are too small, the neighbor graph will only be connected in the local area while the neighbor graph is overly connected if they are too large.

4 Improving the Quality of Output Clusters

After reviewing the spectral clustering, it is clear that the quality of clusters is closely linked to two input parameters. The first one is similarity matrix, which is utilized to represent the whole dataset and it is also the starting point of spectral clustering. The Distance or Similarity Measure plays a key role in clustering. All clustering algorithms group objects that are "close" together and it is important to measure closeness in an appropriate way. Appropriate distance measures lead to more interesting results. A proper similarity matrix is a precondition of finding "correct" clusters. Although Sect. 3 provides some ways to build similarity matrices, it still highly connects to business background and experiences. Thus, constructing a sound similarity matrix is the most direct and efficient way to improve the quality of clusters. Meanwhile, the other input parameter, number of clusters, also has an influence on it. The similarity function is given by

$$s(x_i, x_j) = \exp(-||x_i - x_j||^2 / (2\sigma^2))$$ (3)

where x_i and x_j are two points and σ is a parameter [6] that adjusts the similarity. However, there is a drawback associated with the Gaussian function [8]. This function is very sensitive to the choice of σ. The significance of σ is that it controls the portion of area that serves as the neighborhood of the points in a dataset. When σ is small, the data needs to be closer to zero in order to have a higher similarity value. On the other hand, a very large σ results in all of the similarity values being at or very near to one.

More importantly, unlike K-means, this idea is suitable to spectral clustering. K-means is a typical example of this occasion as it is sensitive to the starting point but the starting point changes after each time sampling [14]. Intuitively, getting more supports from hardware level is also another way to achieve this purpose. For instance, spectral clustering can achieve parallel calculation through deploying on distributed systems [9]. Furthermore, data clustering can gain parallel calculation in one computer through CPU and GPU [15].

5 Experimental Results

The proposed algorithm has been applied to well-known data sets available in UCI machine learning library. We have implemented this Spectral Clustering algorithm in WEKA tool. The sample results have been reported in Figs. 1 and 2. For improving the quality of the output clusters, similarity attribute performs a major role. In Spectral Clustering algorithm the first and foremost step is to build the similarity matrix from the given data set which is the key step for producing the good clusters. In building the similarity matrix, all the researchers are using Euclidean distance measure as a similarity

64   S.V. Suryanarayana et al.

measure. But, all the attributes of the available data sets are not in a suitable format to support the Euclidean distance measure. All data has a specific best geometry representation, which is application dependent, allowing for an accurate calculation of distances between points of the data set. So, we applied Manhattan distance parameter as an affinity object in building the affinity matrix. With this Manhattan distance measure, we are getting the desired number of output clusters.

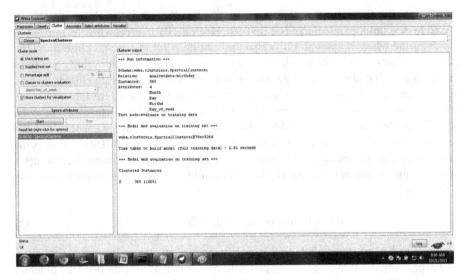

Fig. 1. Spectral Clustering result with Euclidean distance measure for birthday data set

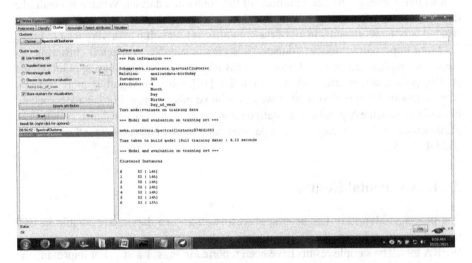

Fig. 2. Spectral Clustering result with Manhattan distance measure for birthday dataset

From the Fig. 1, it is observed that for the case of birthday data set (which contains 7 clusters according to the day of the week) the spectral clustering algorithm implemented with Euclidean distance measure produces only one cluster which is not correct. Whereas for the same dataset, our proposed spectral clustering algorithm produces 7 output clusters which are the desired output as shown in the Fig. 2. Similar results we can observe for other datasets like iris,contact-lens etc., are summarized in Table 1. From this Table 1, we can compare the performance of Spectral Clustering algorithm results with the use of Euclidean distance measure and Manhattan distance measure. It is clearly observed from the values in this table, the spectral clustering algorithm results with Manhattan distance measure are optimized. For example, if we consider the case of Iris dataset, the spectral clustering with Euclidean distance measure performs only one cluster which is not correct. If we consider the Manhattan distance measure for the same data set we are getting three clusters each cluster corresponds to setosa, verginica and versicolor category which is the desired output.

Table 1. Comparison of spectral clustering results with Euclidean and Manhattan measures

S No.	Name of the Data set	Spectral clustering with Euclidean distance measure		Spectral clustering with Manhattan distance measure	
		Number of clusters	Number of instances in each cluster	Number of clusters	Number of instances in each cluster
1	BIRTHDAY	1	360(100%)	7	52(14%)
					52(14%)
					52(14%)
					52(14%)
					52(14%)
					52(14%)
					53(15%)
2	AIDS	1	50(100%)	4	5(10%)
					5(10%)
					20(40%)
					20(40%)
3	IRIS	1	360(100%)	3	50(33%)
					50(33%)
					50(33%)
4	CONTACT LENS	1	24(100%)	4	2(8%)
					13(54%)
					4(17%)
					5(21%)

From Table 1, we can conclude that the quality of the output clusters formed by the spectral clustering are improved by varying the similarity measure which is used in the process of building the similarity matrix. Table 2 shows the number of clusters and the number of instances with percentage within each cluster corresponding to given σ value.

Table 2. Spectral clustering results for different data sets corresponding to σ value

Dataset	σ = 1.0		σ = 1.2		σ = 1.4		σ = 1.6	
	No. of clusters	No. of instances in a cluster	No. of clusters	No. of instances in a cluster	No. of clusters	No. of instances in a cluster	No. of clusters	No. of instances in a cluster
Diabetes	2	500(65%)	1	768(100%)	1	768(100%)	1	768(100%)
		268(35%)						
Glass	5	2(1%)	4	184(86%)	4	184(86%)	4	184(86%)
		26(12%)		27(13%)		27(13%)		27(13%)
		1(0%)		1(0%)		1(0%)		1(0%)
		1(0%)		2(1%)		2(1%)		2(1%)
		184(86%)						
Iris	3	50(33%)	2	50(33%)	2	50(33%)	1	150(100%)
		50(33%)		100(67%)		100(67%)		
		50(33%)						
Contact-lenses	4	2(8%)	3	15(63%)	3	15(63%)	3	15(63%)
		13(54%)		4(17%)		4(17%)		4(17%)
		4(17%)		5(21%)		5(21%)		5(21%)
		5(21%)						

From the below Table 2, we can clearly observe that the value of σ plays a significant role on the performance of the spectral clustering. We observed the results for σ = 1, σ = 1.2, σ = 1.4, σ = 1.6 and σ = 2.

6 Conclusion

The Spectral clustering algorithm is recently evolved clustering technique, which obtained from the field of graph splitting. The input, in this case, is a similarity matrix, constructed from the pair-wise affinity among data points. The algorithm uses the eigen-values and eigenvectors of a normalized affinity matrix to segment the data. The pair-wise affinity among the data points is calculated from the proximity (e.g. similarity or distance) measures. In any clustering task, the proximity measure generally performs a crucial role. In fact, one of the early and fundamental steps in a clustering process is the choice of a suitable affinity magnitude. A number of such measures may be used for this task. However, the success of a clustering algorithm partially depends on the selection of the proximity measure. In this paper, we perform a comparative and exploratory analysis on several existing proximity measures to evaluate their performance when applying the spectral clustering algorithm to a number of diverse data sets. Furthermore, we experienced that the choice for σ might have significant impacts on the performance of the spectral clustering algorithm. When performing the different experiments, the best value for σ has been chosen from a visual perspective by trying out different values and then selecting the one that produced the best clustering results.

References

1. Ding, S.: Research on spectral clustering algorithms and prospects. In: 2nd International Conference on Computer Engineering and Technology (ICCET 2010), Chengdu, vol. 6, pp. 149–153 (2010)
2. Ding, C., Zha, H.: Spectral Clustering, Ordering and Ranking: Statistical Learning with Matrix Factorizations. Springer, Heidelberg (2011)
3. Scott, J.: Social Network Analysis: A Handbook, 2nd edn. Athenaeum Press Limited, London (2005)
4. Spielmat, D.A., Shang-Hua, T.: Spectral partitioning works: planar graphs and finite element meshes. In: 37th Annual Symposium on Foundations of Computer Science, Burlington, pp. 96–105 (1996)
5. Spielman, D.A., Teng, S.H.: Spectral partitioning works: planar graphs and finite element meshes. In: 37th Annual Symposium on Foundations of Computer Science. IEEE Press (1996)
6. Chen, C.H.: Handbook of Pattern Recognition and Computer Vision, 4th edn. World Scientific Publishing Co. Pte. Ltd., Singapore (2010)
7. Bapat, R.B.: Graphs and Matrices. Springer, London (2010)
8. Ulrike, V.L.: A tutorial on spectral clustering. Stat. Comput. **17**, 395–416 (2007)
9. Palma, J.M.L.M.: High Performance Computing for Computational Science – VECPAR 2010. Springer, Heidelberg (2011)
10. Spall, J.C.: Introduction to Stochastic Search and Optimization: Estimation, Simulation and Control. Wiley, Hoboken (2003)
11. Rudnick, J.A., Gaspari, G.D.: Elements of the Random Walk: An Introduction for Advanced Students and Researchers. Cambridge University Press, Cambridge (2004)
12. Han, J.: Data Mining: Concepts and Techniques. Elsevier, Amsterdam, Boston. Morgan Kaufmann, San Francisco (2006)
13. Hagen, L., Kahng, A.: Fast spectral methods for ratio cut partitioning and clustering. In: 1991 IEEE International Conference on Computer-Aided Design, pp. 10–13. IEEE Press (1991)
14. Chen, G., Han, B.: Improve k-means clustering for audio data by exploring a reasonable Sampling rate. In: 2010 Seventh International Conference on Fuzzy Systems and Knowledge Discovery (FSKD 2010), Yantai, Shandong, pp. 1639–1642 (2010)
15. Karantasis, K.I.: Accelerating data clustering on GPU-based clusters under shared memory abstraction. In: 2010 IEEE International Conference on Cluster Computing Workshops and Posters (CLUSTER WORKSHOPS 2010), pp. 1–5. IEEE Press (2010)
16. Trivedi, S., Pardos, Z., Sarkozy, G., Heffernan, N.: Spectral clustering in educational data mining. Accessed 26 Feb 2013
17. Dhillon, I.S., Guan, Y., Kulis, B.: Kernel k-means, spectral clustering and normalized cuts. Accessed 20 Feb 2013

Optimal Tuning of PID Controller for Coupled Tank Liquid Level Control System Using Particle Swarm Optimization

Sanjay Kumar Singh[1(✉)] and Nitish Katal[2]

[1] Amity University, Jaipur, Rajasthan, India
sksingh.mnit@gmail.com
[2] Department of Electrical Engineering, PEC University of Technology,
Chandigarh, India

Abstract. In this paper, particle swarm optimization (PSO) algorithm is presented for determining the optimal gains for a proportional-integral-derivative (PID) controller; which is implemented on a coupled tank control system to maintain the liquid level. Coupled tank system is a non-linear system and finds a wide application in industrial systems; and the quality of control directly affects the quality of uniform products, safety and cost. Initially, Ziegler-Nichols method has been used for the PID tuning but the method does not work perfectly due to lack of precision, long run time and lack of stability. The simulation results indicate that better performance has been obtained for the PSO tuned PID controller as compared with those obtained from GA & ZN tuned controllers. As per the results obtained in the paper, the proposed method is more effective in improving the time domain characteristics such as rise time, settling time, maximum overshoot and steady state error.

Keywords: PID controller tuning · Industrial systems · Robust time response · Genetic algorithm · Particle swarm optimization

1 Introduction

Coupled tank liquid level system, the flow of liquid between tanks is controllers so that the optimum levels are maintained in both of the tanks and such systems are widely used in industries [1, 2]. The use of PID controllers ensures that the liquid level is maintained at desired level and disturbance rejection. Ziegler Nichols classical PID tuning rules has been used for initially tuning the controllers. Then time domain performance index of ISE i.e. integral square error has been used for tuning the controllers using genetic algorithms and particle swarm optimization. From the results obtained, considerably better results have been obtained in case of Particle Swarm Optimization Algorithm optimized PID controllers in step response on the system.

© Springer Nature Singapore Pte Ltd. 2017
K. Deep et al. (eds.), *Proceedings of Sixth International Conference on Soft Computing for Problem Solving*, Advances in Intelligent Systems and Computing 547,
DOI 10.1007/978-981-10-3325-4_8

2 Mathematical Modeling of Couple Tank Liquid Level Control System

Considering the coupled tank system is in Fig. 1. The dynamic equations of the system, by considering the flow balances for each tank, the equations for rate of change of fluid volume in tanks are as [3, 10, 20]:

For tank 1:

$$Q_i - Q_1 = A\frac{dH_1}{dt} \tag{1}$$

For tank 2:

$$Q_1 - Q_0 = A\frac{dH_2}{dt} \tag{2}$$

where H_1, H_2 = Height of tank 1 & 2; A = Cross sectional area of tank 1 & 2; Q_1, Q_2 = Flow rate of the fluid and Q_i = Pump flow rate.

The transfer function of the coupled tank system can be given as:

$$G(s) = \frac{\frac{1}{k_2}}{\left(\frac{A^2}{k_1 k_2}\right).s^2 + \left(\frac{A(2k_1 + k_2)}{k_1 k_2}\right) \cdot s + 1} = \frac{\frac{1}{k_2}}{(sT_1 + 1)(sT_2 + 1)} \tag{3}$$

where:

$$T_1 T_2 = \frac{A^2}{k_1 k_2}$$

$$T_1 + T_2 = \frac{A(2k_1 + k_2)}{k_1 k_2}$$

$$k_1 = \frac{\alpha}{2\sqrt{H_1 - H_2}} \ \& \ k_2 = \frac{\alpha}{2\sqrt{H_2 - H_3}}$$

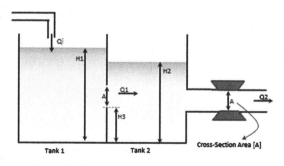

Fig. 1. Schematic representation of coupled tank system.

Using; $H_1 = 18$ cm, $H_2 = 14$ cm, $H_3 = 6$ cm, $\alpha = 9.5$ (constant for coefficient of discharge), $A = 32$; the transfer function can be obtained in Eq. 4.

$$G(s) = \frac{0.002318}{s^2 + 0.201.s + 0.00389} \tag{4}$$

3 Design and Optimization of PID Controllers

PID controllers are the most widely used controllers in the industrial control processes [4] and 90% of the controllers today used in industry are alone PIDs. The general equation for a PID controller can be given by Eq. 5.

$$C(s) = Kp.e(t) + Ki. \int e(t).dt + Kd\frac{de(t)}{dt} \tag{5}$$

Where Kp, Ki and Kd are the controller gains, C(s) is output signal, R(s) is the difference between the desired output and output obtained [5].

3.1 Tuning of PID Controller Using Ziegler Nichols

Ziegler Nichols is the most operative method for tuning the PID controllers. But this method is limited for application since an oscillatory response is generated [3]. Initially, unit step response is derived followed by the computation of the PID gains as suggested by Ziegler-Nichols as in Table 1. Figure 2 shows the closed loop.

Table 1. PID parameters obtained by Ziegler Nichols

PID parameter	Value
Kp	28.214
Ki	4.155
Kd	47.89

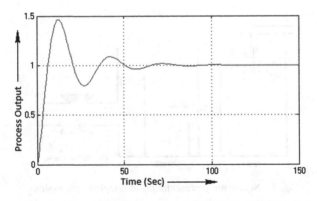

Fig. 2. Closed loop response of the ZN-PID Controllers.

3.2   PID Tuning Using Genetic Algorithm

Genetic algorithms are one of the most robust optimization algorithms and can satisfy any constraint thus boosting their wider applicability [7]. Optimization of the PID controllers with deals with search of the three gain parameters [Kp, Ki, Kd] by minimizing the objective function [9]. For the optimal tuning of the controller, the minimization of the ISE (integral square error) [8] has been carried out.

$$ISE = \int_0^T e^2(t)dt$$

The optimization a population of 40 and tournament-based selection has been considered. The PID gains obtained by optimal tuning using GA are represented in Table 2 and Fig. 3 shows the closed loop response of the GA optimized controllers (Fig. 4).

Table 2. PID parameters obtained by genetic algorithm

PID parameter	GA (ISE)
Kp	28.214
Ki	4.155
Kd	47.89

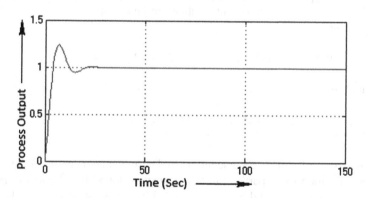

Fig. 3. Closed loop response of GA PID controllers.

3.3   PID Tuning Using PSO

Particle swarm optimization algorithm is meta-heuristic optimization algorithm that is inspired by social behavior of bird flocking or fish schooling. The algorithm is based on the social interaction, which by adjusting the trajectories of the "particles" in the space, tries to minimize the objective. There is vast literature available on variants of PSO algorithms exists [12–19]. The generic constrained PSO algorithm is given under as:

[Step 1]:   Initialization of Swarm

[Step 2]:   Evaluation of fitness of individual particles

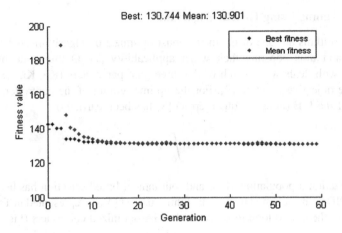

Fig. 4. Best and mean fitness values obtained by optimization using GA (ISE).

[Step 3]:   Modifying gbest, pbest & velocity
[Step 4]:   Modify the position of each particle
[Step 5]:   Move to [Step 2] till the desired convergence is obtained.

In this paper, the optimization has been carried out using PSO with constrained factor [13]; mathematically the velocity and position are given as;
Velocity update:

$$v_i(k) = \chi[v_i(k) + \gamma_{1i} \times (p_i - x_i(k)) + \gamma_{2i} \times (G - x_i(k))]$$

Position update:

$$x_i(k+1) = x_i(k) + v_i(k+1)$$

Where:
χ – Constriction factor; i– Particle index; k – Discrete time index; v – Velocity of ith particle; x – Position of ith particle; p – Best position found by i^{th} particle (personal best); G – Best position found by swarm (global best, best of personal bests); $\gamma_{1,2}$ – Random numbers on the interval [0, 1] applied to i^{th} particle.

Table 3 gives the PID gains obtained by optimization using PSO with constrained factor and Fig. 5 gives the plot for the closed loop response of the controller with PSO optimized gains.

Table 3. PID parameters obtained by PSO

PID parameter	PSO (ISE)
Kp	9.813
Ki	0.318
Kd	2.766

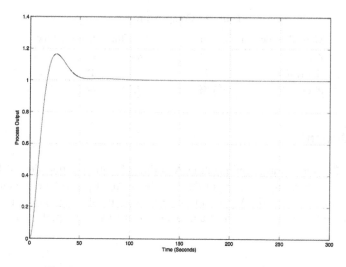

Fig. 5. Closed loop response of PSO PID controllers.

4 Results and Discussion

In this paper, initially the gains of the PID have been estimated using Ziegler Nichols rules which give an oscillatory response, followed by the optimization by genetic algorithm and particle swarm optimization algorithm. The computed parameters are implemented for obtaining the closed-loop response of the system. Figure 6 shows the compared closed loop step response graph, clearly indicating that better results are obtained in case of Particle Swarm Optimization algorithm optimized PID controller with decreased over-shoot percentage, rise and settling time values. Table 4 represents the numerical data of the results obtained.

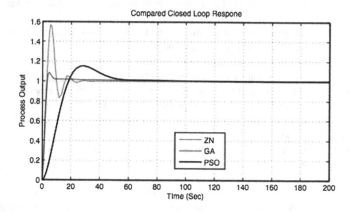

Fig. 6. Compared closed loop response of ZN, GA & PSO optimized PID controllers.

Table 4. Compared results

Method of design	Overshoot %age	Rise time	Settling time
ZN	46.4%	4.83	62.4
GA (ISE)	23.7%	2.93	18.5
PSO (ISE)	8.59%	17.3	54.2

5 Conclusion

The use of Particle Swarm Optimization Algorithm for the optimization of PID controller offers better results in terms of decreased overshoot percentage as compared to Ziegler Nichols and Genetic Algorithm tuned PIDs; thus offering better operation for the coupled tank liquid level control and better plant safety and performance.

References

1. Bhuvaneswari, N.S., Uma, G., Rangaswamy, T.R.: Adaptive and optimal control of a non-linear process using intelligent controllers. Appl. Soft Comput. **9**(1), 82–190 (2009). Elsevier
2. Capón-García, E., Espuña, A., Puigjaner, L.: Statistical and simulation tools for designing an optimal blanketing system of a multiple-tank facility. Chem. Eng. J. **152**(1), 122–132 (2009)
3. Seth, B., J, D.S.: Liquid level control. In: Control System Laboratory (ME413). IIT Bombay 2006-07
4. Åström, K.J., Albertos, P., Quevedo, J.: PID control. Control Eng. Pract. **9**, 159–1161 (2001)
5. Nise, N.S.: Control System Engineering, 4th edn. (2003)
6. Goodwin, G.C., Graebe, S.F., Salgado, M.E.: Control System Design. Prentice Hall Inc., New Jersey (2001)
7. Larbes, C., Aït Cheikh, S.M., Obeidi, T., Zerguerras, A.: Genetic algorithms optimized fuzzy logic control for the maximum power point tracking in photovoltaic system. Renew. Energy **34**(10), 2093–2100 (2009). Elsevier Ltd.
8. Jean-Pierre, C.: Process Control: Theory and Applications, pp. 132–133. Springer, London (2004)
9. Grefenstette, J.J.: Optimization of control parameters for genetic algorithms. IEEE Trans. Syst. Man Cybern. SMC **16**(1), 122–128 (1986)
10. Singh, S.K., Katal, N., Modani, S.G.: Multi-objective optimization of PID controller for coupled-tank liquid-level control system using genetic algorithm. In: Babu, B.V., Nagar, A., Deep, K., Pant, M., Bansal, J.C., Ray, K., Gupta, U. (eds.) Proceedings of the Second International Conference on Soft Computing for Problem Solving (SocProS 2012), December 28–30, 2012. AISC, vol. 236, pp. 59–66. Springer, New Delhi (2014). doi:10.1007/978-81-322-1602-5_7
11. Eberhart, R.C., Kennedy, J.: A new optimizer using particle swarm theory. In: Proceedings of the Sixth International Symposium on Micromachine and Human Science, Nagoya, Japan, pp. 39–43 (1995)
12. Kennedy, J., Eberhart, R.C.: A discrete binary version of the particle swarm algorithm. In: Proceedings of the World Multiconference on Systemics, Cybernetics and Informatics 1997, Piscataway, NJ, pp. 4104–4109 (1997)

13. Kennedy, J., Spears, W.: Matching algorithms to problems: an experimental test of the particle swarm and some genetic algorithms on the multimodal problem generator. In: Proceedings of IEEE Congress on Evolutionary Computation (CEC 1998), Anchorage, Alaska, USA (1998)
14. Clerc, M., Kennedy, J.: The particle swarm-explosion, stability, and convergence in a multidimensional complex space. IEEE Trans. Evol. Comput. **6**(1), 58–73 (2002)
15. Blanks, A., Vincent, J., Anyakoha, C.: A review of particle swarm optimization. Part I: background and development. Nat. Comput. **6**(4), 467–484 (2007)
16. Blanks, A., Vincent, J., Anyakoha, C.: A review of particle swarm optimization. Part II: hybridisation, combinatorial, multicriteria and constrained optimization, and indicative applications. Nat. Comput. **7**(1), 109–124 (2008)
17. Blum, C., Li, X.: Swarm intelligence in optimization. In: Blum, C., Merkle, D. (eds.) Swarm Intelligence. Natural Computing Series, pp. 43–85. Springer, Heidelberg (2008)
18. Hayder, A.H., Ceylan, H., Ayvaz, M.T., Gurarslan, A.: PSOLVER: a new hybrid particle swarm optimization algorithm for solving continuous optimization problems. Expert Syst. Appl. **37**(10), 6798–6808 (2010)
19. Rameshkumar, K., Rajendran, C., Mohanasundaram, K.M.: A novel particle swarm optimisation algorithm for continuous function optimisation. Int. J. Oper. Res. **13**(1), 1–21 (2012)
20. Katal, N., Narayan, S.: Multi-objective optimization-based design of robust fractional-order $PI^\lambda D^\mu$ controller for coupled tank systems. In: Pant, M., Deep, K., Bansal, J.C., Nagar, A., Das, K.N. (eds.) Proceedings of Fifth International Conference on Soft Computing for Problem Solving. AISC, vol. 437, pp. 27–38. Springer, Singapore (2016). doi:10.1007/978-981-10-0451-3_4

Review on Inertia Weight Strategies for Particle Swarm Optimization

Ankush Rathore$^{(\boxtimes)}$ and Harish Sharma

Rajasthan Technical University, Kota, India
ankushrathore777@gmail.com

Abstract. In the category of swarm intelligence based algorithms, Particle Swarm Optimization (PSO) is an effective population-based meta-heuristic used to solve complex optimization problems. In PSO, global optima is searched with the help of individuals. For the efficient search process, individuals have to explore whole search space as well as have to exploit the identified search area. Researchers are continuously working to balance these two contradictory properties i.e. exploration and exploitation and have been modified the PSO in many different ways to improve its solution search capability in the search space. In this regard, incorporation of inertia weight strategy in PSO is a significant modification and after that many researchers have been developed different inertia weight strategies to improve the solution search capability of PSO. This paper presents an analysis of the developed inertia weight strategies in respect to problem-solving capability and their effect in the solution search process of PSO. The effect of 30 recent inertia weight strategies on PSO is measured while comparing over ten well known test functions of having different degree of complexity and modularity.

Keywords: Soft computing · Optimisation · Inertia weight · Swarm intelligence · Nature inspired algorithms

1 Introduction

Particle Swarm Optimization (PSO) algorithm was developed by Eberhart and Kennedy in 1995 [1]. It is inspired by the intelligent behaviour of bird in search of food. The PSO algorithm is used to solve the different complex optimization problems including economics, engineering, complex real-world problems, biology and industry [2]. PSO can be applied to non-linear, non-differentiable, huge search space problems and gives better results with good accuracy [3].

For n-dimensional search space, the velocity and position of the ith particle represents as: $V_i = (v_{i1}, v_{i2}, ..., v_{id})^T$ and $X_i = (x_{i1}, x_{i2}, ..., x_{id})^T$ respectively. Where, v_{id} and x_{id} is the velocity and position of ith particle in d-dimension respectively. The velocity of the swarm (particle) is defined as follows:

© Springer Nature Singapore Pte Ltd. 2017
K. Deep et al. (eds.), *Proceedings of Sixth International Conference on Soft Computing for Problem Solving*, Advances in Intelligent Systems and Computing 547,
DOI 10.1007/978-981-10-3325-4_9

$$v_{id}(new) = v_{id}(old) + c_1 r_1 (p_{id} - x_{id}) + c_2 r_2 (p_{gd} - x_{id}) \tag{1}$$

$$x_{id}(new) = x_{id}(old) + v_{id}(new) \tag{2}$$

where, d = 1, 2, ..., n presents the dimension and i = 1, 2, ..., N represents the particle index, N is the size of the swarm, c_1 and c_2 are called social scaling and cognitive parameters respectively that determines the magnitude of the random force in the direction of particle's previously best visited position (p_{id}) and best particle (p_{gd}) and r_1, r_2 are the uniform random variable between $[0, 1]$. The maximum velocity (V_{max}) assists as a constraint to control the position of the swarms within the solution search space.

Further, Shi and Eberhart [4] was developed the concept of an inertia weight (IW) in 1998 to ensure an optimal tradeoff between exploration and exploitation mechanisms of the swarm population. This inertia weight strategy was to be able to eliminate the need of maximum velocity (V_{max}). Inertia weight controls the particles movement by maintaining its previous memory. The velocity update equation is considered as follows:

$$v_{id}(new) = w * v_{id}(old) + c_1 r_1 (p_{id} - x_{id}) + c_2 r_2 (p_{gd} - x_{id}) \tag{3}$$

This paper discusses the 30 different inertia weight strategies on 10 benchmark functions for PSO algorithm. A comprehensive review on 30 inertia weight strategies have been presented in next section.

2   A Review on Different Inertia Weight Strategies for PSO

Inertia weight plays an important role in the process of providing a trade-off between diversification and intensification skills of PSO algorithm. When the inertia weight strategy is implemented to PSO algorithm, the particles move around while adjusting their velocities and positions according to Eqs. (1) and (2) in the search space.

In 1998, first time Shi and Eberhart [4] proposed the concept of constant inertia weight. A small inertia weight helps in explore the search space while a large inertia weight facilitates in exploit the search space. Eberhart and Shi [5] proposed a random inertia weight strategy and enhances the performance and efficiency of PSO algorithm.

The linearly decreasing strategy [6] increases the convergence speed of PSO algorithm in early iterations of the search space. The inertia weight starts with some large value and then linearly decreases to some smaller value. The inertia weight provides the excellent results from 0.9 to 0.4. In global-local best inertia weight [7], the inertia weight is based on the global best and local best of the swarms in each generation. It increases the capabilities of PSO algorithm and neither takes a linearly decreasing time-varying value nor a constant value.

Fayek et al. [8] introduced a particle swarm simulated annealing technique (PSOSA). This inertia weight strategy is optimized by using simulated annealing and improves its searching capability.

Chen et al. [9] present two natural exponent inertia weight strategies as e1-PSO and e2-PSO, which are based on the exponentially decreasing the inertia weight. Experimentally, these strategies become a victim of premature convergence, despite its quick convergence speed towards the optimal positions at the early stage of the search process.

Using the merits of chaotic optimization, chaotic inertia weight has been proposed by Feng et al. [10] and PSO algorithm becomes better global search ability, convergence precision and quickly convergence velocity.

Malik et al. [11] presented a sigmoid increasing inertia weight (SIIW) and sigmoid decreasing inertia weight (SDIW). These strategies provide better performance with quick convergence ability and aggressive movement narrowing towards the solution region.

Oscillating Inertia Weight [12] provides a balance between diversification and intensification waves and concludes that this strategy looks to be competitive and, in some cases, better performs in terms of consistency.

Gao et al. [13] proposed a logarithmic decreasing inertia weight with chaos mutation operator. The chaos mutation operator can enhance the ability to jump out the premature convergence and improve its convergence speed and accuracy.

To overcome the stagnation and premature convergence of the PSO algorithm, Gao et al. [14] proposed an exponent decreasing inertia weight (EDIW) with stochastic mutation (SM). The stochastic mutations (SM) is used to enhance the diversity of the swarm while EDIW is used to improve the convergence speed of the individuals (Table 1).

Linearly decreasing inertia weight have been proposed by Shi and Eberhart [4] and greatly improved the accuracy and convergence speed. A large inertia weight facilitates at the inceptive phase of search space while later linearly decreases to a small inertia weight.

Adewumi et al. [25] proposed the swarm success rate random inertia weight (SSRRIW) and swarm success rate descending inertia weight (SSRDIW). These strategies use swarm success rates as a feedback parameter. Further, it enhances the effectiveness of the algorithm regarding convergence speed and global search ability.

Shen et al. [18] proposed the dynamic adaptive inertia weight and used to solve the complex and multi-dimensional function optimization problems. This strategy can timely adjust the particle speed, jump out of a locally optimal solution and improve the convergence speed.

Ting et al. [24] proposed the exponent inertia weight. There exist two important parameters as a local attractor (a) and global attractor (b). This method controls the population diversity by adaptive adjustment of local attractor (a) and global attractor (b).

Chatterjee and Siarry [22] proposed nonlinear decreasing inertia weight strategy with nonlinear modulation index. This strategy is quite effective as well as avoid premature issues. Lei et al. [17] proposed adaptive inertia weight. It furnishes with automatically harmonize global and local search ability and obtained the global optima.

Table 1. Inertia weight strategies

S.no.	Name of inertia weight	Formula of inertia weight
1	Logarithm decreasing inertia weight [13]	$w = w_{max} + (w_{min} - w_{max}).log_{10}(a + \frac{10t}{T})$
2	Exponent decreasing inertia weight [14]	$w = (w_{max} - w_{min} - d_1).exp(\frac{1}{1+\frac{d_2 t}{T}})$
3	Natural exponent inertia weight strategy(e2 -PSO) [9]	$w = w_{min} + (w_{max} - w_{min}).e^{-[\frac{t}{(\frac{T}{4})}]^2}$
4	Natural exponent inertia weight strategy(e1 -PSO) [9]	$w = w_{end} + (w_{start} - w_{end}).e^{[\frac{-t}{(\frac{T}{10})}]}$
5	Global-local best inertia weight [7]	$w = [1.1 - \frac{gbest_i}{pbest_i}]$
6	Simulated annealing inertia weight [8]	$w = w_{min} + (w_{max} - w_{min}).\lambda^{k-1}$
7	Oscillating inertia weight [12]	$w = (\frac{w_{min}+w_{max}}{2} + \frac{w_{max}-w_{min}}{2}cos(\frac{2\Pi t}{T}))$, where $T = \frac{2S_1}{3+2k}$
8	Chaotic random inertia weight [10]	$z = 4 * z * (1 - z)$, $w = 0.5 * rand + 0.5 * z$
9	The Chaotic inertia weight [10]	$w = (w_{max} - w_{min}) * (\frac{T-t}{T}) + w_{min} * z$, where, $z = 4 * z * (1 - z)$
10	Linear decreasing inertia weight [6]	$w = w_{max} - (w_{max} - w_{min})(\frac{t}{T})$
11	Sigmoid decreasing inertia weight [11]	$w = \frac{(w_{max}-w_{min})}{(1+e^{-u(k-n*gen)})} + w_{min}$, $u = 10^{log((gen)-2)}$
12	Sigmoid increasing inertia weight [11]	$w = \frac{(w_{max}-w_{min})}{(1+e^{u(k-n*gen)})} + w_{min}$, $u = 10^{log((gen)-2)}$
13	Random inertia weight [5]	$w = 0.5 + 0.5 * rand$
14	Constant inertia weight [4]	$w = c$, where $c = 0.2$ (considered for experiments)
15	Chaotic adaptive inertia weights (CAIWS-D) [15]	$w = [(w_{max} - w_{min})(\frac{T-t}{T}) + w_{min}] * z$, where, $z = 4*SR*(1-SR)$
16	Chaotic adaptive inertia weights (CAIWS-R) [15]	$w = (0.5*SR + 0.5)*z$, where $z = 4*SR*(1-SR)$
17	Decreasing exponential function inertia weight (DEFIW) [15]	$w = t^{-(\sqrt[t]{t})}$
18	Fixed inertia weight (FIW)[16]	$w = \frac{1}{2ln(2)}$
19	Adaptive inertia weight strategy [17]	$w = [\frac{1-(\frac{t}{T})}{(1+S\frac{t}{T})}]$
20	Dynamic adaptive inertia weight [18]	$w = w_{min} + (w_{max} - w_{min})F(t)\Psi(t)$, $\Psi(t) = exp(-\frac{t^2}{2\sigma^2})$ and $\sigma = T/3$
21	Decreasing inertia weight (DIW) [19]	$w = w_{init} * u^{-t}$
22	Inertia weight strategy [20]	$w = \frac{(w_{init}-0.4)(g_{size}-i)}{(g_{size}+0.4)}$
23	Double exponential dynamic inertia weight [2]	$w = exp(-exp(-R))$, where $R = \frac{(T-t)}{T}$
24	Tangent decreasing inertia weight (TDIW) [21]	$w = (w_{max} - w_{min}) * tan(\frac{7}{8}(1 - \frac{t}{T})^k)$
25	Nonlinear decreasing inertia weight (NDIW) [22]	$w = (w_{max} - w_{min})(\frac{T-t}{T})^n + w_{min}$
26	Linear or non-linear decreasing inertia weight [23]	$w = (\frac{2}{t})^{0.3}$
27	Exponent inertia weight [24]	$w = w_0 e^{-a(\frac{t}{T})}$
28	Swarm success rate random inertia weight (SSRRIW) [25]	$w = 0.5 * rand + 0.5 * ssr_{t-1}$
29	Swarm success rate descending inertia weight (SSRDIW) [25]	$w = (w_{max} - w_{min})(\frac{T-t}{T}) + w_{min} * ssr_{t-1}$
30	Descending inertia weight [25]	$w = w_{min} + (w_{max} - w_{min})(\frac{T-t}{T})$

J. asoc. [23] proposed the linear or non-linear decreasing inertia weight. This strategy has global search ability and also helpful to find a better optimal solution. It overcomes the weakness of premature convergence and converges faster at the early stage of the search process. Jiao et al. [19] proposed the decreasing inertia weight (DIW). This strategy provides the algorithm with dynamic adaptability and controls the population diversity by adaptive adjustment of inertia weight.

Li et al. [21] proposed the tangent decreasing inertia weight (TDIW) based on tangent function (TF). This strategy is to increase the diversity of swarm for more exploration of the search space at initial iterations while later exploit the search area. So that this approach provides better results with accuracy.

Chauhan et al. [2] proposed the double exponential dynamic inertia weight (DEDIW). The inertia weight is calculated for whole swarm iteratively by using gompertz function, and it is capable of providing a stagnation free environment with better accuracy. Peram et al. [20] proposed a new inertia weight that provides the less susceptible to premature convergence and less likely to be stuck in local optima. Sheng-Ta Hsieh et al. [16] introduced fixed inertia weight (FIW). It provides better convergence speed and less computational efforts.

Table 2. Test problems, D: Dimension, AE: Acceptable Error

Test problem	Objective function	Search range	Optimum value	D	AE
Sphere	$f_1(x) = \sum_{i=1}^{D} x_i^2$	[−5.12 5.12]	$f(0) = 0$	30	$1.0E - 05$
De Jong f4	$f_2(x) = \sum_{i=1}^{D} i.(x_i)^4$	[−5.12 5.12]	$f(0) = 0$	30	$1.0E - 05$
Ackley	$f_3(x) = -20 + e + exp(-\frac{0.2}{D}\sqrt{\sum_{i=1}^{D} x_i^3})$	[−30, 30]	$f(0) = 0$	30	$1.0E - 05$
Alpine	$f_4(x) = \sum_{i=1}^{D} |x_i \sin x_i + 0.1x_i|$	[−10, 10]	$f(0) = 0$	30	$1.0E - 05$
Michalewicz	$f_5(x) = -\sum_{i=1}^{D} \sin x_i (\sin (\frac{i.x_i^2}{\pi})^{20})$	[0, π]	$f_{min} = -9.66015$	10	$1.0E - 05$
Cosine Mixture	$f_6(x) = \sum_{i=1}^{D} x_i^2 - 0.1(\sum_{i=1}^{D} \cos 5\pi x_i) + 0.1D$	[−1, 1]	$f(0) = -D \times 0.1$	30	$1.0E - 05$
Exponential	$f_7(x) = -(exp(-0.5\sum_{i=1}^{D} x_i^2)) + 1$	[−1, 1]	$f(0) = -1$	30	$1.0E - 05$
brown3	$f_8(x) = \sum_{i=1}^{D-1} (x_i^{2(x_{i+1})^2+1} + x_{i+1}^{2x_i^2+1})$	[−1 4]	$f(0) = 0$	30	$1.0E - 05$
Beale	$f_9(x) = [1.5-x_1(1-x_2)]^2 + [2.25 - x_1(1 - x_2^2)]^2 + [2.625 - x_1(1 - x_2^3)]^2$	[−4.5,4.5]	$f(3, 0.5) = 0$	2	$1.0E - 05$
Colville	$f_{10}(x) = 100[x_2 - x_1^2]^2 + (1 - x_1)^2 + 90(x_4 - x_3^2)^2 + (1 - x_3)^2 + 10.1[(x_2 - 1)^2 + (x_4 - 1)^2] + 19.8(x_2 - 1)(x_4 - 1)$	[−10,10]	$f(1) = 0$	4	$1.0E - 05$

The decreasing exponential function inertia weight (DEFIW) [15] decreases the value of inertia weight iteratively as the algorithm approaches equilibrium state and furnishes the superiority to the competitors in fitness quality.

Arasomwan et al. [15] Proposed chaotic adaptive inertia weights as CAIWS-D and CAIWS-R. These strategies simply combine chaotic mapping with the swarm success rate as a feedback parameter to harness together chaotic and adaptivity characteristics. These approaches provide more refine accuracy, faster convergence speed as well as global search ability.

3 Experimental Results

To evaluate the performance of the inertia weight strategy, it is tested over 10 different benchmark functions (F_1 to F_{10}) as given in Table 2.

3.1 Parameter Settings

Following experimental settings are adopted:

- $G_0 = 100$ and $\alpha = 20$ [26],
- Number of runs = 30,

Table 3. Average number of function evaluations of different inertia weight strategies for different benchmark functions

Inertia Weight	Sphere	De Jong f4	Ackley	Alpine	Michalewicz	Cosine Mixture	Exponential	brown3	Beale	Colville
1	09851.67	08540.00	15218.33	45575.00	49731.67	50100.00	9611.67	10270.00	2771.667	26260.00
2	22313.33	19670.00	32858.33	30181.67	50100.00	42166.67	22125.00	23351.67	6585.00	36578.33
3	25801.66	24028.33	35703.33	33166.67	48105.00	41483.33	25651.67	26728.33	09263.33	39870.00
4	18846.67	16835.00	29221.67	26516.67	48733.33	40691.67	18248.33	19665.00	06175.00	40780.00
5	50100.00	49196.67	50100.00	26536.67	48840.00	50100.00	50100.00	50100.00	00885.00	47108.33
6	13663.33	11920.00	23591.67	21495.00	50100.00	44928.33	13378.33	14915.00	3565.00	39290.00
7	29278.33	23780.00	50100.00	46190.00	47988.33	45698.33	26725.00	32083.333	3526.67	43920.00
8	39623.33	35648.33	50100.00	50100.00	47638.33	49888.33	35870.00	41895.00	3230.00	46575.00
9	32738.33	30196.67	42308.33	39525.00	48946.67	42490.00	31791.67	33640.00	04591.67	36876.67
10	44853.33	42535.00	50100.00	50096.67	48795.00	49615.00	43596.67	45206.67	10890.00	46073.33
11	13218.33	11451.67	23440.00	19735.00	49196.67	46841.67	12383.33	14135.00	02711.67	43118.33
12	25103.33	23176.67	34655.00	32036.67	46930.00	46618.33	24230.00	25821.67	12451.67	41675.00
13	15840.00	13016.67	29701.67	29538.33	49053.33	48191.67	15398.33	17851.67	03503.33	31575.00
14	8433.33	7330.00	18216.67	15083.33	49228.33	50100.00	07688.33	08975.00	02385.00	45735.00
15	26308.33	23300.00	47571.67	38541.67	48951.67	39000.00	24606.67	28170.00	2968.33	41273.33
16	26348.33	22995	47791.67	37485.00	49750.00	39408.33	24390.00	28135.00	4603.33	40158.33
17	11821.67	14613.33	28691.67	25045.00	50100.00	30740.00	11070.00	12343.33	4011.67	49971.67
18	11116.67	10255.00	20148.33	17286.67	49011.67	47786.67	10798.33	12016.67	01056.67	38490.00
19	21593.33	20201.67	28485.00	27230.00	49003.33	42706.67	21200.00	22236.67	07161.67	39731.67
20	13108.33	11238.33	44071.67	20541.67	46168.33	39801.67	36930.00	14003.33	1121.67	38113.33
21	10585.00	09818.33	17485.00	17256.67	47970.00	47661.67	10198.33	11496.67	03293.33	41001.67
22	18141.67	15730.00	26410.00	23940.00	50100.00	37963.33	17363.33	18841.67	06580.00	45256.67
23	32145	29100	44275.00	40775.00	49000.00	37770.00	31460.00	33501.67	03953.33	34638.33
24	17571.67	15713.33	26353.33	24056.67	49713.33	36716.67	16858.33	19021.67	06556.67	41691.67
25	42213.33	39358.33	50041.67	48980.00	49168.33	47785.00	41083.33	42838.33	10196.67	43158.33
26	10585.00	9818.33	17485.00	17256.67	47970.00	47661.67	10198.33	11496.67	03293.33	41131.67
27	24753.33	22180.00	35250.00	31966.67	49148.33	38595.00	23735.00	25720.00	07821.67	37138.33
28	23265.00	19488.33	40460.00	33003.33	49821.67	35823.33	21410.00	24258.33	02921.67	45111.67
29	26200.00	24011.67	36238.33	33400.00	47765.00	37798.33	25305.00	27360.00	07193.33	38271.67
30	45070.00	42221.67	50100.00	50085.00	49010.00	49023.33	43833.33	45480.00	08773.33	40638.33

- Number of populations = 50,
- Maximum number of iterations (T) = 1000,
- Value of c_1 and c_2 are 2.0 [25].

3.2 Results and Discussion

In this section, 30 different inertia weight strategies are analyzed on 10 benchmark problems in terms of average number of function evaluations (AFE's), mean error (ME) and standard deviation (SD). The AFE's, ME and SD are presented in Tables 3, 4 and 5 respectively. Boxplot of AFE's, ME and SD are shown in Figs. 1, 2 and 3 respectively.

It is clear from the reported results that most of the Inertia weight strategies produce poor results in case of michalewicz function (F_5). It clear from Fig. 1 that constant inertia weight and linearly decreasing inertia weight (LDIW) is best and worst strategy respectively in terms of AFE's. It is observed from Fig. 2 that the mean error taken by chaotic random inertia weight strategy and global local best inertia weight strategy are minimum and maximum in terms of mean error respectively compared to the other inertia weight strategies.

Table 4. Mean error value of different inertia weight strategies for different benchmark functions

Inertia Weight	Sphere	De Jong f4	Ackley	Alpine	Michalewicz	Cosine Mixture	Exponential	brown3	Beale	Colville
1	8.90E-06	8.29E-06	5.03E-04	9.14E-04	8.79E-01	4.81E-06	8.40E-06	8.66E-06	4.09E-06	1.50E-03
2	9.15E-06	8.91E-06	9.59E-06	9.57E-06	4.52E-01	1.23E-01	9.23E-06	9.16E-06	3.99E-02	1.57E-03
3	9.32E-06	8.97E-06	9.55E-06	9.64E-06	4.47E-01	1.63E-01	9.17E-06	9.13E-06	3.99E-02	1.34E-03
4	9.28E-06	8.74E-06	9.50E-06	9.54E-06	4.48E-01	1.58E-01	9.28E-06	9.18E-06	3.99E-02	1.22E-03
5	4.54E-01	3.14E-01	5.21E-01	1.15E-04	8.86E-01	2.15E+00	2.19E-01	8.26E-06	5.55E-06	4.84E-03
6	9.08E-06	9.13E-06	9.52E-06	9.25E-06	8.45E-01	2.27E-01	9.15E-06	9.24E-06	2.00E-02	1.18E-03
7	9.04E-06	8.78E-06	3.96E-05	1.37E-05	4.07E-01	1.08E-02	8.77E-06	9.36E-06	2.00E-02	6.33E-03
8	9.44E-06	8.87E-06	4.51E-04	5.51E-05	2.67E-01	5.41E-01	9.33E-06	9.51E-06	2.00E-02	5.13E-03
9	9.25E-06	8.97E-06	9.61E-06	9.13E-06	3.92E-01	5.91E-02	9.28E-06	9.19E-06	2.00E-02	1.84E-03
10	9.26E-06	8.96E-06	1.78E-04	6.26E-05	4.00E-01	3.45E-02	9.30E-06	9.43E-06	3.99E-02	4.95E-03
11	9.15E-06	8.91E-06	9.67E-06	9.56E-06	4.90E-01	1.76E-01	9.20E-06	9.15E-06	2.00E-02	3.64E-03
12	9.22E-06	8.61E-06	9.58E-06	9.40E-06	4.26E-01	2.22E-01	9.11E-06	9.30E-06	4.57E-06	1.54E-03
13	9.17E-06	9.26E-06	9.58E-06	8.69E-06	6.41E-01	4.58E-01	9.13E-06	9.05E-06	2.00E-02	9.58E-04
14	8.96E-06	8.77E-06	1.00E-01	8.07E-06	9.71E-01	6.50E-01	9.23E-06	8.98E-06	2.00E-02	2.55E-03
15	9.13E-06	8.74E-06	1.27E-05	9.31E-06	3.52E-01	4.94E-03	9.14E-06	9.54E-06	2.00E-02	3.33E-03
16	9.43E-06	9.05E-06	1.39E-05	9.62E-06	3.55E-01	9.43E-06	9.26E-06	9.44E-06	3.99E-02	2.86E-03
17	8.52E-06	8.88E-06	8.69E-06	5.52E-04	9.13E-01	4.93E-03	8.08E-06	8.46E-06	3.99E-02	2.88E-01
18	9.06E-06	8.69E-06	9.59E-06	9.31E-06	8.54E-01	3.60E-01	8.94E-06	9.28E-06	5.66E-06	1.32E-03
19	9.12E-06	8.56E-06	9.46E-06	9.36E-06	5.09E-01	1.82E-01	8.96E-06	8.90E-06	2.00E-02	2.06E-03
20	9.32E-06	9.22E-06	9.61E-06	9.38E-06	3.54E-01	1.53E-01	9.18E-06	9.23E-06	4.88E-06	1.09E-03
21	8.84E-06	8.84E-06	9.54E-06	7.78E-06	6.34E-01	3.50E-01	8.64E-06	9.01E-06	2.00E-02	4.69E-03
22	8.98E-06	9.09E-06	9.62E-06	9.38E-06	6.00E-01	1.28E-01	9.12E-06	9.13E-06	5.99E-02	2.13E-03
23	9.32E-06	9.19E-06	9.68E-06	9.52E-06	4.05E-01	9.86E-03	9.26E-06	9.35E-06	2.00E-02	2.02E-03
24	9.24E-06	9.12E-06	9.61E-06	9.63E-06	4.94E-01	1.23E-01	9.19E-06	9.01E-06	5.99E-02	1.94E-03
25	9.11E-06	8.62E-06	3.85E-05	1.29E-05	4.16E-01	6.90E-02	9.33E-06	9.35E-06	3.99E-02	2.93E-03
26	8.84E-06	8.84E-06	9.54E-06	7.78E-06	6.34E-01	3.50E-01	8.64E-06	9.01E-06	2.00E-02	2.01E-03
27	9.17E-06	9.04E-06	9.54E-06	9.50E-06	6.36E-01	7.88E-02	9.49E-06	9.05E-06	5.99E-02	1.18E-03
28	9.26E-06	9.23E-06	9.61E-06	9.67E-06	3.73E-01	2.96E-02	9.43E-06	9.21E-06	2.00E-02	3.46E-03
29	9.27E-06	8.97E-06	9.49E-06	9.65E-06	2.95E-01	5.91E-02	9.27E-06	9.29E-06	5.99E-02	1.54E-03
30	9.31E-06	9.07E-06	1.93E-04	4.63E-05	4.02E-01	1.97E-02	9.52E-06	8.96E-06	2.00E-02	3.62E-03

Table 5. Standard deviation value of different inertia weight strategies for different benchmark functions

Inertia Weight	Sphere	De Jong f4	Ackley	Alpine	Michalewicz	Cosine Mixture	Exponential	brown3	Beale	Colville
1	1.54E-06	9.56E-07	2.66E-03	2.05E-03	6.06E-01	1.96E-01	1.86E-06	1.38E-06	2.38E-06	1.53E-03
2	8.26E-07	1.08E-06	3.58E-07	3.59E-07	3.24E-01	1.15E-01	6.84E-07	8.14E-07	1.49E-01	1.10E-03
3	6.19E-07	9.40E-07	3.49E-07	2.86E-07	3.96E-01	1.76E-01	7.40E-07	6.89E-07	1.49E-01	8.07E-04
4	6.45E-07	9.15E-07	4.10E-07	3.84E-07	3.13E-01	1.62E-01	5.22E-07	7.74E-07	1.49E-01	3.59E-04
5	4.34E-01	1.05E+00	5.36E-01	4.54E-04	6.88E-01	1.00E+00	1.64E-01	7.43E-01	3.02E-06	4.00E-03
6	8.16E-07	1.19E-06	3.28E-07	7.43E-07	5.70E-01	1.65E-01	5.50E-07	7.21E-07	1.07E-01	7.74E-04
7	7.26E-07	1.17E-06	2.10E-05	9.87E-06	3.15E-01	3.70E-02	1.09E-06	4.36E-07	1.07E-01	6.96E-03
8	5.53E-07	1.19E-06	2.74E-04	3.25E-05	2.29E-01	2.65E-02	6.58E-07	6.79E-07	1.07E-01	4.58E-03
9	6.61E-07	7.83E-07	3.54E-07	1.10E-06	2.92E-01	8.18E-02	4.59E-07	6.97E-07	1.07E-01	1.23E-03
10	6.61E-07	7.48E-07	1.23E-04	8.33E-05	3.60E-01	6.25E-02	6.82E-07	3.49E-07	1.49E-01	3.99E-03
11	7.83E-07	1.07E-06	4.02E-07	3.87E-07	4.01E-01	1.43E-01	5.79E-07	8.31E-07	1.07E-01	4.99E-03
12	5.26E-07	1.20E-06	4.27E-07	5.60E-07	3.77E-01	1.61E-01	6.72E-07	6.50E-07	3.30E-06	8.53E-04
13	8.54E-07	6.95E-07	4.37E-07	1.86E-07	5.43E-01	2.30E-01	8.55E-07	7.03E-07	1.07E-01	7.45E-05
14	1.02E-06	1.37E-06	3.75E-01	2.89E-06	5.38E-01	2.40E-01	4.98E-07	1.17E-06	1.07E-01	2.14E-03
15	8.31E-07	9.18E-07	5.80E-06	5.50E-07	2.90E-01	2.65E-02	5.92E-07	5.28E-07	1.07E-01	3.27E-03
16	6.17E-07	9.55E-07	1.22E-05	3.57E-07	2.81E-01	6.24E-07	6.58E-07	5.21E-07	1.49E-01	2.54E-03
17	1.74E-06	1.50E-06	1.63E-06	2.31E-03	5.80E-01	2.65E-02	2.04E-06	1.77E-06	1.49E-01	1.41E+00
18	9.20E-07	1.10E-06	5.99E-07	6.01E-07	6.15E-01	2.25E-01	8.30E-07	5.78E-07	2.85E-06	6.36E-04
19	7.09E-07	1.23E-06	5.04E-07	6.34E-07	4.16E-01	1.60E-01	9.73E-07	1.15E-06	1.07E-01	1.85E-03
20	6.24E-07	9.46E-07	3.94E-07	5.39E-07	2.94E-01	1.35E-01	5.31E-07	5.89E-07	2.98E-06	3.50E-04
21	8.07E-07	8.84E-07	3.55E-07	3.06E-06	5.18E-01	1.93E-01	1.36E-06	8.22E-07	1.07E-01	4.41E-03
22	7.53E-07	7.83E-07	4.40E-07	5.57E-07	4.41E-01	1.46E-01	1.36E-06	7.62E-07	1.80E-01	1.56E-03
23	6.08E-07	9.61E-07	3.55E-07	4.06E-07	3.60E-01	3.69E-02	6.05E-07	5.29E-07	1.07E-01	2.07E-03
24	6.71E-07	1.13E-06	3.58E-07	3.92E-07	3.38E-01	1.43E-01	6.37E-07	9.25E-07	1.80E-01	1.19E-03
25	9.37E-07	1.17E-06	2.93E-05	7.08E-06	3.60E-01	9.90E-02	4.63E-07	5.80E-07	1.49E-01	2.84E-03
26	8.07E-07	8.84E-07	3.55E-07	3.06E-06	5.18E-01	1.93E-01	1.36E-06	8.22E-07	1.07E-01	1.25E-03
27	6.81E-07	7.45E-07	4.11E-07	4.33E-07	5.58E-01	9.90E-02	4.22E-07	9.90E-02	1.80E-01	4.49E-04
28	6.41E-07	7.43E-07	3.61E-07	3.08E-07	3.01E-01	7.04E-02	5.76E-07	6.86E-07	1.07E-01	2.49E-03
29	6.65E-07	8.10E-07	3.12E-07	3.45E-07	2.30E-01	8.18E-02	5.80E-07	7.32E-07	1.80E-01	1.03E-03
30	5.71E-07	8.76E-07	9.32E-05	3.83E-05	3.18E-01	5.02E-02	3.12E-07	7.68E-07	1.07E-01	4.22E-03

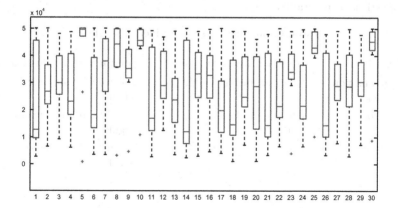

Fig. 1. Boxplots for average number of function evaluations of 30 different inertia weight strategies on 10 benchmark functions as per Table 3

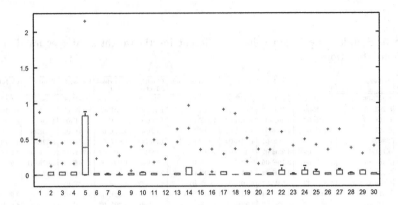

Fig. 2. Mean error value of 30 different inertia weight strategies on 10 benchmark functions as per Table 4

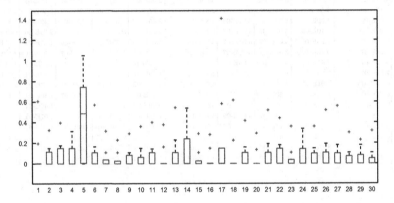

Fig. 3. Standard deviation value of 30 different inertia weight strategies on 10 benchmark functions as per Table 5

Table 6. Summary results for inertia weight

Criterion	Best inertia weight strategy	Worst inertia weight strategy
Average function evaluation	Constant inertia weight	Linear decreasing inertia weight
Mean error	Chaotic random inertia weight	Global-local best inertia weight
Standard deviation	Chaotic random inertia weight	Global-local best inertia weight

If the comparison is made through standard divisions (SD's)the chaotic random inertia weight produces near optimal solutions in comparison to other inertia weight strategies as shown in Fig. 3. The summary results of inertia weight strategies are shown in Table 6.

4   Conclusion

This paper presents the significance of inertia weight strategies in the solution search process of particle swarm optimization (PSO). Here, total 30 inertia weight strategies in PSO are analyzed in terms of efficiency, reliability and robustness while testing over 10 complex test functions. Through boxplots and success rate, it is found that the chaotic random inertia weight is better in terms of accuracy while constant inertia weight performs better in terms of efficiency of PSO among the considered inertia weight strategies.

References

1. Eberhart, R.C., Kennedy, J., et al.: A new optimizer using particle swarm theory. In: Proceedings of the Sixth International Symposium on Micro Machine and Human Science, New York, NY, vol. 1, pp. 39–43 (1995)
2. Chauhan, P., Deep, K., Pant, M.: Novel inertia weight strategies for particle swarm optimization. Memetic Comput. 5(3), 229–251 (2013)
3. Bansal, J.C., Singh, P.K., Saraswat, M., Verma, A., Jadon, S.S., Abraham, A.: Inertia weight strategies in particle swarm optimization. In: Third World Congress on Nature and Biologically Inspired Computing (NaBIC), pp. 633–640. IEEE (2011)
4. Shi, Y., Eberhart, R.: A modified particle swarm optimizer. In: The 1998 IEEE International Conference on IEEE World Congress on Computational Intelligence Evolutionary Computation Proceedings, pp. 69–73. IEEE (1998)
5. Eberhart, R.C., Shi, Y.: Tracking and optimizing dynamic systems with particle swarms. In: Proceedings of the Congress on Evolutionary Computation, vol. 1, pp. 94–100. IEEE (2001)
6. Xin, J., Chen, G., Hai, Y.: A particle swarm optimizer with multi-stage linearly-decreasing inertia weight. In: International Joint Conference on Computational Sciences and Optimization, CSO 2009, vol. 1, pp. 505–508. IEEE (2009)
7. Arumugam, M.S., Rao, M.V.C.: On the performance of the particle swarm optimization algorithm with various inertia weight variants for computing optimal control of a class of hybrid systems. Discrete Dyn. Nat. Soc. 2006, 1–17 (2006)
8. Al-Hassan, W., Fayek, M.B., Shaheen, S.I.: PSOSA: an optimized particle swarm technique for solving the urban planning problem. In: The 2006 International Conference on Computer Engineering and Systems, pp. 401–405. IEEE (2006)
9. Chen, G., Huang, X., Jia, J., Min, Z.: Natural exponential inertia weight strategy in particle swarm optimization. In: The Sixth World Congress on Intelligent Control and Automation, WCICA 2006, vol. 1, pp. 3672–3675. IEEE (2006)
10. Feng, Y., Teng, G.-F., Wang, A.-X., Yao, Y.-M.: Chaotic inertia weight in particle swarm optimization. In: Second International Conference on Innovative Computing, Information and Control, ICICIC 2007, pp. 475–475. IEEE (2007)
11. Malik, R.F., Rahman, T.A., Hashim, S.Z.M., Ngah, R.: New particle swarm optimizer with sigmoid increasing inertia weight. Int. J. Comput. Sci. Secur. 1(2), 35–44 (2007)
12. Kentzoglanakis, K., Poole, M.: Particle swarm optimization with an oscillating inertia weight. In: Proceedings of the 11th Annual conference on Genetic and evolutionary computation, pp. 1749–1750. ACM (2009)

13. Gao, Y.-l., An, X.-h., Liu, J.-m.: A particle swarm optimization algorithm with logarithm decreasing inertia weight and chaos mutation. In: International Conference on Computational Intelligence and Security, CIS 2008, vol. 1, pp. 61–65. IEEE (2008)

14. Li, H.-R., Gao, Y.-L.: Particle swarm optimization algorithm with exponent decreasing inertia weight and stochastic mutation. In: Second International Conference on Information and Computing Science, ICIC 2009, vol. 1, pp. 66–69. IEEE (2009)

15. Arasomwan, M.A., Adewumi, A.O.: On adaptive chaotic inertia weights in particle swarm optimization. In: 2013 IEEE Symposium on Swarm Intelligence (SIS), pp. 72–79. IEEE (2013)

16. Hsieh, S.-T., Sun, T.-Y., Liu, C.-C., Tsai, S.-J.: Efficient population utilization strategy for particle swarm optimizer. IEEE Trans. Syst. Man Cybern. Part B Cybern. 39(2), 444–456 (2009)

17. Lei, K., Qiu, Y., He, Y.: A new adaptive well-chosen inertia weight strategy to automatically harmonize global and local search ability in particle swarm optimization. In: 1st International Symposium on Systems and Control in Aerospace and Astronautics, ISSCAA 2006, p. 4. IEEE (2006)

18. Shen, X., Chi, Z., Yang, J., Chen, C.: Particle swarm optimization with dynamic adaptive inertia weight. In: 2010 International Conference on Challenges in Environmental Science and Computer Engineering (CESCE), vol. 1, pp. 287–290. IEEE (2010)

19. Jiao, B., Lian, Z., Xingsheng, G.: A dynamic inertia weight particle swarm optimization algorithm. Chaos, Solitons Fractals 37(3), 698–705 (2008)

20. Peram, T., Veeramachaneni, K., Mohan, C.K.: Fitness-distance-ratio based particle swarm optimization. In: Proceedings of the 2013 IEEE Swarm Intelligence Symposium, SIS 2003, pp. 174–181. IEEE (2003)

21. Li, L., Xue, B., Niu, B., Tan, L., Wang, J.: A novel particle swarm optimization with non-linear inertia weight based on tangent function. In: Huang, D.-S., Jo, K.-H., Lee, H.-H., Kang, H.-J., Bevilacqua, V. (eds.) ICIC 2009. LNCS (LNAI), vol. 5755, pp. 785–793. Springer, Heidelberg (2009). doi:10.1007/978-3-642-04020-7_84

22. Chatterjee, A., Siarry, P.: Nonlinear inertia weight variation for dynamic adaptation in particle swarm optimization. Comput. Oper. Res. 33(3), 859–871 (2006)

23. Fan, S.-K.S., Chiu, Y.-Y.: A decreasing inertia weight particle swarm optimizer. Eng. Optim. 39(2), 203–228 (2007)

24. Ting, T.O., Shi, Y., Cheng, S., Lee, S.: Exponential inertia weight for particle swarm optimization. In: Tan, Y., Shi, Y., Ji, Z. (eds.) ICSI 2012. LNCS, vol. 7331, pp. 83–90. Springer, Heidelberg (2012). doi:10.1007/978-3-642-30976-2_10

25. Adewumi, A.O., Arasomwan, A.M.: An improved particle swarm optimiser based on swarm success rate for global optimisation problems. J. Exp. Theoret. Artif. Intell. 28, 1–43 (2014)

26. Rashedi, E., Nezamabadi-Pour, H., Saryazdi, S.: GSA: a gravitational search algorithm. Inf. Sci. 179(13), 2232–2248 (2009)

Information Retrieval in Web Crawling Using Population Based, and Local Search Based Meta-heuristics: A Review

Pratibha Sharma[✉], Jagmeet Kaur, Vinay Arora, and Prashant Singh Rana

Computer Science and Engineering Department,
Thapar University, Patiala, Punjab, India
pratibhasharma941@gmail.com

Abstract. The exponential growth and dynamic nature of the world wide web has created challenges for the traditional Information Retrieval (IR) methods. Both issues are the imperative source of problems for locating the information on web. The crawlers expedite web based information retrieval systems by following hyperlinks in web pages to automatically download new and updated content. The web crawlers systematically traverse the web pages, and fetch the information *viz.* nature of the web content, hyper-links present on the web page, *etc.* This paper reviews and compares the meta-heuristic approaches like population based, evolutionary algorithms, and local search used for IR in web crawling. This paper reviews how these techniques has been developed, enhanced and applied.

Keywords: Web crawling · Meta-heuristics · Information Retrieval

1 Introduction

The Internet online service has become an indispensable part of our routine life. Now, people are able to develop their own website or participate in various other online web services provided by the host owner. Moreover, general public can add new web contents or edit the available web content dynamically any time and anywhere. These competencies expedite various web sites to transform their web design from static to dynamic. The Web has undergone an exponential growth in the past few years. It has been found from the survey that approximately 15–20 billion web pages are present on the Web and recently this count exceeded to 1 trillion. The key causes of the success of the World Wide Web are its rapid growth and its dynamic nature. The lack of a centralized control over the web contents is also the main issue. In the World Wide Web, earlier Information Retrieval systems are challenged to traverse the most relevant pages on the web. Today, search engines are the main gateways of information access on the WWW. Web search engines can take together and handle hundreds of millions of queries everyday. They give responses to various user queries by accessing a local

© Springer Nature Singapore Pte Ltd. 2017
K. Deep et al. (eds.), *Proceedings of Sixth International Conference on Soft Computing for Problem Solving*, Advances in Intelligent Systems and Computing 547,
DOI 10.1007/978-981-10-3325-4_10

repository that mirrors the Web. Crawlers facilitate this by following hyperlinks in Web pages to download automatically latest and updated Web pages. While some systems rely on spiders that crawl the Web exhaustively, others incorporate focus within their crawlers to yield application-specific collections or topic-specific collections. Web crawling makes it easier for search engines to give the most imperative results to users queries. The Crawlers traverse through a website and index each web page accordingly. At each stage of the crawling process a URL is removed from the pending queue and the analogous page is retrieved. URLs are extracted from the retrieved page and some or all of these URLs are pushed back into the pending queue for subsequent processing. For performance, crawlers often use asynchronous input/output to allow more than one pages to be downloaded simultaneously or can be structured as multi threaded programs so that each thread executes the steps of the crawling process concurrently with the others. The working model of a web crawler is shown in Fig. 1.

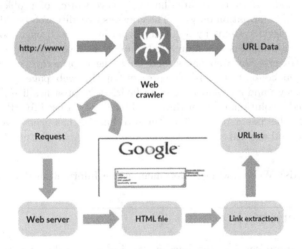

Fig. 1. Working model of web crawler

In order to optimize the web crawling process [1], several meta-heuristic techniques such as population based, evolutionary algorithms, and local search has been used. Their optimization would have a prominent effect on enhancing the searching efficiency and crawling performance. The area of meta-heuristics [2] has been expanded immensely in the last two decades as a panacea to real-world optimisation problems. These techniques work well in certain situations where exact optimisation methods fail to provide adequate results. For intricate kind of problems (Nondeterministic polynomial time-hard problems), these techniques are able to generate the best solution in relatively much lesser time than traditional optimization techniques. Meta-heuristics [3] find applications in a wide ambit of areas including finance, planning, scheduling and engineering design. This paper presents a review of various meta-heuristic techniques and their methodology. In this paper, web crawling using various meta-heuristic techniques are reviewed and compared.

2   Crawling Using Meta-heuristics

Web crawlers [4] are programs that utilise the graph structure of the World Wide Web to link one page to another. Crawling the world wide web is delusively easy: the fundamental algorithm is (a) Fetch a page, (b) Parse the page to extract all linked URLs, (c) For all the unseen URLs, repeat steps (a) to (c). The Internet and web are becoming more dynamic as compared to earlier times in terms of their content as well as their usage. Hence, the rapid growth of the web and its increasing rate of change is moving from a trivial programming approach to a significant algorithmic and system design challenge. Moreover, information retrieval efforts also aim to match with this dynamic scenario by designing intelligent systems. In the last 10–12 years or so, number of researchers from various fields have been engaged in developing intelligent and adaptive systems for web-based Information Retrieval. Meta-heuristics techniques [5] are emerging as typical examples for such systems. Meta-heuristics are formally defined as an iterative generation process that guides a subordinate heuristic by combining intelligently different concepts for exploring and exploiting the search space. Learning strategies are used to structure information in order to find efficiently near-optimal solutions (Osman & Laporte, 1996). Metaheuristics developed an equilibrium between intensification and diversification of the search space. In the past 1–2 decades, several meta-heuristic techniques have been proposed such as genetic algorithm, genetic programming, simulated annealing, ant colony optimisation, particle swarm optimisation, and differential evolution. This paper reviews how these techniques has been used in crawling the web content.

3   Population Based Techniques

3.1   Web Crawling Using Evolutionary Techniques

This section outlines the work related to evolutionary algorithms [6] as a tool for enhancing the capabilities of crawlers as an optimizer or learner. According to the latest trends many researchers are using Genetic Algorithm (GA) both to solve IR problems and to optimise web crawling (Fig. 2).

Using Genetic Algorithms. Until 1980 there were no significant attempts in the field of applying genetic algorithms in web crawling and information retrieval. Raghavan et al. were the first ones who initiated the topic by trying to optimize document clustering using genetic algorithms [7]. Then, Gordon et al. enhanced the document descriptors by proposing a new approach [8]. The fashion was then followed by Yang et al. who improved the weights of keywords related to the specific document topic [9]. Petry et al. also used genetic algorithms to

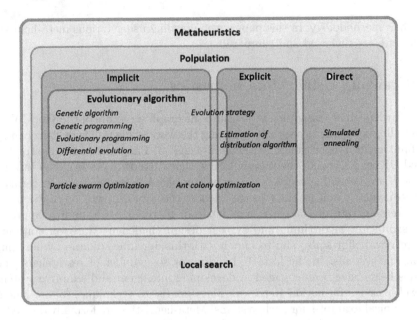

Fig. 2. Metaheuristic techniques

Table 1. Various metaheuristic techniques explored

Metaheuristics	Sub-techniques	Related research paper
Evolutionary algorithms	Genetic Algorithm	[7–28]
	Genetic Programming	[29–38]
	Differential evolution	[39,40]
	Evolutionary programming	[41–48]
	Evolutionary strategy	[49,50]
Implicit	Particle Swarm Optimization	[51–57]
	Ant Colony Optimization	[58–61]
Direct	Simulated Annealing	[62–64]
Local Search	-	[65–74]

boost the process of retrieving content from a collection of weighted indexed documents, by changing the weights of queries [10]. Genetic algorithm (GA) has been used by Chen, et al. [11] to develop a global and personal spider. A set of standard experiments are performed to compare the performance and efficiency of the Best first search (BFS) and GA algorithms based spiders. Jaccard's function has been used to calculate score function for the web pages. Mutation is a process used in GA to find the relevant pages. Author examined the final Jaccards scores of homepage suggestions and noted their CPU times and system times. High Jaccard's score shows high relevance of user's query. A metasearch engine is developed by Z.N. Zacharis et al. [12] that post queries and collects the results from different search engines. An intelligent agent is proposed using metagenetic algorithm which gathers and suggests web pages according to users interest (Table 1).

Using metagenetic algorithm is an easy to learn and adaptive approach. Judy Johnson et al. [13] used genetic algorithm in emerging scenarios by evaluating the previous search queries made by the user. The resultant queries are criticized on the basis of rank function that combines text contents and hyperlinks of the query. This results in better performance compared to earlier used best first strategies. An Intelligent crawler with online processing facility is proposed by Milad shokouhi et al. [14]. He introduced an intelligent agent that uses genetic algorithm named Gcrawler. Gcrawler finds the best path for crawling and expands the keyword by using genetic algorithm. Superiority of this algorithm is there is no need to interact with users. Ibrahim Kushchu et al. [15] presented applications of Evolutionary and Adaptive Systems (EAS) in Web-based (IR). Research is done in two phases: In first phase, researchers often use GA or GP as an optimizer in the contexts of clustering, query improvement, or keyword selection. In second phase, adaptive intelligent agents are employed and evolutionary methods are used as learning mechanisms. EAS approach is more domain independent and flexible to create dynamic web.

Jialun Qin et al. [16] used a genetic algorithm with focused crawling for improving its crawling performance. An advanced crawling technique is proposed to build domain-specific collections for web search engines that consolidate a global search algorithm. The proposed work apply a genetic algorithm to the focused crawling process to find more relevant Web pages in order to overcome the drawbacks of traditional focused crawlers. Web site clustering consists in finding meaningful groups of related web sites. Esteban Meneses et al. [17] analyzed two models and four associated algorithms: vector models are analyzed with k-means and self-organizing maps (SOM), graphs are analyzed with simulated annealing and genetic algorithms and these algorithms are tested by clustering some web sites. The results for clustering this web site collection using both models are compared and show what kind of clusters each one produces. An improved search engine has been proposed by M. Koorangi et al. [18] to increase the efficiency of the web searches. Stationary and mobile agents have been used collectively to enhance the performance. The meta search engine gives required documents needed by user and collects the results obtained by search engine which results in more relevant documents and reduced network traffic. Efficiency has been improved by expunging traffic from the network. In focussed crawling, semantic instances in a web space are given to multiple crawlers to access them and further send these instances to a centralized meta-crawler for integration.

Huilian Fan et al. [19] worked on a strategy to improve the search efficiency of focused web crawler. Taking the combination of advantages of hyperlink structure and web content strategies, he has derived a new strategy which uses hyperlinks as genetic individuals and Vector Space Model (VSM) to calculate individual fitness. To apply crossover and mutation, new URLs are imported. URLs having same prefix are considered as vocation. Gives higher precision and recall as compared to BFS and HITS algorithms. Yuan Xu et al. [20] used Standard TF-IDF for weighting method. Only the average of parents with higher fitness has been used to generate offsprings. The chromosomes discarded are half of the total chromosomes in a generation. Nguyen Quoc Nhan et al. [21] used focused crawler to traverse the web by selecting relevant pages. They proposed a crawling system in order to improve the performance of crawler. Along with searching the best path to follow, the system also traverses its initial keywords using GA. A hybrid approach (consists automata and POS tagging techniques) is used to crawl Vietnamese web pages. Approach used here gives better performance than BFS crawler.

Hui Ning et al. [22] analyzed URL analysis models used in existing focused crawler and proposed a new model using improved GA to collect topic pages. Two existing models has been considered: URL Analysis Model based on content analysis and URL Analysis Model based on link analysis. A vector space model has been introduced for formal representation of words and the count of their occurrences. The proposed algorithm improves accuracy rate, recall rate and other quotas effectively and handles problems of getting into the local optimal solution. Focused crawler intends to explore the Web conform to a specific topic. Banu Wirawan Yohanes et al. [23] discussed the problems caused by local searching algorithms. Crawler can be trapped within a limited Web community and overlook suitable web pages outside its track. A genetic algorithm as a global searching algorithm is modified to address the problems. The genetic algorithm is used to optimize web crawling and to select more suitable web pages to be fetched by the crawler. The utilization of genetic algorithm had empowered focused crawler to traverse the Web comprehensively, despite it relatively small collections. It brought up a great potential for building an exemplary collections compared to traditional focused crawling methods.

Somya Ravi et al. [24] shows how search engines work and evolutionary Algorithm (EA) is used to improve the searching technique by considering the previous searches. Genetic algorithm is a heuristic technique used to search web pages and indexing them. Here EA use previous results and computes new results based on it has learnt from previous searches. Colin G. Johnson et al. [13] explored the idea of applying evolutionary algorithms to search spaces that are defined rationally (i.e. by listing every item in the space). Metaphors of mutation and crossover are defined when the search spaces are within a function that returns similar elements as the given key. Huynh Thi Thanh Binh et al. [25] used Genetic algorithm to create one child from multiple parents. Given a set of keywords and the crawler crawls the set and extract the suitable keywords using genetic algorithm. Different approaches are used to select parents (i.e. roulette wheel selection, rank selection, random selection, levenshtein distance). GA includes five components: chromosome representation, fitness function formation, parent selection, crossover operator and mutation operator. In order to enhance crawler performance, author has used different recombination approaches to select multiple parents and to create child.

Vikas Thada et al. [26] used same set of documents for performance analysis of different similarity coefficients and hence selects the best combination of probability of Crossover (proC) and probability of Mutation (proM) to achieve highest relevancy amongst the similarity coefficients, searching becomes easier using RT and BUB. Use of focused crawlers improves the crawling performance. Pathak et al. [27] used internet service engine for web information retrieval system. It explores new techniques of web information retrieval system and deals with the large size and dynamic nature of web and finds the effective techniques of information retrieval system. Pranali Kale et al. [28] proposed a technique to choose the most promising links in order and tried to maximize the relevancy of a new, unvisited URL. By applying GA for focused web search, optimal results are obtained. Result obtained by this algorithm is more precise than previously used techniques and more relevant links can be searched in less time using genetic algorithm.

Using Genetic Programming. The enlargement in the number of Web pages has led to the advancement and rectification of several search engines. Walker, Reginald L. [29] used search strategies behind the current search engines for the World Wide Web to determine the application of Genetic Programming (GP) to explore the Web documents. This appraisal is applied to the fusion of Web content that have not yet been developed. Results in this approach can be optimized by repeating the search over period of time. Same technique is used by Reginald L. Walker after some period to enhance the performance of system. Search strategies followed by current search engines for the web to determine the application of genetic programming model which results in parallel implementation of pseudo search engine simulator.

Weiguo Fan et al. [30] proposed an information retrieval approach to rank the web pages using GP. Taking into consideration a huge corpus of web links, fitness functions has been discovered that results in better performance. Order-based fitness function based on utility theory is compared with existing order-based fitness functions to find the ranking function. This task has been performed by using GP resulting in upgraded performance. Weiguo Fan et al. [31] generated ranking function using genetic programming in which search engine's performance is calculated by ranking functions. He discovered a ranking function for structural information that works on group of queries. Two experiments are performed to test ranking function: first experiment takes queries from user and the second uses queries relevant to previous searches constructed automatically. Ranking function is working on not only unstructured and semi structured information but also on structural information.

De Almeida, Humberto Mossri, et al. [32] performed experiments with the TREC-8 and WBR99 collections. Results indicate that the use of meaningful components in a GP-based structure leads to elective ranking functions that notably defeat the baselines (standard TF-IDF, BM25 and another GP-based approach). CCA improves the retrieval performance compared to standard TF-IDF, BM25 and other GP-based approaches. It eliminates the overfitted data.

Li Piji et al. [33] has introduced a ranking model, WIRank-Web Image Rank, GP is used to automatically generate ranking function, combining together different types of features of web images. Important element is temporal information (imgFileDate and imgPageDate here), which strongly affects ranking. Comparison is done between ranking results obtained by WIRank with the classical ranking strategy PageRank and other strategies such as VisualRank and ImageRank. Layered implementation of GP (LAGEP) model provided significant improvement in relevancy. Mahdi Bazarganigilani et al. [34] proposed an efficient technique to search relevant pages on web by using GP. GP results in the best possible relevancy among the search engines overcoming accuracy of simple SVM and combination based SVM. Decay methods have been used to score pages on the web and it provides constraint to crawler for searching only relevant pages. To check the strength of GP, DMOZ resources have been used by taking 30% and 10% of resources from it. The performance of GP has been compared with simple SVM and combinatorial SVM resulting best accuracy and efficiency. Aecio S.R. Santos et al. [35] designed a Genetic programming for Crawling (GP4C) framework that has been used to find score function that is used by scheduler to rank the pages crawled by web crawler. GP4C works in two phases: training and validation. In training phase, a set of web pages are trained and in validation phase, the distinct pages so obtained are validated. The best resultant function GP4C is flexible to all the baselines in download cycle. 5-fold cross validation technique has been used by considering four folds for training and validating sets and one fold for testing phase. It seems to be flexible framework to work into and the final score function can be derived by applying new features.

Amir Hosein Keyhanipour, Behzad Moshiri [36] introduced a Layered Multi-Population genetic programming model to the problem of web spam classification. Experiment is done in two phases: first phase is feature selection and second phase is for comparison. Web spam degrades the performance of rank function in search engines. Feature selection is done on data set based on crawls of the UK Web domain and initial population of classifiers has been provided. Two tasks have been accomplished: Computation of evaluation metrics and comparison of the performance of the proposed method with existing algorithms. Genetic programming framework provides a parallel and extensive search to find the near-global optimum results. Genetic Programming framework has been used in order to find freshness and coverage of the content collected by web crawler. Aecio S. R. Santos et al. [37] generated a score function to rank the web pages taking in account the probabilities of being modified. GP4C is evaluated using the ChangeRate and Normalized Discounted Cumulative Gain (NDCG) factors. NDCG is used to maximize the objective function. A GP framework is implemented to automatically generate score function which helps web scheduler to rank web pages. NDCG has been used as fitness function which makes result more robust. GP is used for learning the desired ranking functions. Keyhanipour, Amir Hosein, et al. [38] introduced a new symbolic feature to represent images: secular Information, which is hardly ever exploited in the current information retrieval system. The experimental results show that the proposed algorithms are efficient in acquiring active ranking functions for web image retrieval. Significant enhancement in relevancy is obtained in comparison to some other well-known ranking techniques. GP techniques used in the approach is to find arbitrary target functions. MGP-Rank can utilize such information effectively and out-perform other ranking systems.

Using Differential Evolution. Danushka Bollegala et al. [39] proposed a Rank DE method which uses DE algorithm to rank the pages received by search engine. It is an approach to learn ranking function for information retrieval on LETOR data set. The ranking function is analyzed by taking weights into account. DE performs better than other evolutionary algorithms. A document summarization model has been presented by Rasim M. Alguliev et al. [40] which extracts pri-mary sentences from given corpus while reducing information in the summary. The model is represented as a discrete optimization problem. To solve this dis-crete optimization problem a self-adaptive DE algorithm has been created. The model is implemented on multi-document summarization function. Experiments performed shows that the proposed model is preferred over summarization sys-tems. The resulting summarization system based on the proposed approach is competing on the DUC2002 and DUC2004 data sets.

Using Evolutionary Programming. Reynolds et al. [41] proposed an algo-rithm for solving problems and knowledge extraction. Population and space com-ponents are implemented as agents that interact with web pages, web services and user. Cultural algorithm has been used for search queries and extracting use-ful information. Belief space components act on the data returned by queries, consisting of URLs. It handles additional constraint criteria and retrieve signif-icant and relevant information. Mahdavi et al. [42] has done clustering of web content for searching and extracting knowledge from the web. A novel hybrid harmony search (HS) algorithm has been proposed for clustering the web content that finds a globally ideal partitions of them into a specified number of clusters.

Zhang et al. [43] worked on search engines using topical crawlers [44] by deter-mining seed URLs and downloading the web Pages from the internet. It increased the crawling performance and focuses on semantics but not on keywords. Cec-chini Roco L., et al. [45] considered the problem of context based search as a multi-objective optimization problem and discussed EA-based strategies to search queries. An automated technique is proposed to work on training corpus. The sys-tem attempts to increase the potential of the system to redeem relevant pages while also keeping non-relevant ones. A novel Binary Ant System Harmony Search algorithm (BASHS) has been proposed by Wang, Ling, et al. [46] to extend HS to tackle the binary-coded optimization problems efficiently and effectively. The proposed algorithm is an effective boosted tool in terms of search accuracy and convergence speed and it enhance the global search ability. BASHS has excellent optimization ability and achieves the best performance. Johnson Colin G. [47] pro-posed the methods of integrating evolutionary algorithms to search spaces that are defined by listing data in the search space. Gao et al. [48] introduced the modeling on dynamic web evolution and incremental crawling strategy and concerns about the refresh interval with minimum waiting time.

Using Evolutionary Strategy. Jason J. Jung [49] used meta-crawler to handle the relevant semantic instances and access updated instances in the web space. Meta-crawler used four functions: Knitting, merging, translating and comparing. Evolution strategy is used to efficiently build large web spaces (i.e. local ontologies) with similar structure. Conflicts between the semantic instances obtained from multiple crawlers have been minimized. Rana Forsati et al. [50] presented novel document clustering algorithms based on the Harmony Search (HS) optimization method. An evolution based clustering algorithm has been proposed that finds near-optimal clusters within a reasonable time. Harmony clustering is then integrated with the K-means algorithm in three ways to achieve better results by combining the fundamental power of HS with the refining power of the K-means.

3.2   Web Crawling Using Implicit Metaheuristics

Using Particle Swarm Optimization. Particle Swarm Optimization (PSO) is an estimation method that upgrades a problem by repeatedly trying to enhance a candidate key with observance to a given quality measure. Information Retrieval (IR) techniques are selected to extract interesting documents from large corpus built by high-performance crawlers. Fabio Gasparetti and Alessandro Micarelli [51] presented a robust and modular web search system, which is inspired from ant foraging behavior. Its target is to search autonomous information about distinct topics, in a huge corpus, such as the Web. The algorithm has proven to be vigorous and flexible to users information. Tao Peng et al. [52] discussed an approach to build classifier based on PSO to evaluate the weights of all classifiers generated as PSO finds the best weight combinations. Author has built a focused crawler based on link-contexts supervised by different classifiers to criticize the method using multiple crawls over 37 topics covering hundreds of thousands of pages. In harvest rate and target recall, focused web crawler guided by PSO-based classifier outperforms other several classifiers.

Anna Bou Ezzeddine [53] used a bee hive model for information retrieval. The model implemented shows a multi agent system that uses a bee as a confined agent for retrieving information from Web. An adapted model has been used to trace the searches that are input on the web representing a new approach to information retrieving. Proposed solution using particle swarm method as compared to conventional searching is more flexible and more effective source searching. In search engines, Information retrieval (IR) is one of the most vital components. Due to dynamic nature of web, optimization of IR would have a great effect in improving the efficiency of search engine. Priya I. Borkar and Leena H. Patil [54] presented a model using hybrid Genetic Algorithm-Particle Swarm Optimization (HGAPSO) for retrieving information from the web. HGAPSO works by expanding the search query to expand the new keywords that are related to the user search and also it helps to revolute a user query to improve the results of the corresponding search. Sumathi Ganesan et al. [55] used PSO to identify users with similar interest on the basis of previous searches and cluster them in one cluster. Using PSO technique, clustering of web content is done on the basis of ranks given to search queries. This technique performed better than other techniques like DBSCAN and KMeans.

Shafiq Alam et al. [56] performed four different experiments: (1) detect outliers based on average intra-cluster distance, (2) detect outliers based on maximum intra-cluster distance, (3) detect outliers using the intersection of maximum and average intra-cluster distance, and lastly, (4) identify the synthetically injected web bots and evaluate using traditional precision and recall measures. Proposed approach has successfully identified outliers. The avg. intra-cluster distance, max intra-cluster distance is close to 100% and search space is reduced. In order to solve global optimization problems, James J.Q. Yu and Victor O.K. Li [57] proposed a novel Social Spider Algorithm (SSA). A new social animal foraging strategy model to solve optimization problems has been introduced. Swarm intelligence algorithms which uses Ant Colony Optimization (ACO) and Particle Swarm Optimization (PSO) has been used to find optimal solution. Proposed algorithm has superior performance compared with other state-of-the-art meta-heuristics.

Using Ant Colony Optimization. For hypertext graph crawling, Piotr Dziwinski and Danuta Rutkowska [58] presented a new algorithm based on ants behavior to crawl web. An ant has been used as an agent to extract significant number of irrelevant hypertext documents to extract a given relevant document. During crawling, artificial ants need not to follow a queue to central control crawling process. Proposed algorithm i.e. Focused Ant Crawling Algorithm, for hypertext graph crawling, is better than the conventional crawling algorithm and the best-first search strategy that utilizes a queue for the central control of the crawling process. In order to attain accessibility and performance information Ekachai Jinhirunkul et al. [59] utilized Ant Colony Optimization (ACO) algorithm to analyze an unexplored web site, alignment of its structure and voyage routing. As a sequence, all reachable nodes within the denominated web site can be covered, along with prerequisite performance statistics to emulate near optimal accessible paths to any given node in the web site. By virtue of the simplicity of the Ant Colony Optimization algorithm, some candid mapping techniques were signed to bring about strategic degradation of the proposed algorithm.

Zhen Zhang et al. [60] introduced the computational architecture of Ants, and explained the mechanism of communication and cooperation among them. AntCrawlers architecture consists of a search engine that uses a certain number of crawling agents (called Ants) to search the web for some requested content. These Ants traverse the web by starting from a page and continue by following links to other pages. Ants present a new point of view for further development in the field of focused search. An ACO based algorithm to predict web usage patterns is presented by Trapti Agrawal et al. [61]. It consolidates content, structure as well as web usage data. Taking leverage of ant based optimizing principle, the web users are considered as artificial ants, and use the ant theory as an analogy to model users choice in the website. The proposed algorithm fares amazingly well, provided that a few parameters are suitably tuned. Most of the ants reach to the uppermost threshold most of the time which directly turns into increased prediction correct rate. So it results in increasing the overall efficiency of the learning algorithm.

3.3   Web Crawling Using Direct Metaheuristics

Using Simulated Annealing. C. Yang et al. [62] proposed an approach based on automatic analysis of searches and search agents based on hybrid simulated annealing supporting global search on internet. Comparison is made between the performance of the simulated annealing spider and best-first search spiders. It is operated without human supervision resulting in higher fitness score, higher precision and recall values. Lam et al. [63] used Web information discovery (WID) to discover the relevant pages. Information retrieval techniques to find the relevancy between search query with the content in resulted links. Result depends on relevance score of the web pages. It is an intelligent, fast and automatic web traversal method. Relevance feedback is taken by users to modify the technique. Meneses et al. analyzed two models and four associated algorithms: vector models are analyzed with k-means and self-organizing maps (SOM), graphs are analyzed with simulated annealing and genetic algorithms. These algorithms are tested by clustering some web sites. The results for clustering this web site collection using both models are compared and show what kind of clusters each one produces. Yuan Xu et al. used Standard TF-IDF for weighting method. Only the average of parents with higher fitness are used are generate offspring. The chromosomes discarded are half of the total chromosomes in a generation. Luokai Hu et al. [64] handles the issues with large scale data optimization. It combines the advantages of both simulated annealing and map-reduce. High performance parallel platform outperforms its traditional genetic rivals.

4   Web Crawling Using Local Search

A local Search is Dynamic and memory less algorithm. These modify their objective function during the search process and make the use of information that is collected during the iterative process. Crawling web using local search, Angelaccio et al. [65], described visualization of web that has been used for browsing and viewing search engine results. Local search tool combines the browsing and searching models. Given a prototype description of query model and the corresponding agent architecture for a generic local search tool that includes a system implemented via user interface scripts. It is viewed as a structure useful for many public domain local search tools whose visualisation is often incomplete.

The problems in traditional focused crawler due to local search algorithms were explored by Qin, Jialun et al. [66] and a new domain-specific collection building approach is proposed to address these problems. The approach is used to collect web pages with high precision and high variance. By local search algorithm, best-first search can only traverse the search space by penetrating its neighbour nodes that are already visited. Lovic et al. [67] proposed an expert system to compile expert knowledge and use it in resolving search issues. Performance issues of local search engine has been analyzed and identified by exploring search logs. A set of rules has been derived to fix these search problems resulting in boosted and boomed performances.

Mundluru et al. [68] discussed about crawling and extracting large amount of local data from deep web resources. The challenges faced are extracting local structured data and allocating address to them. Extracted data has been integrated into a local search engine. Deep web crawlers are flexible to extract data from any number of levels forging it mere to be used by local searches. A new hybrid approach to focused crawling is represented by Sun, Yixue, Peiquan Jin, and Lihua Yue. [69] which is based on meta-search and VIPS (VIsion based Page Segmentation) algorithm. In order to obtain better recall and precision, VIPS-based algorithm is used for to compute relevance of a Web page. As local search is increasingly becoming a major focus point of research interest. It is a widely-recognized specialty search with a large application area. Its data is usually aggregated from a variety of sources. Ahlers, Dirk, and Susanne Boll. [70] Examined the application of focused Web crawling to the geo-spatial domain. The approach is described for a geo-aware focused crawling of urban regions with a high building frequency. Focused crawler is able to maintain a spatial focus effectively for long crawl duration.

Redwan et al. [71] presented local search techniques used for designing and implementation of a search engine system for web documents in Amharic language. The proposed search engine allows indexing and searching of documents written in encoding multiple illustrations. A local search engine is a vertical search engine whose subject moves around a certain geographical area. Huitema, et al. [72] described their experiences of developing a crawler for a local search engine for a city in USA. They focused on crawling and indexing a huge number of highly relevant Web pages, and then demonstrated ways in which the search engine is capable to exceed an technical search engine. Using web pages, a large collection of documents in local search engine have been built. Lu, Yiyao, et al. [73] presented an automatic remarking approach that first adjusts the data segments on a result page into different groups such that the data in the same group presents same data. An annotation for the search site is automatically designed and can be used to remark new result pages from the web database. Neunerdt et al. [74] introduced a new type of specific focused crawler using local search techniques, which uses a segregator based on HTML Meta information. A classification component is proposed which leads to high precision and recall values.

5 Conclusion

After going through the procedure of thorough observations, ample overviews, and the extensive literature survey of the web crawling using various metaheuristics, this can be formulated that precision and recall may be used to compare the performance of the algorithms. As scientific research fields and the web are fast evolving, it is really very difficult to develop web collections with both high precision and high diversity by using traditional techniques. By exploring various metaheuristic approaches to optimize the traditional web crawling methods, it is found that the intelligent systems give better performance. There is still a lot of work to do for improving the efficiency of the algorithms. Although the results by applying metaheuristics to web crawling are encouraging, there is still a long way to achieve the outstanding crawling efficiency. Through extensive studies it is found that genetic algorithm is the fastest technique to find relevant web pages.

As future work, exploring metaheuristic techniques for crawling focused web is a broad area to find the relevant pages by prioritizing the crawl frontier and exploring the hyperlinks.

References

1. Yang, X.-S.: Nature-Inspired Optimization Algorithms. Elsevier, Amsterdam (2014)
2. Dréo, J., Petrowski, A., Siarry, P., Taillard, E.: Metaheuristics for Hard Optimization: Methods and Case Studies. Springer, Heidelberg (2006)
3. Crainic, T.G., Toulouse, M.: Parallel Strategies for Meta-heuristics. Springer, Berlin (2003)
4. Lawrence, S.: Context in web search. IEEE Data Eng. Bull. **23**(3), 25–32 (2000)
5. Senvar, O., Turanoglu, E., Kahraman, C.: Usage of metaheuristics in engineering: a literature review. In: Meta-heuristics Optimization Algorithms in Engineering, Business, Economics, and Finance, pp. 484–528 (2013)
6. Peña, J.M., Robles, V., Larrañaga, P., Herves, V., Rosales, F., Pérez, M.S.: GA-EDA: hybrid evolutionary algorithm using genetic and estimation of distribution algorithms. In: Orchard, B., Yang, C., Ali, M. (eds.) IEA/AIE 2004. LNCS (LNAI), vol. 3029, pp. 361–371. Springer, Heidelberg (2004). doi:10.1007/978-3-540-24677-0_38
7. Raghavan, V.V., Agarwal, B.: Optimal determination of user-oriented clusters: an application for the reproductive plan. In: Genetic Algorithms and Their Applications: Proceedings of the Second International Conference on Genetic Algorithms, 28–31 July 1987, at the Massachusetts Institute of Technology, Cambridge, MA. L. Erlhaum Associates, Hillsdale (1987)
8. Gordon, M.: Probabilistic and genetic algorithms in document retrieval. Commun. ACM **31**(10), 1208–1218 (1988)
9. Yang, J., Korfhage, R.R., Rasmussen, E.: Query improvement in information retrieval using genetic algorithms: a report on the experiments of the trec project. In: Proceedings of the 1st Text Retrieval Conference, pp. 31–58 (1993)

10. Petry, F.E., et al.: Fuzzy information retrieval using genetic algorithms and relevance feedback. In: Proceedings of the ASIS Annual Meeting, vol. 30, pp. 122–125. ERIC (1993)
11. Hsinchun, C., Yi-Ming, C., Ramsey, M., Yang, C.C.: An intelligent personal spider (agent) for dynamic internet/intranet searching. Decis. Support Syst. **23**(1), 41–58 (1998)
12. Zacharis, Z.N., Panayiotopoulos, T.: A metagenetic algorithm for information filtering and collection from the world wide web. Expert Syst. **18**(2), 99–108 (2001)
13. Johnson, J., Tsioutsiouliklis, K., Lee Giles, C.: Evolving strategies for focused web crawling. In: ICML, pp. 298–305 (2003)
14. Shokouhi, M., Chubak, P., Raeesy, Z.: Enhancing focused crawling with genetic algorithms. In: International Conference on Information Technology: Coding and Computing, ITCC 2005, vol. 2, pp. 503–508. IEEE (2005)
15. Kushchu, I.: Web-based evolutionary and adaptive information retrieval. IEEE Trans. Evol. Comput. **9**(2), 117–125 (2005)
16. Qin, J., Chen, H.: Using genetic algorithm in building domain-specific collections: an experiment in the nanotechnology domain. In: Proceedings of the 38th Annual Hawaii International Conference on System Sciences, HICSS 2005, p. 102b. IEEE (2005)
17. Meneses, E.: Vectors and graphs: two representations to cluster web sites using hyperstructure. In: Fourth Latin American Web Congress, LA-Web 2006, pp. 172–178. IEEE (2006)
18. Koorangi, M., Zamanifar, K.: A distributed agent based web search using a genetic algorithm. Int. J. Comput. Sci. Netw. Secur. **7**(1), 65–76 (2007)
19. Fan, H., Zeng, G., Li, X.: Crawling strategy of focused crawler based on niche genetic algorithm. In: Eighth IEEE International Conference on Dependable, Autonomic and Secure Computing, DASC 2009, pp. 591–594. IEEE (2009)
20. Xu, Y., Deli, Y., Yu, L.: Efficient annealing-inspired genetic algorithm for information retrieval from web-document. In: Proceedings of the First ACM/SIGEVO Summit on Genetic and Evolutionary Computation, pp. 1017–1020. ACM (2009)
21. Nhan, N.Q., Son, V.T., Binh, H.T.T., Khanh, T.D.: Crawl topical vietnamese web pages using genetic algorithm. In: 2010 Second International Conference on Knowledge and Systems Engineering (KSE), pp. 217–223. IEEE (2010)
22. Ning, H., Wu, H., He, Z., Tan, Y.: Focused crawler url analysis model based on improved genetic algorithm. In: 2011 International Conference on Mechatronics and Automation (ICMA), pp. 2159–2164. IEEE (2011)
23. Yohanes, B.W., Handoko, H., Wardana, H.K.: Focused crawler optimization using genetic algorithm. TELKOMNIKA (Telecommun. Comput. Electron. Control) **9**(3), 403–410 (2013)
24. Ravi, S., Ganesan, N., Raju, V.: Search engines using evolutionary algorithms. Int. J. Commun. Netw. Secur., ISSN: 2231-1882
25. Binh, H.T.T., Long, H.M., Khanh, T.D.: Recombination operators in genetic algorithm–based crawler: study and experimental appraisal. In: Nguyen, N., Trawiński, B., Katarzyniak, R., Jo, G.S. (eds.) Advanced Methods for Computational Collective Intelligence. SCI, vol. 457, pp. 239–248. Springer, Heidelberg (2013). doi:10.1007/978-3-642-34300-1_23
26. Thada, V., Jaglan, V.: Performance analysis of similarity coefficients in web information retrieval using genetic algorithm
27. Pathak, P., Gordon, M., Fan, W.: Effective information retrieval using genetic algorithms based matching functions adaptation. In: Proceedings of the 33rd Annual Hawaii International Conference on System Sciences, p. 8. IEEE (2000)

28. Kale, P., Dahiwale, P.: Design and implementation of focused web crawler using genetic algorithm
29. Walker, R.L.: Assessment of the web using genetic programming. In: Proceedings of the Genetic and Evolutionary Computation Conference, San Francisco, pp. 1750–1755 (1999)
30. Fan, W., Fox, E.A., Pathak, P., Wu, H.: The effects of fitness functions on genetic programming-based ranking discovery for web search. J. Am. Soc. Inf. Sci. Technol. 55(7), 628–636 (2004)
31. Fan, W., Gordon, M.D., Pathak, P., Xi, W., Fox, E.A.: Ranking function optimization for effective web search by genetic programming: an empirical study. In: Proceedings of the 37th Annual Hawaii International Conference on System Sciences, p. 8. IEEE (2004)
32. de Almeida, H.M., Gonçalves, M.A., Cristo, M., Calado, P.: A combined component approach for finding collection-adapted ranking functions based on genetic programming. In: Proceedings of the 30th Annual International ACM SIGIR Conference on Research and Development in Information Retrieval, pp. 399–406. ACM (2007)
33. Piji, L., Jun, M.: Learning to rank for web image retrieval based on genetic programming. In: 2nd IEEE International Conference on Broadband Network & Multimedia Technology, IC-BNMT 2009, pp. 137–142. IEEE (2009)
34. Bazarganigilani, M., Syed, A., Burki, S.: Focused web crawling using decay concept and genetic programming. Int. J. Data Min. Knowl. Manag. Process (IJDKP) 1(1), 1–12 (2011)
35. Santos, A.S.R., Ziviani, N., Almeida, J., Carvalho, C.R., de Moura, E.S., da Silva, A.S.: Learning to schedule webpage updates using genetic programming. In: Kurland, O., Lewenstein, M., Porat, E. (eds.) SPIRE 2013. LNCS, vol. 8214, pp. 271–278. Springer, Cham (2013). doi:10.1007/978-3-319-02432-5_30
36. Keyhanipour, A.H., Moshiri, B.: Designing a web spam classifier based on feature fusion in the layered multi-population genetic programming framework. In: 2013 16th International Conference on Information Fusion (FUSION), pp. 53–60. IEEE (2013)
37. Santos, A.S.R., de Carvalho, C.R., Almeida, J.M., de Moura, E.S., da Silva, A.S., Ziviani, N.: A genetic programming framework to schedule webpage updates. Inf. Retrieval J. 18(1), 73–94 (2015)
38. Keyhanipour, A.H., Moshiri, B., Oroumchian, F., Rahgozar, M., Badie, K.: Learning to rank: new approach with the layered multi-population genetic programming on click-through features. Genet. Program. Evolvable Mach., 1–28 (2016)
39. Bollegala, D., Noman, N., Iba, H., et al.: Rankde: learning a ranking function for information retrieval using differential evolution—nova. The university of newcastle's digital repository (2011)
40. Aliguliev, R.M., Aliguliyev, R.M., Isazade, N.R.: Multiple documents summarization based on evolutionary optimization algorithm. Expert Syst. Appl. 40(5), 1675–1689 (2013)
41. Reynolds, R.G., Stefan, J.M.: Web services, web searches, and cultural algorithms. In: IEEE International Conference on Systems, Man and Cybernetics, vol. 4, pp. 3982–3987. IEEE (2003)
42. Mahdavi, M., Chehreghani, M.H., Abolhassani, H., Forsati, R.: Novel metaheuristic algorithms for clustering web documents. Appl. Math. Comput. 201(1), 441–451 (2008)
43. Zhang, H., Jing, L.: Sctwc: an online semi-supervised clustering approach to topical web crawlers. Appl. Soft Comput. 10(2), 490–495 (2010)

44. Menczer, F., Pant, G., Srinivasan, P.: Topical web crawlers: evaluating adaptive algorithms. ACM Trans. Internet Technol. (TOIT) **4**(4), 378–419 (2004)
45. Cecchini, R.L., Lorenzetti, C.M., Maguitman, A.G., Brignole, N.B.: Multiobjective evolutionary algorithms for context-based search. J. Am. Soc. Inf. Sci. Technol. **61**(6), 1258–1274 (2010)
46. Wang, L., Zhou, P., Fang, S., Niu, Q.: A hybrid binary harmony search algorithm inspired by ant system. In: 2011 IEEE 5th International Conference on Cybernetics and Intelligent Systems (CIS), pp. 153–158. IEEE (2011)
47. Johnson, C.G.: Search-based evolutionary operators for extensionally-defined search spaces: applications to image search. In: 2012 IEEE Congress on Evolutionary Computation (CEC), pp. 1–7. IEEE (2012)
48. Gao, K., Wang, W., Gao, S.: Modelling on web dynamic incremental crawling and information processing. In: 2013 Proceedings of International Conference on Modelling, Identification & Control (ICMIC), pp. 293–298. IEEE (2013)
49. Jung, J.J.: Using evolution strategy for cooperative focused crawling on semantic web. Neural Comput. Appl. **18**(3), 213–221 (2009)
50. Forsati, R., Mahdavi, M., Shamsfard, M., Meybodi, M.R.: Efficient stochastic algorithms for document clustering. Inf. Sci. **220**, 269–291 (2013)
51. Gasparetti, F., Micarelli, A.: Swarm intelligence: agents for adaptive web search. In: ECAI, vol. 16, p. 1019 (2004)
52. Peng, T., Zuo, W., He, F.: Svm based adaptive learning method for text classification from positive and unlabeled documents. Knowl. Inf. Syst. **16**(3), 281–301 (2008)
53. Ezzeddine, A.B.: Web information retrieval inspired by social insect behaviour. Inf. Sci. Technol. Bull. ACM Slovakia **3**, 93–100 (2011)
54. Borkar, P.I., Patil, L.H.: A model of hybrid genetic algorithm-particle swarm optimization (hgapso) based query optimization for web information retrieval. IJRET **2**, 59–64 (2013)
55. Ganesan, S., Sivaneri, A.I.U., Selvaraju, S.: Evolving interest based user groups using pso algorithm. In: 2014 International Conference on Recent Trends in Information Technology (ICRTIT), pp. 1–6. IEEE (2014)
56. Alam, S., Dobbie, G., Koh, Y.S., Riddle, P.: Web bots detection using particle swarm optimization based clustering. In: 2014 IEEE Congress on Evolutionary Computation (CEC), pp. 2955–2962. IEEE (2014)
57. James, J.Q., Li, V.O.K.: A social spider algorithm for global optimization. Appl. Soft Comput. **30**, 614–627 (2015)
58. Dziwiński, P., Rutkowska, D.: Ant focused crawling algorithm. In: Rutkowski, L., Tadeusiewicz, R., Zadeh, L.A., Zurada, J.M. (eds.) ICAISC 2008. LNCS (LNAI), vol. 5097, pp. 1018–1028. Springer, Heidelberg (2008). doi:10.1007/978-3-540-69731-2_96
59. Jinhirunkul, E., Sophatsathit, P.: Web navigation analysis and simulation using ant colony optimization. In: 4th International Conference on Autonomous Robots and Agents, ICARA 2009, pp. 595–600. IEEE (2009)
60. Zhang, Z., Du, Y., Li, C.: Antcrawlers: focused crawling agents based on the idea of ants. In: IITA International Conference on Control, Automation and Systems Engineering, CASE 2009, pp. 250–253. IEEE (2009)
61. Srivastava, S., Mathur, A.: An efficient prediction based on web user simulation approach using modified ant optimization model and hierarchical clustering. In: 2013 International Conference on Machine Intelligence and Research Advancement (ICMIRA), pp. 191–198. IEEE (2013)

62. Yang, C.C., Yen, J., Chen, H.: Intelligent internet searching engine based on hybrid simulated annealing. In: Proceedings of the Thirty-First Hawaii International Conference on System Sciences, vol. 4, pp. 415–422. IEEE (1998)

63. Lam, W., Wang, W., Yue, C.-W.: Web discovery and filtering based on textual relevance feedback learning. Comput. Intell. **19**(2), 136–163 (2003)

64. Hu, L., Liu, J., Liang, C., Ni, F.: A mapreduce enabled simulated annealing genetic algorithm. In: 2014 International Conference on Identification, Information and Knowledge in the Internet of Things (IIKI), pp. 252–255. IEEE (2014)

65. Angelaccio, M., Buttarazzi, B.: A visualisation system for web local search. In: Proceedings of the IEEE International Conference on Information Visualization, pp. 474–478. IEEE (2000)

66. Qin, J., Zhou, Y., Chau, M.: Building domain-specific web collections for scientific digital libraries: a meta-search enhanced focused crawling method. In: Proceedings of the 2004 Joint ACM/IEEE Conference on Digital Libraries, pp. 135–141. IEEE (2004)

67. Lovic, S., Lu, M., Zhang, D.: Enhancing search engine performance using expert systems. In: 2006 IEEE International Conference on Information Reuse and Integration, pp. 567–572. IEEE (2006)

68. Mundluru, D., Xia, X.: Experiences in crawling deep web in the context of local search. In: Proceedings of the 2nd International Workshop on Geographic Information Retrieval, pp. 35–42. ACM (2008)

69. Sun, Y., Jin, P., Yue, L.: A framework of a hybrid focused web crawler. In: Second International Conference on Future Generation Communication and Networking Symposia, FGCNS 2008, vol. 2, pp. 50–53. IEEE (2008)

70. Ahlers, D., Boll, S.: Urban web crawling. In: Proceedings of the First International Workshop on Location and the Web, pp. 25–32. ACM (2008)

71. Redwan, H., Mindaye, T., Atnafu, S.: Search engine for amharic web content. In: AFRICON 2009, pp. 1–6. IEEE (2009)

72. Huitema, P., Fizzano, P.: A crawler for local search. In: Fourth International Conference on Digital Society, ICDS 2010, pp. 86–91. IEEE (2010)

73. Yiyao, L., He, H., Zhao, H., Meng, W., Chu, Y.: Annotating search results from web databases. IEEE Trans. Knowl. Data Eng. **25**(3), 514–527 (2013)

74. Neunerdt, M., Niermann, M., Mathar, R., Trevisan, B.: Focused crawling for building web comment corpora. In: 2013 IEEE Consumer Communications and Networking Conference (CCNC), pp. 685–688. IEEE (2013)

Parameter Optimization for H.265/HEVC Encoder Using NSGA II

Saurav Kumar, Satvik Gupta, Vishvender Singh, Mohit Khokhar,
and Prashant Singh Rana$^{(\boxtimes)}$

Computer Science and Engineering Department, Thapar University, Punjab, India
psrana@gmail.com

Abstract. High Efficiency Video Coding (H.265/HEVC) is the latest technology standard proposed by Joint Collaborative Team on Video Coding (JCT-VC). There are quite a few parameters for this encoder required to accomplish this goal. If a single standard configuration file is used for all genres of videos that may not maintain the optimal quality in all encoded videos. This is because every video has objects with unlike speeds of movement. Therefore, encoding factors must be customized in the most favorable way for each video separately. The work propose here is to use NSGA II for multi-objective optimization in order to find out the respective personalized encoding parameters to obtain higher Compression Ratio and Peak Signal-to-Noise Ratio (PSNR). Experiments on six QCIF videos with resolution 176×144 and different configuration files have been performed. Results demonstrate that the proposed technique gives enhanced video compression quality. Test videos and code used in the research is available as supplement at http://bit.ly/HEVC-NSGA-II.

Keywords: H.265/HEVC parameter · NSGA II · H.265/HEVC configuration file · Video compression

1 Introduction

H.265/HEVC is the latest video compression standard set by Joint Collaborative Team on Video Coding (JCT-VC) in January 2013 [1–3]. It delivers 50% superior coding efficiency as compared to H.264/AVC [4]. Its standard has been designed to address essentially all the existing applications of H.264/MPEG-4 AVC developed during 1999–2003 and extended in 2003–2009 [5]. The first edition of H.265/HEVC was officially release in January 2013. Final aligned specification was approved by ITU-T as H.265 and MPEG-4 Part 2 by ISO/IEC [6] respectively. H.265 compression standard is extensively used for a wide range of applications such as broadcasting of High Definition TV signals over cable, internet, satellite/terrestrial transmission systems, and video transmission over mobile network to name a few [5]. As compared with previous conventional standards, it provides better compression ratio. This encoder requires tuning

© Springer Nature Singapore Pte Ltd. 2017
K. Deep et al. (eds.), *Proceedings of Sixth International Conference on Soft Computing for Problem Solving*, Advances in Intelligent Systems and Computing 547, DOI 10.1007/978-981-10-3325-4_11

of several parameters that are fluctuating in nature over the domain of a fixed range of discrete values. Parameter tuning is the key phase to find best possible video compression.

For optimal encoding we seek the lowest possible bit rate without compromising subjective video quality for human perception. Thus the objective function is to compress video, subjected to constraints on benchmark subjective video quality. In H.265/HEVC encoder, a single standard configuration file is used for all videos which may be a primitive assumption about the characteristics of underlying data. In contrast to this static model an agile line of attack to gauge diverse movement attributes and correspondingly set the encoding factors for configuration files are required. Video encoding can provide a power economical architecture which would be of key interest to the battery operated mobile applications. Thus by achieving high resolution and lower resource requirements, we can make best of both worlds [7].

A configuration file of an encoder must attain the parameter values dynamically for each video sequence to achieve high PSNR and low RMSE for encoded video sequence. This thought motivates us to find out the optimal encoder parameters for each testing video sequence. So there is a need to scrutinize the encoder parameters for each video separately. We have proposed an organized and sophisticated approach to solve this parameter optimization problem. The algorithm can trade off encoding delay, compression rate, computational efficiency and robustness while supporting competitive parallel processing techniques with elevated video resolution.

Underlying proposed approach is to use NSGA II for multi-objective optimization and compute optimum parameters for encoder which can provide higher compression ratio and higher PSNR values.

This paper is organized as follow: Section 2 discusses the fundamentals of H.265/HEVC encoder along with its parameters and quantitative metrics. NSGA II is concisely explained in Sect. 3. Section 4 talks about parameter optimization using NSGA II. Final results, conclusion and future projections are discussed in Sect. 5.

2   H.265/HEVC Encoder

2.1   Basics of H.265/HEVC Encoder

H.265/HEVC makes use of a complex course for video encoding and is designed for several purposes such as mobile friendly, data loss resilience, parallel processing architecture. A basic encoder block diagram is describe in Fig. 1.

Hanhart et al. [8] proposed that size of group of pictures (GOP) is set to 8 and for videos having 24, 30, 50, and 60 frames per seconds. The corresponding intra period was taken as 24, 32, 56 and 64 respectively. Kim et al. [9] have proposed HM reference encoder based upon random access configuration. This improves the coding efficiency of low delay configuration encoders anticipated by Horowitz et al. [10]. The coding order is set to 0, 8, 4, 2, 1, 3, 6, 5, 7 and test conditions are selected as in [6]. Additional encoder parameters are taken from

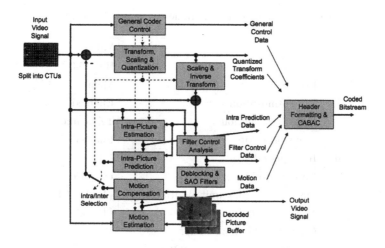

Fig. 1. H.265/HEVC encoder block diagram.

"CFG 16" configuration file, which is presented by Correa et al. [6]. This file was confirmed to be optimal from computational complexity as well as coding efficiency point of view. Table 1 shows the mentioned HM reference software encoder [9] configuration parameters.

2.2   H.265/HEVC Encoder Parameters

There are a number of encoding parameters for H.265/HEVC regarding video compression. These are quantization parameter factor, coding unit's height and width, file I/O, source parameters, profile and level parameters, coding structure parameters, motion estimation parameters, mode decision parameters, slice coding parameters and de-blocking filter parameters. These factors being highly effective for the encoded video quality are described in Table 1 along with their parameter range [5].

– *Maximum and Minimum Coding Unit for Width and Height*: A prediction unit structure depends upon coding unit level. Luma and chroma coding blocks are further subdivided on the basis of prediction type decision. Core of the coding layer in H.265/HEVC is called coding tree unit (CTU). Its size can be larger than traditional macro block, selected by the encoder [5]. The coding tree unit consists of a luma and corresponding chroma coding tree block (CTB) along with syntax elements. Size (L) of luma CTB may be taken as L = 16, 32 or 64 samples. So maximum coding unit (CU) size is selected as $L \times L$ (*Width* $\times$ *Height*) of a luma CTBs having larger size that enables better compression [4].
– Log_2 *of Max/Min Size of Quad Tree Transform Unit:* This unit is represented as a quad tree having the coding unit positioned at root. The luma and chroma sample arrays that exists in a CU are given as coding blocks (CB). Subdivision

108   S. Kumar et al.

Table 1. Configuration parameters of H.265/HEVC.

SN	Parameters	Values
1	Encoder version	HM 15.0
2	Profile	Main
3	RD optimization	Enabled
4	Motion estimation	TZ Search
5	Transform skip	Enabled
6	Intra period	1 S
7	Rate control	Disabled
8	Maximum coding unit Width	{16, 32, 64}
9	Maximum coding unit Height	{16, 32, 64}
10	Quad tree TU Log$_2$ min/max size	{2, 3, 4, 5}
11	Quantization parameter (QP)	[0–51]
12	Quantization parameter factor (QPf)	[0–1]
13	Search range	[2–6]
14	Maximum partition depth	[1–4]

of chroma CTBs of a CTU is always aligned with that of a luma CTB as shown in the Fig. 2 [1]. Luma CB residual can be split into smaller luma Transform Blocks (TB) or same as luma TB, which are then applied to the chroma TBs afterwards [5]. *QuadTreeTU Log$_2$MaxSize* is log of base 2 of luma CTB chosen and QuadTreeTULog2MinSize is log of base 2 of minimum luma CTB. It is generally set to 2 [1].

– *Quantization Parameter (QP) and Quantization Parameter Factor (QPf)*: It is used to determine the quantization of transformed coefficients. Values of QP range from 0 to 51. Upon incrementing QP by 1, the step size increases ≈12% because mapping of QP values to step size is a logarithmic change by 12%. It means the percentage reduction is observed in bit rate i.e. range of QP taken in HM reference software is 20 to 51. Quantization Parameter Factor is the weight assigned during rate distortion optimization. Low value signifies high quality and additional number of bits taken. Its value typically ranges from 0.3 to 1 [1].

– *Search Range*: In motion vector signaling, a motion vector predictor is prepared by using Advanced Motion Vector Prediction (AMVP) scheme. It is based upon motion compensation scheme that exploits both spatial and temporal motion vectors. This is done by creating a candidate set of best predictors from the PU neighbors [11]. To select a motion vector predictor (MVP), a merge scheme is used among a merge candidate set that contains one temporal MVP and four spatial MVPs'. Henceforth, the Search Range is used for motion estimation and is defined around the predictor. Motion vector may acquire values outside the search range. Search Range for QCIF video sequences,

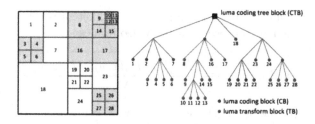

Fig. 2. Coding tree structure.

have maximum coding unit width/height up to 64 $\in \{8, 16, 32, 64\}$. In HM, the reference software search range is set to 8 [1].

- *Maximum Partition Depth*: It defines the depth of coding unit tree and ranges from 1 to 4. In HM reference software, its value is assumed to be 4 [1].

2.3 Quality Assessment Metrics

The quality of digital video is measured on a basis of subjective perception as experienced by humans. Subjective analysis is essential because PSNR may not correlate with the viewers' ratings. So it may be used as a ground truth to develop innovative techniques [12]. Subjective and objective measures are two vital advocates to decide about digital video quality.

2.4 Objective Metrics

Objective assessment provides numerical estimations of the video performance by mathematical models. This calculation is essential because human perception is limited and varies among different viewer for the same video. Moreover objective metrics enable automation and comparison of the output in a consistent manner.

Lately, several objective metrics have been proposed. Among most common ones are Peak Signal-to-Noise Ratio (PSNR) and Mean Squared Error (MSE) between the input versus output video frames as described below:

1. **PSNR:** Peak signal to noise ratio (PSNR) is a measure of signal quality which is computed on logarithm scale. It is defined in terms of MSE between the original picture and distorted picture. In video's context, it is calculated on the picture basis as a weighted sum of PSNR of individual components (Y-PSNR, U-PSNR and V-PSNR) [13]. Generally, Y-PSNR (luma component) is used since it is close to human perception and defined by Eq. 1.

$$PSNR_{dB} = 10 Log_{10} \left(\frac{MAX_I^2}{MSE} \right) \tag{1}$$

Here MAX_I represents the maximum possible pixel value of an image represented using 8 bits per sample. Thus $MAX_I = 255$. As far as video compression is concerned to the human perception, values of PSNR typically lie between 30 dB and 50 dB.

2. **RMSE:** It is defined as root of mean of squared differences between luminance values of the source and reconstructed video frames

$$\text{RMSE} = \sqrt{\frac{\sum_{i=0}^{X-1}\sum_{i=0}^{Y-1}(F_m(x,y) - f_m(x,y))^2}{XY}} \qquad (2)$$

where $F_m(x,y)$ and $f_m(x,y)$ are luminance (Y-component) values of source and encoded pixel of the video frame m, at point (x,y) in 2-dimensional space. X and Y are resolution variables in terms of pixels which may vary according to the size of image (set to 256 for our research).

2.5   Subjective Metrics

Subjective quality measures are time consuming, costly and require human experts sometimes with equipments. It could be a slow and expensive procedure. Nevertheless, it is a crucial ritual for quality assurance and users' satisfaction. PSNR does not consider the saturation effect of human eyes, so underlying nature of artifacts is not fully captured. Subjective quality measure can be calculated using two metrics, Mean Opinion Score (MOS) and Structural Similarity Index as described below:

Mean Opinion Square (MOS): It measures the human quality impression between 1 and 5. This rating is then correlated with PSNR values [12], computed frame by frame basis. Relation between PSNR and MOS is defined in Table 2 [12].

Table 2. Relationship between MOS and PSNR.

Scale	Quality	Impairment	PSNR(dB)
5	Excellent	Imperceptible	>37
4	Good	Perceptible, but not annoying	31–37
3	Fair	Slightly annoying	25–31
2	Poor	Annoying	20–25
1	Bad	Very annoying	<20

Structural Similarity Index: Human preferences towards evaluating visual quality are measured by Structural Similarity Index (SSI) [14]. It depends upon Human Visual system (HVS) and is suitable to extract the structural information from observed scene. Therefore, distortion or structural similarity thoroughly measures the perceptual image quality. In other worlds, SSIM attempts to classify the distortions which have negligible effect on image structure. Sampat et al. [14] have shown that SSIM provides better prediction of image quality for a range of distortions associated with images. Maximum value of SSIM index is 1 if both original and encoded image are identical.

In this work PSNR, RMSE and MOS have been used to quantify the video quality. The computation of encoder parameters may not be represented in terms of single objective function that needs to be optimized. So, the classical optimization technique is not appropriate for computing optimum values of encoder parameters. An alternative modern approach for parameter estimation is soft computing. Non-dominated sorting genetic algorithm - II (NSGA II) has been utilized for this rationale, details of which are given in Sect. 3.

3   Non-dominated Sorting Genetic Algorithm II (NSGA II)

Non-dominated sorting genetic algorithm II (NSGA II) [15] searches a secondary set of solutions using a population for potential solutions, known as chromosomes. Chromosomes are represented by a D-dimensional vector $[t_{i1}, t_{i2}, ..., t_{iD}]$ where i = 1, 2, ..., NP is the population size (number of chromosomes). Each decision variable t_{ij} for j = 1, 2, ..., D in i^{th} chromosome represents j^{th} threshold value. The population is initialized randomly and corresponding objective function values are calculated using Eqs. 3 and 4. Before applying the selection, crossover and mutation operator to generate a new offspring, given population is sorted using a fast non-dominated sorting technique and crowding distance. Fast non-dominated sorting approach assigns the rank to each individual with $O(FNP^2)$ computational complexity, where F signifies number of objectives. Crowding distance is a measure of closeness of an individual relative to its neighbors i.e. the density of solutions. It is calculated using the average of distances from its immediate neighbors along same front in each dimension. Selection of parents from the population is done using binary tournament election method which is based two factors, crowding distance and rank. Now from these parents, offsprings are generated using crossover and mutation operators. This process is repeated until the termination condition is satisfied which is known as number

---

Algorithm 1. Non-dominated sorting genetic algorithm II (NSGA II).

Input: Population size (NP) and number of generations.
Output: A set of best individuals known as pareto-front
1. Initialize the population PP of size NP randomly.
2. Calculate the objective values using equation 3 and equation 4.
3. Arrange the population using non-dominated sorting technique and crowding distance.
4. Select the parents using a binary tournament selection.
5. Apply crossover and Mutation operator on the selected parents
6. Perform selection from the parents and their offsprings.
7. Replace unfit individuals with the fit ones to maintain a constant population size.
8. Repeat the steps 2−7 until termination condition is satisfied.

---

of generations. In this work, binary tournament [16], single point crossover [17], and single point mutation operator [17] are used for NSGA II implementation.

4   Parameter Optimization for H.265/HEVC Using NSGA II

There are several parameters used by H.265/HEVC encoder in which some remain constant while others vary within a specified range. The literature reveals that there are total six parameters that affect the video quality and compression ratio. These parameters are described in Table 1 along with their range of values. There are two objectives for video encoding. First is to maximize the Y-PSNR (YPSNR(x)) and second is to minimize the file size (FileSize(x)) for a given set of parameters. Both the objective functions are defined as:

$$Objective\ Function1 = maximum(YPSNR(x)) \tag{3}$$

$$Objective\ Function2 = minimum(FileSize(x)) \tag{4}$$

The parameters set is represented as $\mathbf{x} = (x_1, x_2, x_3, x_4, x_5, x_6)$ where, $x_1 \rightarrow$ maximum coding unit width, $x_2 \rightarrow$ maximum coding unit height, $x_3, \rightarrow$ quad tree TU log2 maximum size, $x_4 \rightarrow$ quad tree TU log2 minimum size, $x_5 \rightarrow$ Maximum partition depth, $x_6 \rightarrow$ QP. The detailed steps for estimating above parameters are given in Algorithm 1. This algorithm returns a set of best individuals known as pareto-optimal front. Initial population is generated randomly and sample of the population is shown below:

x_1	x_2	x_3	x_4	x_5	x_6
32	32	4	2	3	33

In this work, the target quality is set to 35 dB as it comes under 'good' quality range indicated in Table 2.

5   Results and Discussion

This section discuses about environment setup, NSGA II implementation, its parameter setting, test video sequence files, reference parameters for H.265/HEVC encoder and results.

5.1   Environment Setup

Simulation were carried out by using HM (version 15.0) software for H.265/HEVC encoder, which uses a GOP structure of IBBBBBBB with Hierarchical-B structure enabled. Details about machine configuration and software used are listed in the Table 3.

Table 3. Environment used for simulation.

Category	Model/Configuration
Motherboard	DV-6000 AMD
Processor	Intel® Core™ i5-M370, 3.2 GHz
Graphics card	ATI Radeon Fire Pro V8800
RAM	2×4 GB PC3-10700
HDD (Storage)	Western Digital 2×2 TB
Operating system	Linux (Ubuntu 14.10)
Video player	VLC/Media Player Classic 64 bit
HM	Version 16.0
NSGA II implementation	Octave 3.6.4

5.2 Parameter Setting for NSGA II

NSGA II is implemented in Octave (Version 16.0), an open source software licensed under GNU GPL. Parameters are set with Crossover Rate (CR) = 0.9, Mutation Rate (MR) = 0.01, binary tournament selection [16], single point crossover [17], and single point mutation operator [17]. The population size is set to 50 and the maximum number of generation is set to 200. Code is available in the supplement information.

5.3 Test Video Sequences

Videos composed of less than 10 frames per seconds are sometimes used for low bit rate (<64 kbps) video communication. Typically, 10 to 20 frames per seconds are considered for low bit rate video communication. Standard television signals are generally sampled at 25 to 30 frame per seconds, while 50 to 60 frames per seconds exhibit smooth apparent motion [18]. RGB format is used to capture and display colored images while YCbCr format is more efficient for compression. Here, Y represents pixel brightness (luminance), Cb and Cr are chrominance components of the pixel. A standard video supports several sampling patterns for Y, Cb, Cr format. Some typical patterns are 4:4:4, 4:2:2 and 4:2:0. The number in ratio N1:N2:N3 represents relative sampling rate in horizontal direction. Here N1 = number of Y samples in both odd and even rows, N2 = number of Cb and Cr samples in odd rows and N3 = number of Cb and Cr samples in even rows. Also Cb = U and Cr = V represent color components of YUV color space. On the basis of resolution, some commonly used 4:2:0 YUV formats are specified below according to the number of pixels [12]:

In this work, six QCIF format videos with resolution of the order 176×144 have been used and corresponding details are specified in the Table 4. Test videos under consideration have diverse speeds of the moving objects. For example, very fast, moderately paced, slow and sluggish. We recommend to use different

Video format	Resolution
Sub Quarter Common Intermediate Format (SQCIF)	128×96 pixels
Quarter Common Intermediate Format (QCIF)	176×144 pixels
Common Intermediate Format (CIF)	352×288 pixels
Source Intermediate Format (SIF)	352×240 pixels

Table 4. Sample test sequences [18,19]

Test video	Resolution	Total no. of frames	No. of frames to be encoded
AkiyoQcif.yuv	176×144	75	30
SuzieQcif.yuv	176×144	75	30
ForemanQcif.yuv	176×144	75	30
FootballQcif.yuv	176×144	75	30
CoastguardQcif.yuv	176×144	75	30
BusQcif.yuv	176×144	75	30

configuration files for video encoding in order to get the optimal results. All the test videos are made available as a supplement information to this paper.

5.4   Results and Discussion

The above mentioned videos are implemented on the HM 15.0. Experiment was carried out for the parameters through which encoder has some default setting. Subsequently, the optimal values of parameters are computed through proposed technique. List of default reference parameters are given below:

$$\begin{array}{cccccc} x_1 & x_2 & x_3 & x_4 & x_5 & x_6 \\ 64 & 64 & 5 & 2 & 4 & 32 \end{array}$$

Table 5 shows the performance comparison of various encoded test video sequences using reference and optimized parameters. It has been observed that by using reference parameters, Y-PSNR of the test videos demonstrates unreliable quality. For BusQcif and FootballQcif, Y-PSNR goes down to 28.97 dB and

Table 5. Performance comparison of encoded test video sequences using optimized parameters and reference parameters.

Test video	Encoding using Reference Parameters*			Encoding using Optimized Parameters									
	Y-PSNR	Encoded File Size (bytes)	Compression Ratio	Y-PSNR	Encoded File Size (bytes)	Compression Ratio	Optimized Parameters					
							x_1	x_2	x_3	x_4	x_5	x_6
AkiyoQcif	36.477	2542	448.755	35.0215	1935	589.527	64	64	4	2	4	34
SuzieQcif	34.4189	2382	478.898	35.1142	2521	452.493	64	64	4	2	4	32
ForemanQcif	32.7594	5966	191.206	34.9189	7279	156.716	64	64	3	2	3	30
FootballQcif	28.6437	33944	33.606	34.9989	92650	12.31	64	64	2	2	4	25
CoastguardQcif	31.3543	7709	147.974	35.2879	19764	57.717	64	64	3	2	4	26
BusQcif	28.9785	26342	43.304	34.8586	73454	15.529	64	64	4	2	3	25

* x_1=64 x_2=64 x_3=5 x_4=2 x_5=4 x_6=32

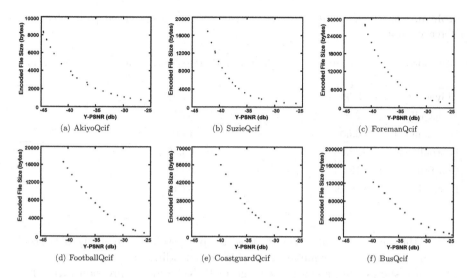

Fig. 3. Pareto optimal graph between encoded file size and Y-PSNR for test video sequences.

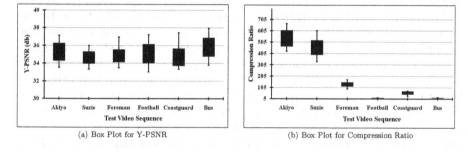

Fig. 4. Box plots for Y-PSNR and compression ratio for test video sequences.

28.64 dB respectively while having somewhat annoying quality. In both videos, most of the objects are movable. Therefore, motion estimation takes large size and the quality is reduced. It is evident from Table 5 that PSNR value is maintained approximately at 35 dB. This ensures that the quality of videos encoded is promising when our proposed technique is employed. For AkiyoQcif and SuzieQcif, Y-PSNR is obtained to 36.477 dB and 34.418 dB respectively. Thus, it provides a good quality reference model as well within the proposed model. Most of the frames contained in these videos are sluggish in movement. Therefore it is established that the proposed technique behaves quite effectively for videos having variety of movement related activities. Figure 3 shows the pareto optimal graphs that are generated for each test video. Here, x-axis represents Y-PSNR (dB) and y-axis represents compression ratio. It is observed that for the video sequences SuzieQcif, ForemanQcif, FootballQcif, CoastguardQcif and BusQcif, Y-PSNR is maintained high with reasonably good compression ratio.

Table 6. Change in encoded test video sequences using optimized parameters and reference parameters.

Test video	Percentage change in Y-PSNR	Percentage change in compression ratio
AkiyoQcif	−3.99	+31.37
SuzieQcif	+2.02	−5.51
ForemanQcif	+6.60	−18.03
FootballQcif	+22.18	−63.36
CoastguardQcif	+12.54	−60.99
BusQcif	+20.29	−64.14

This is the case especially when videos are processed by involving optimized encoding parameters as compared to reference encoding parameters.

Table 6 shows the percentage change in PSNR and compression ratio after using proposed algorithm. It has been observed that for all videos under testing, Y-PSNR increases within a range between [2%, 20%] with reasonably good compression ratio, except AkiyoQcif video. Above results justify the need to have different video encoding parameters for smart encoding. To validate the performance of proposed methodology, 30 simulations are carried out for each video sequence. Figure 4(a) and (b) present the box-plot for Y-PSNR and compression ratio. It is found that for all the test videos (Fig. 4(a)) Y-PSNR is maintained near to 35 dB with ±2 error. Likewise, another reflection is that (Fig. 4(b)) BusQcif and FootballQcif have extremely low compression ratio because these videos contain fast moving objects. In support of AkiyoQcif and SuzieQcif, the compression ratio is incredibly high because videos having awfully slow moving objects. These quantitative results substantiate the fact that different videos require encoding parameters based upon their inherent characteristics.

6   Conclusion

In this work, we have carried out the parameter optimization of H.265/HEVC encoder using NSGA II algorithm. It maintains the proper balance between Y-PSNR and compression ratio while encoding a video. Two objectives are formulated for video encoding: first is to maximize Y-PSNR and second is to minimize encoded file size for the set of parameters. At this point, six parameters are chosen for optimization. These are, maximum coding unit width, maximum coding unit height, quad tree TU Log_2 min/max size, quantization parameter (QP), quantization parameter factor (QPf) and search range. NSGA II is used for multi-objective optimization to find out the optimal parameters for encoder which can provide sound quality compressed video file. For testing, six YUV videos files have been used with 176 × 144 resolutions and comprising of 30 frames. Results show that the unlike videos entail distinct configuration profiles for fine quality and compression ratio. Test operations included range of videos from extremely swift to leisurely moving objects, which demanded unique treatment for compression. To authenticate the performance of proposed methodology, 30 simulations were performed for each video sequences. It was found that (i) Y-PSNR ≈ 35 dB ± 2 error for each video (ii) Compression Ratio $\propto 1/($Object Movements) i.e. if the video has lethargic movements then it can be compressed to a greater ratio.

Lastly, reference parameters cannot maintain Y-PSNR at a good level for all the videos because they possess varying speed objects. So there arises a need to develop some new self dependable approach which can implicitly optimize H.265/HEVC video encoder parameters according to the picture's motion. Although, NSGA II is time-consuming for video encoding, yet it can be implemented in parallel to speedup the encoding process.

Supplement Information

The test videos and code used in the study is available as supplement at http://bit.ly/HEVC-NSGA-II.

References

1. Sze, V., Budagavi, M.: A comparison of cabac throughput for HEVC/H. 265 VS. AVC/H. 264. In: IEEE Workshop on Signal Processing Systems (SiPS), pp. 165–170 (2013)
2. Henot, J.-P., Ropert, M., Le Tanou, J., Kypréos, J., Guionnet, T.: High efficiency video coding (HEVC): replacing or complementing existing compression standards? In: IEEE International Symposium on Broadband Multimedia Systems and Broadcasting, pp. 1–6 (2013)
3. Grois, D., Marpe, D., Mulayoff, A., Itzhaky, B., Hadar, O.: Performance comparison of H. 265/MPEG-HEVC, VP9, and H. 264/MPEG-AVC encoders. In: PCS, vol. 13, pp. 8–11 (2013)
4. Zhao, T., Wang, Z., Kwong, S.: Flexible mode selection and complexity allocation in high efficiency video coding. IEEE J. Sel. Top. Sign. Process. **7**(6), 1135–1144 (2013)
5. Sullivan, G.J., Ohm, J., Han, W.-J., Wiegand, T.: Overview of the high efficiency video coding (HEVC) standard. IEEE Trans. Circ. Syst. Video Tech. **22**(12), 1649–1668 (2012)
6. Correa, G., Assuncao, P., Agostini, L., da Silva Cruz, L.A.: Performance, computational complexity assessment of high-efficiency video encoders. IEEE Trans. Circ. Syst. Video Technol. **22**(12), 1899–1909 (2012)
7. He, G., Zhou, D., Li, Y., Chen, Z., Zhang, T., Goto, S.: High-throughput power-efficient VLSI architecture of fractional motion estimation for ultra-hd HEVC video encoding (2015)
8. Hanhart, P., Rerabek, M., De Simone, F., Ebrahimi, T.: Subjective quality evaluation of the upcoming HEVC video compression standard. In: SPIE Optical Engineering + Applications, p. 84990-99. International Society for Optics and Photonics (2012)

9. Kim, I.-K., McCann, K., Sugimoto, K., Bross, B., Han, W.-J.: Hm10: high efficiency video coding (HEVC) test model 10 encoder description. In: Proceedings of the Joint Collaborative, Team Video Coding (2013)
10. Horowitz, M., Kossentini, F., Mahdi, N., Xu, S., Guermazi, H., Tmar, H., Li, B., Sullivan, G.J., Xu, J.: Informal subjective quality comparison of video compression performance of the HEVC and H. 264/MPEG-4 AVC standards for low-delay applications. In: SPIE Optical Engineering Applications, pp. 84990–84999. International Society for Optics and Photonics (2012)
11. Mora, E.G.: Multiview video plus depth coding for new multimedia services. Ph.D. thesis, Telecom ParisTech (2014)
12. Klaue, J., Rathke, B., Wolisz, A.: EvalVid – a framework for video transmission and quality evaluation. In: Kemper, P., Sanders, W.H. (eds.) TOOLS 2003. LNCS, vol. 2794, pp. 255–272. Springer, Heidelberg (2003). doi:10.1007/978-3-540-45232-4_16
13. Zhang, F.: Quality of experience-driven multi-dimensional video adaptation. Ph.D. thesis, Universität München (2014)
14. Sampat, M.P., Wang, Z., Gupta, S., Bovik, A.C., Markey, M.K.: Complex wavelet structural similarity: a new image similarity index. IEEE Trans. Image Process. **18**(11), 2385–2401 (2009)
15. Deb, K., Pratap, A., Agarwal, S., Meyarivan, T.: A fast and elitist multiobjective genetic algorithm: NSGA-II. IEEE Trans. Evol. Comput. **6**(2), 182–197 (2002)
16. Miller, B.L., Goldberg, D.E.: Genetic algorithms, selection schemes, and the varying effects of noise. Evol. Comput. **4**(2), 113–131 (1996)
17. Shin, S.-Y., Lee, I.-H., Kim, D., Zhang, B.-T.: Multiobjective evolutionary optimization of DNA sequences for reliable DNA computing. IEEE Trans. Evol. Comp. **9**(2), 143–158 (2005)
18. Rao, K.R., Kim, D.N., Hwang, J.J.: Video Coding Standards: AVS China, H. 264/MPEG-4 PART 10, HEVC, VP6, DIRAC and VC-1. Springer Science & Business Media, Dordrecht (2013)
19. Yuv data set. http://trace.eas.asu.edu/YUV/

Circumferential Temperature Analysis of One Sided Thermally Insulated Parabolic Trough Receiver Using Computational Fluid Dynamics

Yogender Pal Chandra[1(✉)], Arashdeep Singh[1], S.K. Mohapatra[1],
and J.P. Kesari[2]

[1] Mechanical Engineering Department, Thapar University,
Patiala 147004, Punjab, India
yogender027mae@gmail.com, arashdeep012@gmail.com,
skmohapatra@thapar.edu
[2] Department of Mechanical Engineering, Delhi Technological University,
New Delhi 110042, India
drjpkesari@gmail.com

Abstract. Low temperature industrial thermal applications like process heating involving solar thermal technology renders the usage of inexpensive air filled annuli receivers despite they are below par in thermal performance. This work is cantered around the air filled receiver system and more importantly try to assess both conventional and modified air filled annulus system using computational fluid dynamics (CFD) in terms of their performance parameters. For modification purpose, conventional receiver was fitted with thermal insulation in non-concentrating half section of receiver which is actually short of concentrated sun's radiation. Finally it was simulated for significantly reduced circumferential temperature distribution (CTD) around the absorber and was compared with conventional air filled annulus receiver. This comparison could be supposed to serve as a means of advancement for the development of small scale solar thermal based heat producing plants.

Keywords: Parabolic trough collectors · Air filled annulus receiver · Computational fluid dynamics

1 Introduction

Linear concentration technology has taken concentrated solar power (CSP) by storm: there is no definitive focusing point, rather a line. Serious development and evolution of this technology, in fact, came into existence when researchers analyzed step by step and wrap by wrap, the existing forefront point focusing technology and suddenly stumbled upon a common notion—why, point concentrate the heat, when it will be redistributed anyways, in line circuitry of the heat transfer fluid. Parabolic trough based concentrated solar thermal power plants, for the most part consists of parabolic trough solar fields, heat generation system or absorber/receiver system, power block powered with Rankine steam turbine and a temporary or optional power storage system.

© Springer Nature Singapore Pte Ltd. 2017
K. Deep et al. (eds.), *Proceedings of Sixth International Conference on Soft Computing for Problem Solving*, Advances in Intelligent Systems and Computing 547,
DOI 10.1007/978-981-10-3325-4_12

As far as thermal performance of parabolic trough collector is concerned, thermo-physical parameters *i.e.* solar irradiance, wind velocity, mass flow rate and inlet temperature of heat transfer fluid (HTF) holds paramount importance. Presently, most eminent technology used in CSP is vacuumed annulus receiver. These receivers ensure curtailment of all the unfavourable convection losses which are quite a conspicuous problem in high temperature CSP applications. In addition, with regards to their high thermal performance, they are very expensive amounting to not less than 20% of cost of the whole solar parabolic trough collector (PTC) field.

Furthermore, relatively inexpensive *i.e.* air filled annulus receivers are very much suited for low temperature applications like process heating and industrial water heating. A great many of the work on low temperature industrial usage employing CSP could be referred in [1, 2]. As an illustration of simple inclusive design restraint of receivers *viz. de fecto*, only the part that faces the parabolic trough receives the concentrated solar irradiance while the other half is starved of it. This is highly dependent on collector design or more specifically rim angle of the trough. Let's put it this way, due to inhomogeneous heating of conventional and commercially available receivers which is essentially due to its part facing towards and part facing away of the concentrating trough, the whole system is eventually disposed up to the net loss of gained heat by radiation and convection from the non-concentrated side of the receiver.

One of the solutions for the above mentioned problem could be application of thermal insulation material in the upper portion of absorber which would not only serve as to restrain the heat draining from non-concentrating side but also reduce the convection currents instigated by buoyancy effects of confined air. This leads to better homogeneity in absorber tube temperature contrary to conventional receiver resulting in reduced glass tube temperature as well which is followed by reduction in convection losses from ambient air. Further to this, stratification of air is eventually ensued as its consequence *i.e.* hot air is kept in vicinage of absorber while cold air is retained near glass tube. This was not the case with conventional receiver due to practically huge temperature difference in absorber and glass casing which ultimately resulted in recurrent movements of convection currents. Also, these stratified currents are restricted in the lower portion of receiver which also serves as the source of dramatic reduction in convection current losses.

Accordingly, for effective and efficient heat transfer process it is predominantly important that heat transfer fluid (HTF) is subjected to a better temperature distribution. Yet varying the physical parameters such as DNI, HTF velocity and temperature at inlet could possibly affect the CTD to some extent however, biased solar influx impact is most dominant and unfavourable. The result of this sporadic heating is the intermittent temperature rise of HTF *i.e.* higher at bottom while lower at upper portion. In other words, this happens due to non-symmetrical heating up of receiver for the reason of high intensity solar radiation focusing near the bottom. This is explicitly shown in Fig. 2(a) in which practical working conditions of DNI, wind speed, inlet temperature, and velocity of HTF are imitated for typical industrial receiver. Accordingly, This is clear that HTF is non-uniformly heated up due to inhomogeneous solar heating and is in perfect agreement with He et al. [3].

With this intention, this work is dedicated to find the above mentioned glitches of conventional air filled receiver *viz.* irregular heating of absorber and would put forward through numerical simulation approach, the results of modified receiver in the form of more favourable and uniform temperature distribution around absorber. In other words, this paper serves to mitigate heat flux localization with added benefit of decline in convection currents instigated by buoyancy effects of air trapped in annulus [4]. This is actually achieved by application of single side thermal insulation which has the property to withhold the gainful heat getting drained out from non-concentrating region of conventional receiver.

As a matter of fact, glass wool was used for the purpose of thermal insulation and was modelled in ANSYS – Fluent design modeller having perfect simpatico agreement with commercial receivers of PTC.

2   Description of Modified Receiver

Physical parameters and its description are described in Fig. 1 and Table 1. It is designed in ANSYS – Fluent modeller according to the working data given in Table 1. Thermal exchange phenomenon of the model is defined in Fig. 3(a), (b) and (c) in which front and isometric view of current geometry are illustrated. Geometry is modelled having steel as absorber material encapsulated with glass envelope. While sun facing side is fitted with glass wool as insulation, lower trough facing annulus is filled with air at pressure greater than 1 Torr.

Figure 3(a) depicts the thermal exchange phenomena in the receiver accommodating TherminolVp-1 as heat transfer fluid (HTF). As an illustration to the heat exchange phenomena, trough focuses solar irradiance equivalent to (q_{sol}) onto receiver, solar irradiance transmitted by radiation to absorber through glass casing $(q_{sol\text{-}abs})$,

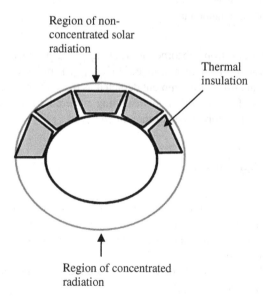

Region of non-
concentrated solar
radiation

Thermal
insulation

Region of concentrated
radiation

Fig. 1. Insulated receiver to stamp down convection and adverse radiation losses.

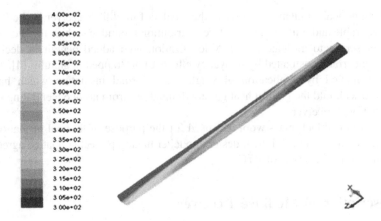

Fig. 2. Temperature distribution in heat transfer fluid.

Table 1. Physical specification of solar field.

Aperture width	5.57 m
Aperture length	12.5 m
Outer diameter of absorber	0.0720 m
Inner diameter of absorber	0.0705 m
Outer diameter of glass envelop	0.1205 m
Inner diameter of glass envelop	0.1115 m
Absorber steel material	321-H stainless steel
Absorber glass material	Silicate glass
Rim angle	70°
Concentration ratio	70

convective heat transfer from absorber to HTF ($q_{a,\ f\text{-}conv.}$), convection heat transfer instigated by buoyancy effect in entrapped air ($q_{a\text{-}conv.}$), radiation thermal exchange from glass casing and absorber to ambient depicted as ($q_{g\text{-}rad.}$) and ($q_{a\text{-}rad.}$) respectively, conduction heat loss from absorber to insulating material ($q_{a\text{-}cond.}$) and lastly, convection losses from glass tube to environment ($q_{g\text{-}conv.}$).

2.1   Governing Equations

- Radiation heat transfer falling in infrared zone equates to zero.
- Overall dimension of receiver is large enough to render any error associated with the idea that solar thermal exchange is defined in terms of heat flux at any level amounts to zero [5, 6].
- Steady sate and incompressible fluid flow is assumed on the grounds of insignificant negative pressure gradient.

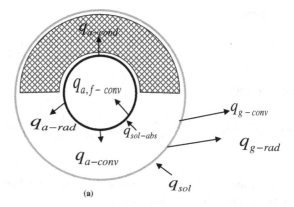

(a)

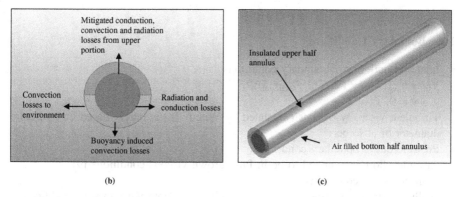

(b)   (c)

Fig. 3. (a) Heat transfer model of receiver (b) Font view of modified receiver (c) Isometric view of modified receiver.

- End losses are assumed to be minimal with adiabatic edge profile of geometry [7].
- Due to finitely long longitudinal span of receiver, flow is assumed to be two dimensional.

Continuity equation, energy equation and momentum equation as given in Eqs. (2), (3) and (4) are among the partial differential equations (PDEs) governing the fluid flow [8]. K-ε turbulence model as used by Wilcox [8] was used in reference to high Reynolds number to define the fluid flow in turbulent region.

Equation for steady state condition and incompressible flow:

$$\frac{\partial p}{\partial t} = \frac{\partial p}{\partial x} = \frac{\partial p}{\partial y} = \frac{\partial p}{\partial z} = 0 \tag{1}$$

Continuity equation:

$$\nabla . u_i = \frac{\partial u}{\partial x} + \frac{\partial v}{\partial y} + \frac{\partial w}{\partial z} = 0 \tag{2}$$

Energy equation

$$\rho C_p u_i \frac{\partial T}{\partial x_i} = (k_f + k_T) \frac{\partial}{\partial x_i} \left(\frac{\partial T}{\partial x_i} \right) \tag{3}$$

Momentum equation:

$$\frac{\partial}{\partial t} (\rho u_i) + \nabla.(\rho u_i u_j) = -\nabla p + \frac{\partial}{\partial x_j} \left[(\mu + \mu_t) \left(\frac{\partial u_i}{\partial x_j} + \frac{\partial u_j}{\partial x_i} \right) - \frac{2}{3} (\mu + \mu_t) \frac{\partial u_i}{dx_i} \delta_{ij} - \frac{2}{3} \rho k_f \delta_{ij} \right] + \rho g_i \tag{4}$$

Where k_T, k_f and μ_t represents respectively turbulent conductivity, thermal conductivity and eddy viscosity as given by Wilcox [8].

2.2 Boundary Conditions

- Inlet and outlet flow boundary conditions:
  $V_{in} = u_{avg} = w = (V/A_{cross \ sectional})$;
  $u = v = 0; T = T_{in}$
  Turbulent viscosity (k_{in}) as calculated by Wilcox [8] is purely based on hydraulic diameter of absorber.
- Pressure outlet boundary conditions:
  At outlet, flow is assumed to be in fully grown viscous conditions [9].
- Wall boundary conditions:
  Walls are strictly assumed hydro – dynamically smooth with no slip conditions.
- Receiver tube is subjected to concentrated solar irradiance and is simulated as heat flux around the absorber this implies that absorber tube is exposed to non – zero heat flux conditions [10]. Direct normal irradiance (DNI) around absorber is presumed to be constant.

2.3 Numerical Method

Table 1 precisely illustrates the data used for geometric modelling. ANSYS academic version 15.0 was used for the whole modelling. After geometric modelling, domains of the current model were assigned and fixed as solid or fluid depending to the inherent nature of subsystem of receiver. Afterwards, meshing operation was performed on the geometry which involves discretization of model into number of smaller finite elements and nodes. This is depicted in Fig. 4(a) and (b) which shows front and isometric views of model under consideration adopting quadrilateral mesh systems. Meshing grids have higher compactness in the interior domain of model due to the complexity of geometry this suggests why it is not spread evenly in the full computational zone.

At first, Energy is enabled in ANSYS workbench for solving heat transfer equations, afterwards k-ε model is initiated to take into account the turbulent nature of fluid flow. Finally all the previously mentioned boundary conditions are invoked and applied to the respective named selection of the geometry and surface to surface (S2S) model is

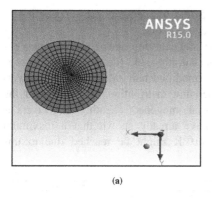 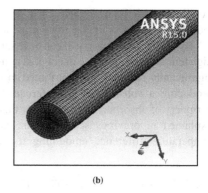

(a)   (b)

Fig. 4. Grid generation for numerical simulation (a) front view and (b) isometric view of receiver tube

Table 2. Thermo physical properties of material used

Material	k (W/m-k)	ρ (kg/m³)	C_p (J/kg-k)	μ (kg/m-s)
TherminolVp1	0.117	1068	2270	0.000457
Air	0.0242	1.125	1006.43	0.00001789
Glass wool	0.04	18	670	—
Silicate glass	1.4	2.5	840	—
Stainless steel	16	7920	444	—

turned on to simulate heat transfer problem. Table 2 illustrates thermo – physical properties of utilised material while assigning the subsystem of the geometry in respective cell zone conditions.

3   Results and Discussions

An explicit explanation in the form of graphical and simulation representation of the simulated model is proposed to demonstrate circumferential temperature distribution (CTD) in both receivers. Part of this is elucidated in Fig. 5 which shows the isothermal zones or temperature profiles of both conventional and modified air filled receiver system. Again, modified receiver sustains glass wool as thermal insulation in its non-concentrating side of half annulus. Figure 5 explains two temperature counters namely temperature contour – 1 and temperature contour – 2. Temperature contour – 1 defines isotherm for conventional receiver while temperature contour – 2 defines it for modified receiver. Also, respective steady state temperature difference and time taken by boundary conditions to achieve this state for both receivers are shown here. In other words, both the receivers along with the fluid inside are assessed on the basis of circumferential temperature distribution (CTD) along with steady state temperature difference and time taken for aforementioned boundary conditions to achieve the same.

To illustrate, it is apparently clear from Fig. 5 that temperature gain accounting to modified receiver is higher in contrast to conventional one.

By preliminary analysis it can be assured that concentrating side of both receivers has high temperature influence in contrast to non-concentrating side, (though this influence is significantly reduced and distributed in case of modified half side insulated receiver) producing a substantial circumferential temperature distribution (CTD) around the periphery of absorber. The reason for this being inhomogeneous heating of receivers as explained in previous sections. The results have already shown that a maximum temperature difference amounting to nearly 110 K might be reached due to this

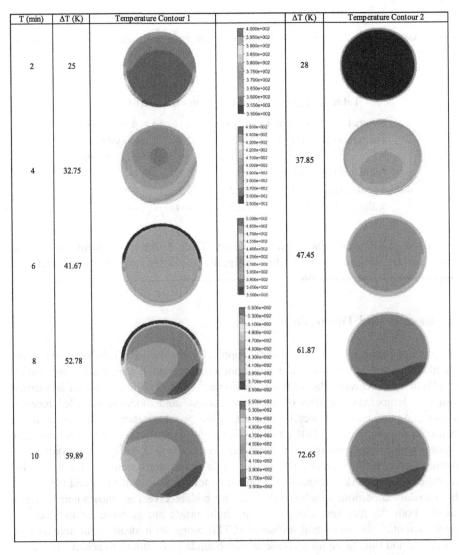

Fig. 5. CTD profile of both absorbers along with isothermal zones in HTF.

inhomogeneous heating. This huge temperature difference is highly unfavourable and affects thermal performance very severely in terms of net temperature gain and unwanted fluid stratification. Furthermore, it is apparently evident from the temperature profiles, that isotherms of modified receivers are more uniform and symmetrical across absorber as compared to conventional one. This leads to improved thermal performance and an increase in temperature gain of HTF. This is further corroborated by the upsurge in temperature of the non-concentrating side of modified receiver's absorber which is due to the fact of mitigated conduction and convection losses. Correspondingly, increase in temperature of non-concentrating side of absorber of thermally insulated receiver results in improved temperature distribution around the absorber thus achieving enhanced heat transfer mechanism.

As a matter of fact, simulated boundary conditions have shown that heat flux around the conventional receiver could vary from a lowest value of few W/m^2 to the highest value of 45 W/m^2 in circumferential direction. This is however not the case with one side insulated receiver which is subjected to substantially reduced heat flux difference as compared to conventional one (refer Fig. 5). More specifically, Fig. 6(a) and (b) compares the circumferential temperature distribution (CTD) around the absorber tube for both receivers. Graphical representation suggests that temperature at locations $\varphi = 0°$ and $\varphi = 360°$ is comparatively lower than the temperature at $\varphi = 180°$ for both the receivers. The reason again belongs to the fact that positions $\varphi = 0°$ and $\varphi = 360°$ belongs to the non-concentrating side of receiver while $\varphi = 180°$ belongs to concentrating side thus receiving only sparse value of sunlight. To quantify, it can be most significantly deduced that surface temperature variation or circumferential temperature distribution (CTD) profile of both receivers is not similar in any case of physical size or magnitude. In other words, inhomogeneous heating of conventional receiver caused steeper temperature profile while it is dispersed comparatively low. This can be easily evaluated by the sudden change in temperature in radial location $\varphi = 90°$ to $\varphi = 180°$ (refer Fig. 6(a)). Consequently, conventional receiver distributes temperature spanning for the full range of 420–550 K. On the other hand, reasonably more gradual increase in temperature from 460 to 510 K took place in case of half insulated modified receiver (refer Fig. 6(b)). In conclusion, both receivers are

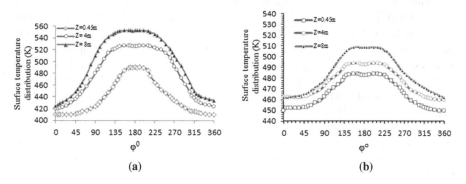

Fig. 6. Circumferential temperature distribution (CTD) profile of (a) conventional receiver, and (b) modified receiver.

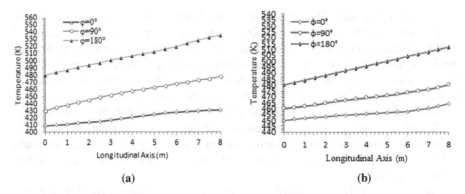

Fig. 7. Temperature distribution profile along the length of the absorber for different circumferential location of (a) conventional receiver (b) modified receiver.

compared on the basis of circumferential temperature distribution (CTD) at different longitudinal location and it was inferred that one side insulated receiver renders more favourable CTD and increased heat transfer contrary to conventional one.

Furthermore, CTD becomes more intense as longitudinal distance is approached, In other words CTD has substantially higher value at Z = 4 m then at Z = 0.45 m. Figure 7(a) and (b) illustrates the difference in longitudinal variation of CTD for both the receivers at various locations of receiver i.e. at z = 0.45 m, z = 4 m and z = 8 m. Contrasting feature of both figures suggest that CTD is mitigated quite significantly even in longitudinal direction in case of modified receiver.

4   Conclusion

In order to calculate and assess the impact of inhomogeneous heating of parabolic trough collector's conventional air filled annuli receiver system, numerical simulation approach of computational fluid dynamics (CFD) is used. Modified receiver in the form of one side insulation is also modelled in ANSYS – Fluent and compared with conventional one. Comparison was made on the grounds of circumferential temperature distribution (CTD) around the absorber. CTD in conventional receiver was found tremendously and unfavourably high. However, off-limit design of modified receiver in the form of Sun facing half annulus thermal insulation instilled an improved, more dispersed and consistent angular temperature distribution in absorber tube which clearly rendered an enhanced thermal exchange in HTF. This is also evident from the fact that CTD in conventional receiver varied in the full range of 420–550 K while, it varied in comparatively lower range i.e. 460–510 K in case of modified receiver.

Acknowledgement. All the authors bestow a profound gratitude to Mr. Ram Chandra for his invaluable feedback and suggestions. Along with this, work of all the authors used in references is highly appreciated.

Nomenclature

Symbols

A	Area (m^2)
C_p	Specific heat at constant pressure (kJ/kg K)
D	Diameter (m)
k	Thermal conductivity (kW/m K)
k_{in}	Turbulent intensity (%)
k_T	Turbulent conductivity (kW/m K)
$\dot{m}$	Mass flow rate (kg/s)
P	Pressure (N/m^2)
q	Heat flux (W/m^2)
T	Temperature (K)
t	Time (s)
u, v, w	Velocities in x, y, z direction
V	Velocity (m/s)
x, y, z	Cartesian coordinates

Greek

ε	Turbulent energy dissipation or emissivity
ρ	Density (kg/m^3)
μ	Dynamic viscosity (Pa s)
μ_t	Turbulent eddy viscosity (Pa s)
Δ	Increment value
φ	Circumferential angle

Subscripts

a	Absorber interaction point
a-cond	Conduction losses from absorber
a-conv	Buoyancy induced convective heat transfer
a-f, conv	Heat transfer from absorber to fluid *via.* convection
a-rad	Radiation losses from absorber
avg	Average
f	Heat transfer fluid
g-conv	Convection losses from glass to ambient
g	Condition pertaining to glass envelope
g-rad	Radiation losses from glass envelope
in	Condition at inlet
i, j	Pertaining to nodes i, j
o	Condition at outlet
sol	Solar incidence
sol-abs	Solar radiation transmitted through glass envelop to absorber *via.* radiation from absorber to trapped air in annulus

References

1. Sozen, A., Altiparmak, D., Usta, H.: Development and testing of a prototype of absorption heat pump system operated by solar energy. Appl. Therm. Eng. **22**, 1847–1859 (2002)
2. El-Fadar, A., Mimet, A., Azzabakh, A., Perez-Garci, M., Castaing, J.: Study of new solar absorption refrigeration powered by a parabolic trough collector. Appl. Therm. Eng. **29**, 1267–1270 (2009)
3. He, Y., Xiao, J., Cheng, Z., Tao, Y.: A MCRT and FVM coupled simulation method for energy conversion process in parabolic trough solar collector. Renew. Energy **36**, 976–985 (2011)
4. Al-Ansari, H., Zeitoun, O.: Numerical study of conduction and convection heat losses from a half-insulated air-filled annulus of the receiver of a parabolic trough collector. Sol. Energy **85**, 3036–3045 (2011)
5. Ozisik, M.N.: Radiative Transfer and Interaction with Conduction and Convection. Wiley, New York (1973)
6. Modest, M.: Radiative heat transfer, 2nd edn. Academic Press, Burlington (2003)
7. Xu, C., Chen, Z., Li, M., Zhang, P., Ji, X., Luo, X., Liu, J.: Research on the compensation of the end loss effect for parabolic trough solar collectors. Appl. Energy **115**, 128–133 (2014)
8. Wilcox, D.C.: Turbulence modelling for CFD. DCW Industries Inc (1998)
9. Tao, W.Q.: Numerical Heat Transfer, 2nd edn. Xi'an Jiaotong University Press, Xi'an (2001)
10. Cheng, Z.D., He, Y.L., Xiao, J., Tao, Y.B., Xu, R.J.: Three-dimensional numerical study of heat transfer characteristics in the receiver tube of parabolic trough solar collector. Int. Commun. Heat Mass Transfer **37**(7), 782–787 (2010)
11. Barlev, D., Vidu, R., Stroeve, P.: Innovation in concentrated solar power. Sol. Energy Mater. Sol. Cells **95**, 2703–2725 (2011)
12. Arasu, A.V., Sornakumar, T.: Design, manufacture and testing of fibreglass reinforced parabolic trough for parabolic trough solar collector. Sol. Energy **81**(10), 1273–1279 (2007)

Optimization of Wind Turbine Rotor Diameters and Hub Heights in a Windfarm Using Differential Evolution Algorithm

Partha P. Biswas[1]([⊠]), P.N. Suganthan[1], and Gehan A.J. Amaratunga[2]

[1] School of Electrical and Electronic Engineering,
Nanyang Technological University, Singapore, Singapore
parthapr001@e.ntu.edu.sg, epnsugan@ntu.edu.sg
[2] Department of Engineering, University of Cambridge, Cambridge, UK
gaja1@hermes.cam.ac.uk

Abstract. In this paper some of the optimized wind turbine layouts in a wind-farm, as presented by many authors, have been chosen as basis for further evaluation and study. The objective of most of earlier studies was to minimize cost per kW of power produced. This paper focuses from different perspective of optimization of turbine rotor diameters and hub heights to improve overall wind-farm efficiency at a reduced cost. Differential Evolution (DE) algorithm is employed to optimize rotor diameters and hub heights of turbines in a wind-farm.

Keywords: Windfarm layout · Wind turbine rotor diameter · Hub height · Differential evolution (DE) algorithm · Efficiency · Cost function

1 Introduction

Windfarm layout is highly complex and several factors are involved in optimizing the layout in order to achieve maximum power at minimum cost. After Mosetti *et al.* [3] publication on wind turbine layout optimization in a windfarm, many papers have been published by various authors in pursuit of improving the results. Among various publications, layouts proposed by Grady *et al.* [4] and Mosetti *et al.* [3] are considered here for further examination. The total power produced and published with the layouts proposed in these papers will be achieved, however with a better efficiency of the wind-farm with optimized turbine rotor diameters and hub heights.

By optimizing wind turbine rotor diameters and hub heights in a staggered manner, we are practically considering effect of wakes in downstream of a row of wind turbines. Due to effect of several wakes by upstream turbines, capacity of rotors at downstream is not fully utilized. The concept presented in this paper is – get maximum out of the wind energy by the upstream turbine row and install turbine of smaller dimension at the under-utilized downstream row. This optimization in turn increases the overall efficiency

© Springer Nature Singapore Pte Ltd. 2017
K. Deep et al. (eds.), *Proceedings of Sixth International Conference on Soft Computing for Problem Solving*, Advances in Intelligent Systems and Computing 547,
DOI 10.1007/978-981-10-3325-4_13

of the wind farm and overall cost of installation is reduced. In a closely spaced layout of upstream row, the optimization is more effective.

Rotor blades and Tower are major components for wind turbine contributing part of total cost. A comparative study on overall installation cost has been presented in the subsequent sections of this paper.

2   Windfarm Modelling

As in most of the other papers Jensen wake decay model [1, 2] is employed for consideration of the wind velocity inside the wake region. In the equations below for clarity k-th turbine is considered under the influence of single turbine at m-th location (Fig. 1). Assuming that the momentum is conserved in the wake, the wind speed in the wake region is calculated by–

$$u_k = u_{0k}\left[1 - \frac{2a}{\left(1 + \alpha_m \frac{x_{mk}}{r_{m1}}\right)}\right] \tag{1}$$

$$a = \frac{1 - \sqrt{1 - C_T}}{2} \tag{2}$$

$$r_{m1} = r_m \sqrt{\frac{1 - a}{1 - 2a}} \tag{3}$$

$$\alpha_m = \frac{0.5}{\ln\left(\frac{h_m}{z_0}\right)} \tag{4}$$

Where,

u_{0k} is the local wind speed at k-th turbine without considering the wake effect
x_{mk} is the distance between m-th & k-th turbine
r_m is the radius m-th turbine rotor
r_{m1} is the downstream rotor radius of m-thturbine
h_m is the hub height of the m-thturbine
α_m is the entrainment constant pertaining to m-thturbine
a is the axial induction factor
C_T is the thrust coefficient of the wind turbine, which represents the thrust exerted on the wind rotor by air. C_T is considered same for all turbines.
z_0 is the surface roughness of windfarm.

The wake region is conical for the linear wake model and radius of the wake region is represented by wake influence radius defined as

$$R_{wm} = \alpha_m x_{mk} + r_{m1} \tag{5}$$

Local wind speed at k-th turbine is dependent upon the hub height of the turbine as velocity of wind changes with height from ground. Logarithmic law has been used here to represent the local wind speed.

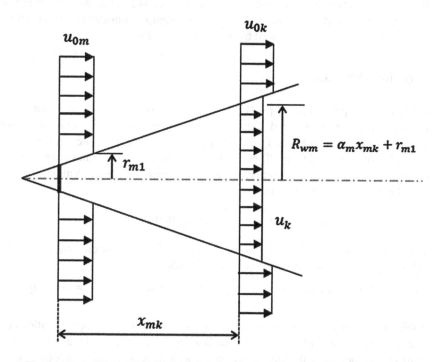

Fig. 1. Linear Wake Model - k-th turbine under the influence of single m-th turbine

$$u_{0k} = u_{ref} \log\left(\frac{h_k}{z_0}\right) \Big/ \log\left(\frac{h_{ref}}{z_0}\right) \tag{6}$$

For comparison with earlier studies performed in [3, 4], reference height and wind speed at reference height are considered as, $h_{ref} = 60$ m & $u_{ref} = 12$ m/s respectively; and z_0 as 0.3.

When one area is inside multiple wake flows, the velocity deficit will be enhanced. If we consider i-th turbine under the influence of multiple wakes, taking into account the wake flow superimposed effect, the wind speed at the position of i-th turbine can be written as [5, 6]

$$u_i = u_{0i}\left[1 - \sqrt{\sum_{j=1}^{N_T}\left(1 - \frac{u_{ij}}{u_{0j}}\right)^2}\right] \tag{7}$$

Where,

u_{0i} & u_{0j} are the local wind speeds (free stream velocity) at i-th & j-th turbine respectively without considering the wake effect

u_{ij} is the wind velocity at i-th turbine under the influence of j-th turbine

N_T is the number of turbines affecting the i-th turbine with wake effects.

It is to be noted here that the variation in rotor diameters and hub heights are not of wide range and the turbine downstream are far enough to consider full wake effect instead of partial wake effect induced by the upstream turbine.

3   Previous Studies

Currently in this paper layouts of [3, 4] for Case-1 i.e. *Constant Wind Speed and Fixed Wind Direction* has been revisited. For power calculation both Mosetti *et al.* & Grady *et al.* made some approximation. Total Power equation for each turbine has been used as $P_i = 0.3u^3$ for rotor radius of 20 m. As rotor radius in this paper has been used as variable the constant factor in power equation cannot be approximated. Without approximation the power equation becomes–

$$P_i = 0.5 * \rho * \pi * r^2 * u^3 * C_P/1000\,\text{kW} \tag{8}$$

For air density $\rho = 1.2254$ kg/m^3, rotor radius $r = 20$ m & rotor efficiency $C_p = 0.4$ $P_i = 0.307976611*u^3$ kW with wind velocity u in m/s.

Turbine thrust coefficient, as in other papers, is considered as $C_T = \dfrac{8}{9}$ i.e. approx. *0.8888.*

Table 1 summarizes the results obtained in previous studies with approximation and without approximation for Case-1: *Constant Wind Speed and Fixed Wind Direction.*

In this paper the same calculated power or higher will be achieved with optimization of rotor diameter and hub height at a better overall efficiency of the wind farm.

4   Differential Evolution

Differential Evolution (DE), introduced by Storn and Price in 1996, is stochastic, population based optimization algorithm where the individuals in the population evolve and improve their fitness through probabilistic operators like recombination and mutation. These individuals are evaluated and those that perform better are selected to compose the population in the next generation. More details of recent updates can be found in paper [8]. Following are the steps involved in Differential Evolution algorithm.

4.1   Initialization

The first step in the DE optimization process is to create an initial population of candidate solutions by assigning random values to each decision vector of the population. Such values must lie inside the feasible bounds (between maximum & minimum) of the decision vector. We may initialize j-th component of the i-th decision vector as

$$x_{i,j}^{(0)} = x_{\min,j} + rand_{ij}[0, 1]\left(x_{\max,j} - x_{\min,j}\right)$$

Where $rand_{ij}[0, 1]$ is a uniformly distributed random number lying between 0 and 1.

4.2 Mutation

After initialization, DE creates a donor/mutant vector $v_i^{(t)}$ corresponding to each population member or target vector $x_i^{(t)}$ in the current iteration through mutation *(the superscript 't' denotes parameter at t-th iteration)*. There are quite a few strategies for mutation. The one used here is:

$$v_i^{(t)} = x_{R_1^i}^{(t)} + F\left(x_{R_2^i}^{(t)} - x_{R_3^i}^{(t)}\right)$$

The indices R_1^i, R_2^i & R_3^i are mutually exclusive integers randomly chosen from the population range. The scaling factor F is a positive control parameter for scaling the difference vectors

4.3 Crossover

Through crossover the donor vector mixes its components with the target vector $x_i^{(t)}$ to form the trial/offspring vector $u_i^{(t)} = (u_{i,1}^{(t)}, u_{i,2}^{(t)}, \dots, u_{i,d}^{(t)})$. Binomial crossover, which is adopted here, operates on each variable whenever a randomly generated number between 0 and 1 is less than or equal to a pre-fixed parameter Cr, the crossover rate. The scheme is expressed as:

$$u_{i,1}^{(t)} = \begin{cases} v_{i,j}^{(t)} & \text{if } j = K \text{ or } rand_{i,j}[0, 1] \leq Cr, \\ x_{i,j}^{(t)} & \text{otherwise} \end{cases}$$

Where K is any randomly chosen natural number in $\{1,2,\dots,d\}$, d being the dimension of real-valued decision vectors.

4.4 Selection

Selection determines whether the target (parent) or the trial (offspring) vector survives to the next iteration i.e. at t = t+1. The selection operation is described as:

$$x_i^{(t+1)} = \begin{cases} u^{(t)} & \text{if } f\left(u^{(t)}\right) \leq f\left(x_i^{(t)}\right), \\ x_i^{(t)} & \text{otherwise} \end{cases}$$

Where f(.) is the objective function to be minimized.

5 Application and Approach Using DE

For all practical purposes it is not possible to have as many variants of rotor diameters and hub heights as the number of turbines. The layouts considered for further evaluation in this paper have 30 & 26 turbines in a 2 km x 2 km windfarm area consisting of many 200 m x 200 m grids.

3 different values of both rotor diameters and hub heights are randomly selected from a defined range of these parameters. The selected values are then evenly distributed and assigned to all the wind turbines in the windfarm. Range of rotor diameter is input in DE as 36 to 44 m, while that of hub height is 54 to 62 m. It may be noted here that the selected values in DE are rounded off to calculate the objective function defined below. Power from each turbine inside wake is calculated using Eqs. (1) to (8) above.

Efficiency of the windfarm is described as,

$$\eta = \frac{\sum_{i=1}^{N} P_i}{\sum_{i=1}^{N} P_{i,\max}} \tag{9}$$

Where, $P_{i,\max}$ is the maximum output from i-th turbine had there been no wake effect. The objective of the optimization: Maximize 'η' subject to

$$\sum_{i=1}^{N} P_i = P_{desired} + \varepsilon \tag{10}$$

Where, $P_{desired}$ is the total calculated power output as detailed in Table 1. ε is a positive error arbitrarily chosen in the range of $0 < \varepsilon \leq 30$ to facilitate the iteration process.

Table 1. Recalculated power of previous layouts [3, 4]

Parameter	Grady et al.		Mosetti et al.	
	Reported	Calculated (without approx.)	Reported	Calculated (without approx.)
No. of Turbines (N)	30	30	26	26
Total Power	14310	14667	12352	12654
Efficiency (%)	92.015	91.867	91.645	91.452

6 Results and Discussion

The DE optimization for 30 turbine layout proposed by Grady *et al.* gives following output for Case 1: *Constant Wind Speed and Fixed Wind Direction.*

Figures 2 *and* 3 below specify the selected rotor diameters and hub heights for the windfarm. Do take note of wind direction indicated by arrow in the figures.

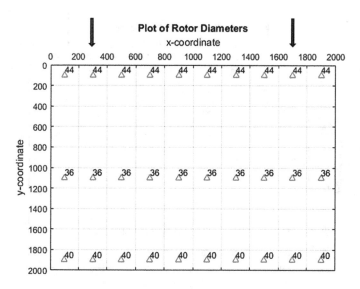

Fig. 2. Layout indicating Optimized Rotor Diameters

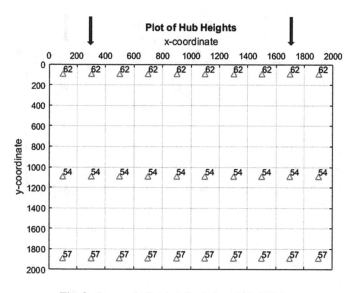

Fig. 3. Layout indicating Optimized Hub Heights

The DE optimization for 26 turbine layout proposed by Mosetti *et al.* gives following output for Case 1: *Constant Wind Speed and Fixed Wind Direction.*

Figures 4 *and* 5 below specify the selected rotor diameters and hub heights for the wind farm proposed by Mosetti *et al.*

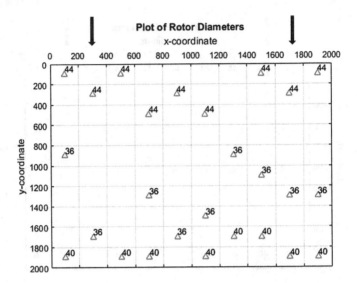

Fig. 4. Layout indicating Optimized Rotor Diameters

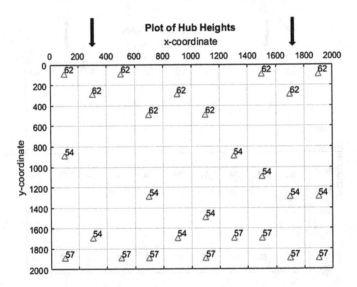

Fig. 5. Layout indicating Optimized Hub Heights

It is obvious from the optimized layout is that the turbines with larger diameter and higher hub height tend to align in a row facing the oncoming wind. This helps to extract maximum power out of the wind. The turbines at the last row (farthest from oncoming wind) experience the least wind speed. Hence to recover power out of the wind, these turbines assume intermediate diameter and hub heights. The turbines in the middle have smallest rotor diameter and shortest hub height.

Adopting different rotor diameters and hub heights essentially increases the overall efficiency of the windfarm as can be seen from Tables 2 and 3. Though some of the rotor radii and hub heights have increased from earlier studies considering fixed rotor radius of 20 m and hub height of 60 m, many have decreased as well to bring down the overall investment cost which has been discussed in subsequent clause here. It is to be noted here that the above solutions are not unique and different combinations of variable radii and hub heights can be obtained by running the optimization program several times.

Table 2. Present layout vs Grady et al. layout [4]

Parameter	Grady et al.		Optimized layout (using differential evolution)
	Reported	Calculated (without approx.)	
No. of turbines (N)	30	30	30
Total power	14310	14667	14692
Efficiency (%)	92.015	91.867	93.064

Table 3. Present layout vs Mosetti et al. layout [3]

Parameter	Mosetti et al.		Optimized layout (using differential evolution)
	Reported	Calculated (without approx.)	
No. of turbines (N)	26	26	26
Total power	12352	12654	12680
Efficiency (%)	91.645	91.452	92.612

7   Cost Model

The cost model introduced by Mosetti *et al.* accounts for only one variable, number of wind turbines. The cost for a windfarm having N turbines can be expressed as –

$$Cost_{base} = N\left(\frac{2}{3} + \frac{1}{3}e^{-0.00174N^2}\right) \tag{11}$$

In a wind turbine, Turbine Blades and Tower account for 17.7% & 21.9% cost respectively [11]. Wind turbine costs about 75% of total installation cost of a windfarm setup [10]. For rotor radius change in the range of $\pm$ 10% from base radius of 20 m, 1% change in rotor radius affects the blade cost by about 3% [9]. Tower cost varies proportional to its height [7].

Considering variations in rotor diameters and hub heights, and to compare the cost with previous studies the modified cost model, taking into account the weight age of rotor cost and tower cost w.r.t. total installation cost, can be represented as:

$$Cost = Cost_{base}\left[1 + \frac{1}{N}\sum_{i=1}^{v_1} 0.0039825x_i N_i + \frac{1}{N}\sum_{j=1}^{v_2} 0.0016425x_j N_j\right] \tag{12}$$

Where,

$Cost_{base}$ = Calculated cost using Eq. (11) above

x_i = % change in rotor radius for i-th variant of turbines from R_{base} = 20 m, '−ve' for reduction, '+ve' for increase in radius

x_j = % change in hub height for j-th variant of turbines from H_{base} = 60 m, '−ve' for reduction, '+ve' for increase in height

R_{base} = Rotor radius of turbine used in previous study = 20 m

H_{base} = Hub height of turbine used in previous study = 60 m

v_1 = Number of variants of turbines having rotor radii different from R_{base}

v_2 = Number of variants of turbines having hub heights different from H_{base}

N_i = Number of turbines of i-th variant, turbine rotor radius different from R_{base}

N_j = Number of turbines of j-th variant, turbine hub height different from H_{base}.

Using Eq. (12) above the Cost/kW of power generation is calculated and tabulated below (Tables 4 and 5).

Table 4. Cost per kW comparison (for layout proposed by *Grady* et al.)

Parameter	Grady et al.		Optimized layout (with different turbine rotor diameters & hub heights)
	Reported	Calculated (without approx.)	
Total power	14310	14667	14692
Cost/kW	1.5436×10^{-3}	1.506×10^{-3}	1.4939×10^{-3}

Table 5. Cost per kW comparison (for layout proposed by *Mosetti* et al.)

Parameter	Mosetti et al.		Optimized layout (with different turbine rotor diameters & hub heights)
	Reported	Calculated (without approx.)	
Total power	12352	12654	12680
Cost/kW	1.6197×10^{-3}	1.5810×10^{-3}	1.5678×10^{-3}

8 Conclusion

The selection of hub heights and rotor diameters in a staggered manner helps improve overall efficiency of the windfarm; also offers an economical advantage when wind speed and direction are mostly fixed. A supposition made in the study is of constant thrust co-efficient for different rotor diameters and this can be achieved with proper design of rotor. Similar to earlier studies wind speed at respective hub height is considered as local wind speed contributing to the power output from that wind turbine. An optimized layout found using Genetic or Greedy or any other algorithms can further be reviewed in terms of selection of rotor diameters and hub heights. The effectiveness of different hub heights and rotor diameters for more practical cases when wind speed and direction are variable, remain topic for further studies.

Acknowledgements. This project is funded by the National Research Foundation Singapore under its Campus for Research Excellence and Technological Enterprise (CREATE) program.

References

1. Jensen, N.O.: A note on wind generator interaction (1983)
2. Katic, I., Højstrup, J., Jensen, N.O.: A simple model for cluster efficiency. In: European Wind Energy Association Conference and Exhibition (1986)
3. Mosetti, G., Poloni, C., Diviacco, B.: Optimization of wind turbine positioning in large windfarms by means of a genetic algorithm. J. Wind Eng. Ind. Aerodyn. **51**(1), 105–116 (1994)
4. Grady, S.A., Hussaini, M.Y., Abdullah, M.M.: Placement of wind turbines using genetic algorithms. Renew. Energy **30**(2), 259–270 (2005)
5. Mittal, A.: Optimization of the layout of large wind farms using a genetic algorithm. Dissertation Case Western Reserve University (2010)
6. Chen, Y., et al.: Wind farm layout optimization using genetic algorithm with different hub height wind turbines. Energy Convers. Manage. **70**, 56–65 (2013)
7. Chen, K., et al.: Wind turbine layout optimization with multiple hub height wind turbines using greedy algorithm. Renew. Energy **96**, 676–686 (2016)
8. Das, S., Mullick, S.S., Suganthan, P.N.: Recent advances in differential evolution–an updated survey. Swarm Evol. Comput. **27**, 1–30 (2016)
9. Fingersh, L.J., Hand, M.M., Laxson, A.S.: Wind turbine design cost and scaling model (2006)
10. European Wind Energy Association: The economics of wind energy. In: EWEA (2009)
11. Jamieson, P.: Innovation in wind turbine design. Wiley, Chichester (2011)

Big Data Analytics Based Recommender System
for Value Added Services (VAS)

Inderpreet Singh[1], Karan Vijay Singh[1(✉)], and Sukhpal Singh[2]

[1] Engineer-Product Development, Mahindra Comviva, Gurgaon, India
inderpreet.singh993@gmail.com, karanvijay.singh93@gmail.com
[2] Computer Science and Engineering Department,
Thapar University, Patiala, Punjab, India
ssgill@thapar.edu

Abstract. The increasing number of services/offers in telecom domain offers more choices to the consumers. But on the other side, these large number of offers cannot be completely looked by the customer. Hence, some offers may pass unobserved even if they are useful for the particular kind of customers. To solve this issue, the usage of recommender systems in telecom sector is growing. So, there is need to notify the customer about the offers which are made on the basis of customer interests. The recommender system is based on demand or interest of consumer. In this paper we proposed a Big Data Analytics based Recommender System for Value Added Services (VAS) in case of telecom organizations so that they could gain profitability in the market by generating customer specific offers.

Keywords: Analytics · Recommendation · Big data · Telecom · Segmentation · Hybrid model

1 Introduction

In recent years, wireless data traffic has been increasing exponentially. This traffic is mainly driven by the usage of smart phones and increasing penetration of Internet services in the world. Nano and Pico cells have become a reality [1] due to increase in demand of smart phones and decrease in size of them. Thus resulting in a rapid increase not only in number and type of smart phones but also in the services needed to track and administer to provide high quality services to the customers. As most of data in the world is now generated by the smart phones and as telecom network is the carrier of this data, so telecom operators are experiencing the explosion of data making it difficult for them to decide about data needed for predicting customer behaviour.

Recommender systems are used to predict the liking that a customer will have for a particular item [2]. They are subclass of information filtering systems i.e. systems which remove unwanted data from the data stream. A recommendation produced by the recommender system is the output of analysis performed on data sets of customer's preferences. Recommendations in case of web services are difficult because of unavailability of customer's exact location and also service history of customer cannot be easily

© Springer Nature Singapore Pte Ltd. 2017
K. Deep et al. (eds.), *Proceedings of Sixth International Conference on Soft Computing for Problem Solving*, Advances in Intelligent Systems and Computing 547,
DOI 10.1007/978-981-10-3325-4_14

defined. But in the case of telecom, this problem can be pacified as smart phone becomes main access point for customer and its usage can be stored and analysed by the company to get ingenious insights about customer behaviour. The availability of customer data in case of telecom domain makes easier for performing different analysis which results in suggesting more worthy services to a target customer.

Using recommendation engines, giving customers the services of their choice and not loading them with the unnecessary services, companies can easily gain customer's confidence and improve customer experience [15, 16]. There are different ways that can be put to use in a recommendation system namely Collaborative filtering, Content Based filtering, and also hybrid approaches. The most commonly used, Collaborative filtering technique makes a prediction for the target customer based on gathering information of (older) customer's with similar interests and activities. Finding likeness between customers effectively is the pivotal part of Collaborative Filtering technique. The challenge occurs when there is only minuscule data available of older customer's or customer's with identical interests leading to cold starting problem [17, 18]. On the other hand, Content based technique uses the distinct characteristics of the product/item that the customer had earlier bought or had shown a liking, so as to give recommendations to the customer of the products with identical characteristics. The combination of the above two leads to hybrid approach that can be executed in several ways like making Content based and collaborative filtering predictions individually and then utilizing results of both the techniques to give final recommendation to the customer. Also this approach can also be used to eliminate cold start problem [19].

The main things playing roles behind the popularity of big data analytics are people skilled in analytics, efficient analytical tools, robust data infrastructure, fact-based decision-making strategy, association between IT approaches, durable committed sponsorship and clear business requirement [15, 16]. An efficient way of analysing and retrieve data in analytics makes it as popular decision oriented data management. In traditional data warehouses and databases, business value cannot produced from stored data. With the help of this new technology, data can create incredible value after storing it efficiently [17, 18]. A new products, techniques and technologies are emerged for data analysis which can be easily used for big data analytics, such as appliances, in-database analytics and in-memory analytics [19].

In this paper, we are giving two recommendation engine models. One, which generates the recommendations on the basis of customer segmentation and metadata details; and other which generates the recommendations on the basis of customer segmentation and service comparison. Former model is used for services like ringtone, games recommendation etc. While the latter is used for services like astrology, cricket, jokes etc.

2 Related Work

Data generated from smart phones is very complex due to which it poses difficulty in analysis. It has problems like heterogeneity and noisiness associated with it. In [3], authors have made a mobile recommender system that helps in providing profitable routes for taxis. It takes present location of taxi, time status and operational status i.e.

with or without passengers. While in [4], authors give example of a recommendation system which suggests appropriate information depending on the customer's situation and interests. Recommendation systems are now utilized in different areas like: In [5], authors discusses about news recommender system that pushes news articles to smart phone customers based on customer's contextual information as well as news content. Where customer's information needs are estimated by the use of Bayesian network technique. While in [14], author has proposed a real-time location-tagged contents recommender system based on smart phone social network. This system locates a customer with the use of GPS (Global Positioning System), and then applies filtering methods. In addition to the use of GPS, the correct location of the customer can also be found with the help of Wi-Fi positioning system (WPS) to get high precision [6]. In [7], authors have developed a smart phone recommender system for indoor shopping based on positioning approach by using received signal patterns of smart phones. It eliminates the disadvantages of existing positioning technologies. Apart from these, there are also recommender systems for movies [8], personalized blog content [9], experts [10], collaborators [11], financial services [12] and Twitter pages [13].

3 Various Types of Data Needed

To generate recommendations, recommendation engine needs certain sets of data. Depending on what type of recommendation needs to be generated, recommendation engine will use specific set. In this paper, we will be discussing about the recommendation engines which will be used for recommending Ringtones and Value added Services (VAS). Consumer application form details are the details that consumer fills while registering for a particular service or set of services provided by the telecom company. It contains details about the consumer's basic information. The set of CAF details which the company has is mentioned in below Table 1.

Table 1. Customer application form (CAF) details

CAF details		
S. no	Data	Purpose of data
1	Name	Basic Information
2	Age	Differentiate customer on the basis of Age
3	Sex	Differentiate customer on the basis of sex
4	Marital Status	Finding whether the customer is Single/Married
5	Address	Get the customer's home address
6	City	To know about the customer's home city

Company level details are the details of consumer transactions which the company stores in their databases (Table 2). Based on these details and metadata of ringtone, recommendation engine generates the ringtone recommendations for the customer. In case of other VAS (Value Added Services), recommendations generated are based on the segment to which consumer profile belongs rather than on the basis of metadata as metadata details are not associated for these kind of services. The set of company level details which the company maintains in their databases is given below in the table.

Table 2. Company level details

Company level details		
S. no	Data	Purpose of data
1	Talk time Value	Total Duration of calls used for generating recommendations on the basis of call plan
2	Data Usage	Details of data used for generating recommendations on basis of data packs
3	No. of SMS Sent/Receive	Details of messages sent/received for generating recommendations on basis of messages packs
4	Customer Location	To know about the customer's present location
5	Most Frequently Called Numbers	To give customer customized calling plans
6	Handset Type	Various types of services supported by customer's handset
7	Metadata of VAS Services	Generate recommendations on the basis of customer's VAS profile
8	Metadata of Customer Ringtone	Generate recommendations on the basis of customer's ringtone

Metadata details of ringtone as shown in Table 3, is used in case of recommending caller tones to the customers. In this case, metadata of customer's past and present caller tones is used to know to which segment the customer belongs, and on the basis of that segment, recommendation is generated for the customer.

Table 3. Metadata details

Metadata details		
S. no	Data	Purpose of data
1	Type	Type of Ringtone e.g. pop, jazz etc.
2	Film/Album	Film/Album to which Ringtone belongs
3	Composer	To generate recommendations on basis of composer
4	Singer	To give recommendations on basis of singer
5	Director	To generate recommendations on basis of music director
6	Language	Language of the ringtone

4   Model

Figure 1 Represents the architecture of the recommendation engine which works on the basis of customer segmentation and metadata comparison.

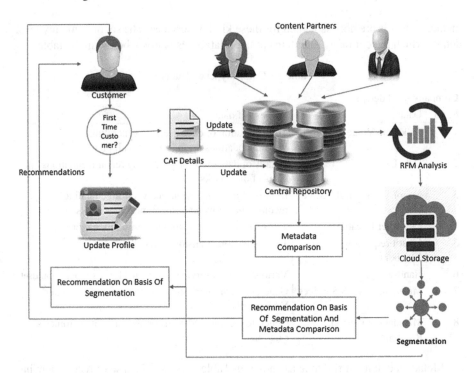

Fig. 1. Architecture of model based on customer segmentation and metadata comparison

4.1 Customer Profile

A new customer profile gets created when the customer registers himself/herself for the services of the telecom company. Customer profile contains basic details about the customer like name, age, gender etc. and these are the details which the customer gives himself by filling consumer application form (CAF). Then, it is checked whether the customer is first time customer or he/she wants to change the caller tone of smartphone. In case of former, recommendation engine generates the recommendations on the basis of customer segmentation i.e. the segment to which the customer's profile belongs. But in case of lateral, recommendations are generated on the basis of metadata and customer profile details.

4.2 Recommendation on Basis of Customer Segment

When the customer register himself for the first time, then there are no metadata details present for that customer, then the recommendations are generated on the basis of matching the customer profile parameters to the profiles of already existing customers in the repository. The profile which matches maximally is used to know the segment to which the customer profile belongs and on basis of that segment, the recommendation is generated.

4.3 Updating Profile and Metadata

If the already registered customer wants to change the caller tone of his/her smartphone, then firstly the metadata details of the customer's record is changed accordingly. So, customer's profile gets updated whenever customer wants to change the caller tone.

4.4 Metadata Comparison

In this step, recommendation engine will compare the metadata details of customer profile with the metadata of the other profiles i.e. we will be calculating the similarity index of the customer metadata details with other customer's metadata details and the profile corresponding to the maximum similarity index will be then looked up to see to which segment it belongs on basis of which recommendations will be made.

The similarity index between two customer's profiles is represented by a number between -1.0 and 1.0. The possibility of customer liking/selecting particular ringtone will be between -1.0 and 1.0. Similarly, in case of not liking/selecting the number will be between -1.0 and 1.0. For finding similarity index, we will have two sets corresponding to each customer. One corresponding to the customer liking/selecting the caller tune and other for not selecting/liking the ringtone. According to Jaccard's formula [20], the similarity index is calculated as follows

$$J(X, Y) = |X \cap Y| \div |X \cup Y|$$

The calculation involves the division of the total number of common elements in both sets by the total number of the elements in both sets (only counted once). The Jaccard index of two similar sets will always be 1, while for two sets with no common elements will always yield 0. Jaccard index for two profiles on the basis of liking of each parameter is,

$$J(X, Y) = |S1 \cap S2| \div |S1 \cup S2|$$

Now as two customers selecting same ringtone is similar, then two customers not selecting the same ringtone are also similar. So, by changing above equation we get,

$$J(X, Y) = (|S1 \cap S2| + |NS1 \cap NS2|) \div (|S1 \cup S2 \cup NS1 \cup NS2|)$$

I.e. instead of considering same selection we also have taken into the account the deselection. In denominator, we have taken the total number of selection/deselection that customer has made. Here, we have considered the customer selection or deselection in independent sort of way. But what if customer likes ringtone but other customer doesn't and vice-versa. To take this thing into account we again have to modify the equation as,

$$J(X, Y) = (|S1 \cap S2| + |NS1 \cap NS2| - |S1 \cap NS2| - |NS1 \cap S2|) \div (|S1 \cup S2 \cup NS1 \cup NS2|)$$

Now this equation will give 1.0 if two customer's profiles have same selection/liking for caller tones and -1.0 if two customers have deselection/disliking for caller tones.

4.5   Recommendation on Basis of Customer Segmentation and Meta-data Comparison

RFM analysis of customer transaction information plus the metadata of profile that matched with the customer's profile gives us the segment to which the customer profile belongs. RFM is method for analysis of market which is used to examine which customers are best ones by examining how recent the customer has made purchase (Recency), how often he/she purchases (Frequency) and how much the customer spends on the purchase (Monetary). It is based on the fact that "80% of business comes from 20% of the customers". Customers are then given ratings on the basis of these three input parameters by the telecom organisation. The RFM score and metadata details matched profile gives the segment to which the customer profile belongs. On the basis of which the recommendations are generated for the customers i.e. Hybrid recommendation systems [21].

4.6   Content Provider

Content providers belonging to different partners are responsible for uploading the content to the telecom company's central repository on which RFM analysis is performed and output is stored on the central cloud storage, so that results can be accessible anytime and from anywhere. Based on these results, customer segmentation is done.

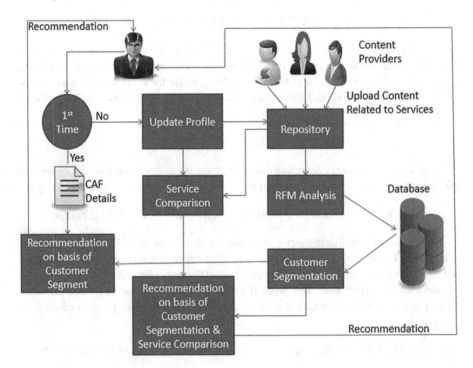

Fig. 2. Architecture of model based on customer segmentation and service comparison

The similar type of model can be used in case of Games i.e. based on customer profile, recommendations can be given whether one likes action, racing, puzzles, sports games etc. While new customer are given recommendations based on similar CAF details of other customers who have already subscribed to games, old customers get recommendations based on metadata of games that they have subscribed along with the segment of customer profile to which they belong. Figure 2 represents the architecture of the recommendation engine which works on the basis of customer segmentation and Service comparison.

In case of other VAS related services, if a new customer wants to subscribe to any service like Astrology, Cricket, Jokes etc., he/she is recommended based on customer profile segmentation of other customers with similar CAF details as we have no idea about the interests of the new customer. Also there is no meta-data related to these type of services. On the other hand, old customers are given recommendations based on their updated profile and RFM analysis of transaction information of the customer which gives the segment to which the customer profile belongs.

5 Conclusion

This paper discussed the approach that how recommendation engine could collect the data, analyse it and generate recommendations on basis of it. In it we have explained two types of recommendation engine. One, which uses the metadata details and customer segmentation to generate recommendations and other which uses services comparison along with customer segmentation. Recommendation not only helps telecom companies to target their customers effectively, but also help the customers in showing the content which is relevant to them.

6 Future Scope

The presented recommendation engines can be used to help the telecom companies in recommending appropriate services to the customer. Hence, gaining competitive edge in the market. Some open issues that need special attention in future research are,

- Approach to integrate above recommendation engine with the systems already used without increasing workload of them.
- To have more input parameters on basis of which recommendation could be given.
- To give real time recommendations to the customer on the basis of location, time etc.

References

1. Ericsson, L.M.: More than 50 billion connected devices. White Paper (2011)
2. Ricci, F., Rokach, L., Shapira, B.: Introduction to Recommender Systems Handbook. Springer, US (2011)
3. Ge, Y., Xiong, H., Tuzhilin, A., Xiao, K., Gruteser, M., Pazzani, M.J.: An energy-efficient mobile recommender system (PDF). In: Proceedings of the 16th ACM SIGKDD International Conference on Knowledge Discovery and Data Mining, pp. 899–908. ACM, New York (2010). Accessed 17 Nov 2011

4. Bouneffouf, D.: Following the customer's interests in mobile context-aware recommender systems: the hybrid-e-greedy algorithm. In: Proceedings of the 2012 26th International Conference on Advanced Information Networking and Applications Workshops (PDF). LNCS, pp. 657–662. IEEE Computer Society (2012). ISBN: 978-0-7695-4652-0 [dead link]
5. Yeung, K.F., Yang, Y.: A proactive personalized mobile news recommendation system. In: 2010 Developments in E-systems Engineering (DESE), pp. 207–212. IEEE (2010)
6. Danalet, A., Farooq, B., Bierlaire, M.: A Bayesian approach to detect pedestrian destination-sequences from WiFi signatures. Transp. Res. Part C Emerg. Technol. **44**, 146–170 (2014). doi:10.1016/j.trc.2014.03.015
7. Fang, B., Liao, S., Xu, K., Cheng, H., Zhu, C., Chen, H.: A novel mobile recommender system for indoor shopping. Expert Syst. Appl. **39**(15), 11992–12000 (2012)
8. Colombo-Mendoza, L.O., Valencia-García, R., Rodríguez-González, A., Alor-Hernández, G., Samper-Zapater, J.J.: RecomMetz: a context-aware knowledge-based mobile recommender system for movie showtimes. Expert Syst. Appl. **42**(3), 1202–1222 (2015)
9. Chiu, P.-H., Kao, G.Y.-M., Lo, C.-C.: Personalized blog content recommender system for mobile phone customers. Int. J. Hum. Comput. Stud. **68**(8), 496–507 (2010)
10. Buettner, R.: A framework for recommender systems in online social network recruiting: an interdisciplinary call to arms. In: 47th Annual Hawaii International Conference on System Sciences, Big Island, Hawaii, pp. 1415–1424. IEEE (2014). doi:10.13140/RG.2.1.2127.3048
11. Chen, H., Gou, L., Zhang, X., Giles, C.: Collabseer: a search engine for collaboration discovery. In: ACM/IEEE Joint Conference on Digital Libraries (JCDL) (2011)
12. Felfernig, A., Isak, K., Szabo, K., Zachar, P.: The VITA financial services sales support environment. In: AAAI/IAAI 2007, Vancouver, Canada, pp. 1692–1699 (2007)
13. Goel, A., Gupta, P., Sirois, J., Wang, D., Sharma, A., Gurumurthy, S.: The who-to-follow system at Twitter: strategy, algorithms, and revenue impact. Interfaces **45**(1), 98–107 (2015)
14. Kwon, H.-J., Hong, K.-S.: Personalized real-time location-tagged contents recommender system based on mobile social networks. In: IEEE International Conference on Consumer Electronics (ICCE), pp. 558–559. Las Vegas (2012)
15. Singh, S., Chana, I.: EARTH: energy-aware autonomic resource scheduling in cloud computing. J. Intell. Fuzzy Syst. **30**(3), 1581–1600 (2016)
16. Singh, S., Chana, I.: Resource provisioning and scheduling in clouds: QoS perspective. J. Supercomput. **72**(3), 926–960 (2016)
17. Singh, S., Chana, I.: QoS-aware autonomic resource management in cloud computing: a systematic review. ACM Comput. Surv. (CSUR) **48**(3), 42 (2016)
18. Singh, S., Chana, I.: QRSF: QoS-aware resource scheduling framework in cloud computing. J. Supercomput. **71**(1), 241–292 (2015). Springer
19. Singh, S., Chana, I., Singh, M., Buyya, R.: SOCCER: self-optimization of energy-efficient cloud resources. Cluster Comput. **19**, 1–14 (2016). doi:10.1007/s10586-016-0623-4. Springer
20. Jaccard, P.: Etude comparative de la distribution florale dans une portion des Alpes et du Jura. Impr. Corbaz (1901)
21. Adomavicius, G., Tuzhilin, A.: Toward the next generation of recommender systems: a survey of the state-of-the-art and possible extensions. IEEE Trans. Knowl. Data Eng. **17**(6), 734–749 (2005). doi:10.1109/TKDE.2005.99

PSO Based Context Sensitive Thresholding Technique for Automatic Image Segmentation

Anshu Singla[1]([⊠]) and Swarnajyoti Patra[2]

[1] Computer Science and Engineering Department,
Thapar University, Patiala 147004, India
`asheesingla@gmail.com`
[2] Department of Computer Science and Engineering,
Tezpur University, Tezpur 784028, India
`swpatra@gmail.com`

Abstract. Image segmentation is the area of research to study the number of homogenous regions present in the image and to analyze the objects present in the image. The set of pixels belong to each object present in the image can be assigned same gray level to visualize certain characteristics. In this article, Particle Swarm Optimizer(PSO) based context sensitive thresholding technique has been presented to detect optimal thresholds present in the image automatically. The main objective behind utilization of the PSO is to demonstrate its effectiveness when applied to context sensitive thresholding technique to determine optimal thresholds of the image to be segmented. Further the results are compared with the two state-of-art thresholding techniques for image segmentation cited in literature. The achieved improvements are validated in terms of quantitative and qualitative parameters on the large dataset of images.

1 Introduction

Image segmentation partitions the image into different regions and each pixel of the image is assigned the region to which it belongs. Each region represents the different real world object [1]. It can be broadly classified into four categories [2] based on: (a) texture, (b) thresholding, (c) clustering and (d) region. Thresholding is one of the effective technique for image segmentation and widely used in image processing applications such as Optical character recognition(OCR), automatic visual inspection of defects, video change detection, moving object segmentation, medical imaging and so on. In literature, there exist many thresholding techniques [3,4] like histogram based [5–15], biologically inspired [16–20]. The mentioned techniques either do not consider the spatial contextual information or unable to detect the number of objects present in the image automatically. The two major challenges for thresholding techniques are: (i) To consider spatial contextual information of the image. (ii) Automatic detection of number of objects present in the image. Recently, one-dimensional energy curve has been proposed [21] which maintains the characteristics of 1D-histogram of

© Springer Nature Singapore Pte Ltd. 2017
K. Deep et al. (eds.), *Proceedings of Sixth International Conference on Soft Computing for Problem Solving*, Advances in Intelligent Systems and Computing 547,
DOI 10.1007/978-981-10-3325-4_15

the image and also includes the spatial contextual information of the image. Another GA based context sensitive technique based on energy curve has been proposed recently [22] which detects the number of objects present in the image automatically.

In this article, both the energy curve and PSO has been employed, to determine the optimal thresholds for automatic image segmentation. To assess the effectiveness of the proposed methodology the results have been compared with two state-of-art techniques: (i) recently developed Genetic algorithm(GA) based context sensitive [22], and (ii) traditional context insensitive [8] thresholding techniques. The image data set considered for experiment consist of images with different number of objects present in it. Experimental results obtained for those images confirmed the effectiveness of the proposed technique.

The rest of this paper is organized as follows: A brief overview of PSO algorithm is presented in Sect. 2. The proposed PSO based technique for image segmentation is presented in Sect. 3. For the considered image dataset, the experimental settings and result analysis has been presented in Sect. 4. Finally, Sect. 5 draws the conclusion of the proposed work and the ways ahead in future.

2   PSO Overview

PSO is a swarm-intelligence based algorithm proposed by kennedy and Eberhart in 1995 [23]. This algorithm simultaneously examines different solutions in different regions of the solution space to find the global optimum solution skipping local optimum traps. PSO algorithm can be employed as GA to solve many same kind of problems [24]. Initially, PSO algorithm has no definition of inertia weight constant but later introduced by Shi and Eberhart [25]. They proposed that small inertia weight favours local search where as larger helps in global search. It is applied to set of particles, where each particle has assigned a randomized velocity. Further each particle is allowed to move towards problem space and has to remember its personal best position/fitness it has passed before and the best solution of its neighbouring particles i.e. global fitness. The personal fitness is known as "p-best" and the global fitness value is known as "g-best". The movement method of particle is presented in the Eq. 1.

$$Vel_{i,t} = \alpha * Vel_{i,t} + M_1 * Rand() * (p\_best_i - CP_i) + M_2 * Rand() * (g\_best_{id} - CP_i) \quad (1)$$

The position of new particle is determined by the Eq. 2 as follows:

$$CP_{i,t} = CP_{i,t} + Vel_{i,t} \quad (2)$$

where Vel_i denotes the velocity of ith particle within the range $[-Vel_{Max}, Vel_{Max}]$ which is predefined by the user. α denotes the inertial weight coefficient that lies in the interval $[0.4, 1.4]$. M_1 and M_2 denote the self confidence and social confidence coefficients respectively. These learning factors of stochastic process determine the impact of the personal best and global best respectively. The Rand() is a random number generator which can generate a random real number between 0 and 1 under normal distribution. CP_i represents the current position of a particle i. The maximum number of iterations for which the PSO is run is given by Max_{iter}. The algorithm 1 presents the pseudocode of PSO.

Algorithm 1. Standard PSO algorithm

```
Input:
      Maximum Number of iterations (Max_iter)
      Acceleration constants (M_1,M_2)
      Velocity constant (Vel_Max)
      inertial weight (α)
Output:
      Global best particle (gbest )
Begin:
Initialize position and velocity of all particle.
itr ← 1
while itr ≠ Max_iter do
    for all swarm particles do
        Calculate fitness value
        if new fitness value is better than pbest then
            pbest ← new calculated fitness value
        end if
    end for
    Choose the particle as gbest among all pbest.
    for all swarm particles do
        Calculate particle velocity according to eq.(1).
        Update particle position according to eq.(2).
    end for
    itr = itr + 1
end while
```

---

3   Proposed Technique

In this study, we focussed on PSO based context sensitive thresholding technique for automatic image segmentation. The methodology used in the proposed technique is presented graphically in the Fig. 1.

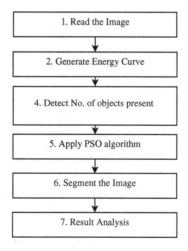

Fig. 1. Methodology

The major steps of proposed technique are explained below:

Step 1: Read the image I of size M*N such that $I = \{l_{ij}, 1 \leq i \leq M, 1 \leq j \leq N\}$. l_{ij} is the gray value of the image I at pixel position (i, j). L is the maximum gray value of the image I.

Step 2: Determine the Energy value at each gray level of the image I using energy function [21] given in Eq. 3.

$$E_l = -\sum_{i=1}^{M}\sum_{j=1}^{N} \sum_{pq \in S_{ij}^2} b_{ij}.b_{pq} + \sum_{i=1}^{M}\sum_{j=1}^{N} \sum_{pq \in S_{ij}^2} c_{ij}.c_{pq} \tag{3}$$

where B_l is the matrix given as $B_l = \{b_{ij}, 1 \leq i \leq m, 1 \leq j \leq n\}$ such that $b_{ij} = 1$ if $l_{ij} > l$; else $b_{ij} = -1$ and C is another matrix defined as $C = \{c_{ij}, 1 \leq i \leq m, 1 \leq j \leq n\}$ such that $c_{ij} = 1\forall(i, j)$. And S_{ij}^2 represents the second-order neighbor pixels of the pixel at spatial position (i, j).

Step 3: The energy curve obtained in the step 2 contain number of peaks and valleys. There may exist potential threshold in every region of two consecutive peaks. The middle points of line joining the consecutive peaks are considered as initial candidate thresholds. Let there exist n initial potential thresholds $th_1, th_2, ..., th_n$, then there exist initially $n + 1$ clusters that represent non-overlapping homogenous regions of the image. The cluster validity measure called Davies Boulding(DB) index [26] is used to calculate DB-index of the initial clusters set say $T_n = [0 - th_1], [th_1 + 1, th_2], ..., [th_n + 1, L]$ obtained for the gray levels $[0, L]$ of the image. DB-index defined as the ratio of the sum of within-object scatter to between object separations. Let $\omega_1, \omega_2, ..., \omega_k$, be the k objects defined by thresholds $t_1 < t_2 < t_3 < .... < t_k$. Then the DB index is defined as:

$$R_{ij} = \frac{\sigma_i^2 + \sigma_j^2}{d_{ij}^2}$$

$$R_i = \max_{\substack{j=1....k \\ i \neq j}} \{R_{ij}\} \tag{4}$$

$$DB = \frac{1}{k}\sum_{i=1}^{k} R_i$$

where σ_i^2 and σ_j^2 are the variances of object ω_i, and ω_j, respectively, and d_{ij}^2 is the distance of object centers ω_i and ω_j. Smaller the DB value, better is the segmentation.

The number of regions obtained initially may be greater than the actual number of regions present in the image. Now iteratively clusters are merged one by one till the number of clusters reduced to two by eliminating thresholds th_1, th_2 and so on. For each combination of clusters DB-index is calculated and the threshold combination for which DB-index is minimum is selected as potential clusters that identifies the object present in the image. The elimination of threshold is a two step process and hence very effective. In the first

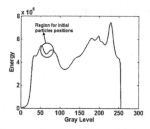

Fig. 2. Initial threshold Region where initial positions are generated for all particles.

step say, from the set of thresholds present in T_n each threshold is eliminated one by one and DB-index for the remaining $n - 1$ thresholds is calculated. Initially the number of such combinations for $n + 1$ clusters will be n. Among those n combinations, the tuple with minimum DB-index is selected. Now the number potential thresholds in the set T_{n-1} will be $n - 1$. Again the same process is applied to obtain the set T_{n-2} containing $n - 2$ thresholds. The process is repeated till the set T_1 with only one threshold is obtained. In the second step, the set from $T_n, T_{n-1}, ...T_1$ for which the DB-index is minimum is selected. Let T_k be the set of thresholds for which the DB-index is minimum. The number of thresholds present in T_k is k ensures that the number of non-overlapping regions present in the image are $k + 1$.

Step 4: The potential thresholds obtained in the step 3 may not be optimal as it is based on the mid-point assumption. To detect the near optimal threshold values for the image segmentation, the potential thresholds set T_k obtained in the step 3 is assumed as the initial best particle to be fed to the PSO algorithm described in 1. The consecutive modes of the energy curve in which the initial threshold/s exist is taken as the region to generate initial positions of all particles. Like for the image Ash, the region where the initial threshold exist is represented using circle in the energy curve as shown in Fig. 2.

Each particle uses k different elements as decision variables with in optimization algorithm. In the proposed technique, a particle is an array consisting of thresholds in the interval of consecutive modes where the potential threshold exists. Randomly generated initial positions of all particles of the image Ash are shown in the Table 1 along with the DB-indices. The velocities of all the particles initialized randomly for the image Ash are shown in the Table 2. For multiple thresholds there will be multiple regions where the initial threshold exist. The swarm size may vary from image to image depending on the size of region in which the initial threshold exists. The proposed technique allows the particles to move to other positions based on the Eqs. 1 and 2, and repeat the steps until the stopping criterion is satisfied. If the stopping criterion is satisfied, then obtain the global best position(g-best) among all the personal best positions(p-best) of all the particles. This global best position gives the near optimal thresholds to segment the image under consideration.

Table 1. Randomly generated initial positions and corresponding DB-indices for each position of all particles for the image Ash.

Particle	x_1	DB-index	x_2	DB-index	x_3	DB-index
1	67	0.1411	68	0.1393	69	0.1375
2	69	0.1375	71	0.1340	73	0.1307
3	70	0.1358	73	0.1307	76	0.1260
4	71	0.1340	75	0.1276	79	0.1215
5	67	0.1411	66	0.1429	65	0.1449
6	65	0.1449	63	0.1488	61	0.1532
7	64	0.1468	61	0.1532	58	0.1605
8	63	0.1488	59	0.1580	55	0.1685

Table 2. Randomly generated initial velocities of all particles for the image Ash.

Particle	x_1	x_2	x_3
1	−1.1071	2.7172	3.0903
2	−1.4888	0.2856	3.6092
3	2.3472	−4.8275	1.2752
4	−4.9255	1.5028	4.8832
5	−4.9438	2.7228	−2.7145
6	−0.4777	−0.6815	−4.9095
7	−4.0936	3.3148	2.0589
8	−1.4173	−0.0898	1.9454

Step 5: Segment the image to analyse the results qualitatively. Let $t_1, t_2, ...t_k$ be the optimal thresholds obtained in the step 4. Then all the pixels of the image with gray level between $[0, t_1]$ will be assigned gray level 0, pixels of the image with gray level between $[t_{i-1}, t_i]$ will be assigned gray level $(i-1) \times (256/k-1)$ and the pixels with gray level greater than t_k will be assigned gray level 255.

4   Implementation

4.1   Description of Experiment

The data set of six gray images with varied number of homogeneous regions has been considered for the experiment as shown in Fig. 3. Images of Ash and Crow shown in Fig. 3(a) and (b) consist of two homogenous regions, image of Peppers shown in Fig. 3(c) consist of three homogenous regions, the image of Lena shown in Fig. 3(d) consist of four, House image shown in Fig. 3(e) of five, and the Flinstones image shown in Fig. 3(f) consist of six homogenous regions respectively. The energy curves of the respective images shown in Fig. 4 consist

(a) Ash   (b) Crow   (c) Peppers

(d) Lena   (e) House   (f) Flinstones

Fig. 3. Original image data set

of peaks and valleys hence discriminate the different homogenous regions present in the image.

To confirm the effectiveness of the proposed technique cluster validity measure DB-index of each image is compared with the GA based context sensitive technique developed recently [22] and the traditional non-meta heuristic Kapur's thresholding technique [8]. As DB-index is involved in the methodology used, for fair comparison another cluster validity measure S_Dbw [27] is taken into account. It is defined on the basis of the cluster compactness in terms of intra-cluster variance and inter-cluster density. For C number of clusters, S_Dbw(C) is defined as:

$$S\_Dbw(C) = Scat(C) + Den(C) \qquad (5)$$

where $Scat(C)$ and $Den(C)$ represent the intra-cluster variance and inter-cluster density, respectively.

The algorithms have been implemented in Matlab(R2012a). For the proposed PSO based technique, the inertial weight coefficient α is taken as 1.4. Both the self confidence M_1 and social confidence M_2 is given 1.5 value. The velocity constant is taken as 5. In case of GA based technique, the population size is taken as 20. Stochastic selection strategy is used to select fittest chromosomes from the mating pool. The crossover and mutation probability is set as 0.8 and 0.01 respectively.

4.2   Results Analysis

Table 3 shows the optimal thresholds, DB-index and S_Dbw index obtained by proposed PSO based and existing GA based context sensitive thresholding technique along with the histogram based Kapur's technique. It has been observed

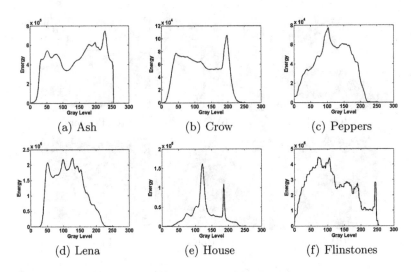

(a) Ash   (b) Crow   (c) Peppers

(d) Lena   (e) House   (f) Flinstones

Fig. 4. Energy curves of the image data set

that for the proposed technique the cluster validity measures DB-index and S_Dbw has been reduced significantly for all the images except for the images Lena and Flinstones, hence produced better segmentation results.

Table 3. Optimal thresholds, DB-index and S_Dbw obtained from the proposed PSO based, existing GA based and Kapur's thresholding technique for considered image dataset.

Images	Using PSO			Using genetic			Kapur's method		
	Threshold	DB-index	S_Dbw	Threshold	DB-index	S_Dbw	Threshold	DB-index	S_Dbw
Ash	98	**0.0995**	**0.3584**	80	0.1200	0.4097	154	0.1187	0.3880
Crow	98	**0.0540**	**0.2079**	118	0.0673	0.2353	148	0.1182	0.3380
Peppers	52, 130	**0.1264**	**0.2382**	51, 122	0.1294	0.2510	79, 149	0.1891	0.3730
Lena	76, 116, 180	**0.1479**	**0.2198**	80, 115, 179	0.1492	0.2291	81, 125, 179	0.1559	0.2313
House	15, 88, 158, 220	**0.0722**	**0.0609**	17, 88, 158, 219	0.0862	0.0675	58, 87, 114, 193	0.2711	1.0891
Flinstones	57, 151, 236, 249, 252	**0.1280**	**0.1701**	48, 147, 230, 241, 249	0.1293	0.1893	0, 54, 93, 134, 188	0.1440	0.1589

The segmented images of the proposed, recently developed GA based, and traditional Kapur's thresholding technique are shown in Figs. 5, 6 and 7. It can be observed from the segmented images that the proposed technique have better visualization as compared to the segmented results of existing one which is more clearly visible for the images of Ash and Crow. The tabular results in Table 3 also shows that cluster validity measures has been reduced significantly for the images Ash and Crow.

To validate the consistency of the proposed approach another experiment was performed where 20 simulations of the meta heuristic thresholding techniques have been compared. Figure 8(a) and (b) present the boxplots of the DB-index for considered images. It can be observed from the figures of boxplot that the variations are very less with the proposed PSO based approach when compared

(a) Ash   (b) Crow   (c) Peppers

(d) Lena   (e) House   (f) Flinstones

Fig. 5. Segmented images using proposed PSO based context-sensitive thresholding technique.

(a) Ash   (b) Crow   (c) Peppers

(d) Lena   (e) House   (f) Flinstones

Fig. 6. Segmented images using GA based existing thresholding technique.

(a) Ash   (b) Crow   (c) Peppers

(d) Lena   (e) House   (f) Flinstones

Fig. 7. Segmented images using Kapur's thresholding technique.

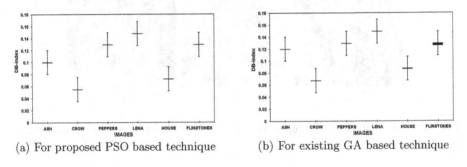

(a) For proposed PSO based technique   (b) For existing GA based technique

Fig. 8. BoxPlot of the image data set

with the GA based approach even if the number of thresholds present in the images increases.

5 Conclusion

PSO has been popular technique used to solve optimization problems due to its high quality of solution and insignificant computational burden. It is based on random numbers and global communication among swarm particles rather than mutation or crossover or pheromone, thus easier to implement as there is no encoding/decoding of parameters into binary strings. This article presents an approach combining context-sensitive thresholding technique with the PSO to detect the optimal thresholds for image segmentation automatically. The main characteristics of this article can be summarized as follows:

First the authors exploit the recently defined energy curve of the image which preserve the features of histogram of the image that is lesser storage space low computational burden.

Second the authors demonstrate the robustness of PSO-model in the area of image segmentation by comparing its accuracy with GA-based model for image segmentation. The authors also presents the flexibility of PSO-model that can be applied to variety of different images irrespective of the number of modes present in the image. The main objective of this article is not to devise a superior thresholding technique to beat all the currently available methods but to exhibit PSO algorithm as an alternative in the area of automatic image segmentation.

Still, there are scopes to explore the following:

1. To study the variants of PSO-model to improve the results of image segmentation. 2. To study various Bio-inspired algorithms to achieve more optimal thresholds for the image segmentation.

References

1. Gonzalez, R.C.: Digital Image Processing. Pearson Education India (2009)
2. Pal, N.R., Pal, S.K.: A review on image segmentation techniques. Pattern Recog. **26**(9), 1277–1294 (1993)
3. Sezgin, M., Sankur, B.: Survey over image thresholding techniques and quantitative performance evaluation. J. Electron. Imaging **13**(1), 146–165 (2004)
4. Sahoo, P.K., Soltani, S., Wong, A.: A survey of thresholding techniques. Comput. Vis. Graph. Image Process **41**(2), 233–260 (1988)
5. Otsu, N.: A threshold selection method from gray level histograms. IEEE Trans. Syst. Man Cybern. **9**, 62–66 (1979)
6. Kittler, J., Illingworth, J.: Minimum error thresholding. Pattern Recogn. **19**(1), 41–47 (1986)
7. Pun, T.: A new method gray-level picture thresholding using the entropy of the histogram. Sig. Process. **2**, 223–237 (1980)
8. Kapur, J.N., Sahoo, P.K., Wong, A.K.C.: A new method for gray-level picture thresholding using the entropy of the histogram. Comput. Vis. Graph. Image Process. **29**(3), 273–285 (1985)
9. Qiaoa, Y., Hua, Q., Qiana, G., Luob, S., Nowinskia, W.L.: Thresholding based on variance and intensity contrast. Pattern Recogn. **40**, 596–608 (2007)
10. Karasulu, B., Korukoglu, S.: A simulated annealing-based optimal threshold determining method in edge-based segmentation of grayscale images. Appl. Soft Comput. **11**, 2246–2259 (2011)
11. Ananthi, V.P., Balasubramaniam, P., Lim, C.P.: Segmentation of gray scale image based on intuitionistic fuzzy sets constructed from several membership functions. Pattern Recogn. **47**, 3870–3880 (2014)
12. Liao, P.S., Chen, T.S., Chung, P.C.: A fast algorithm for multilevel thresholding. J. Inform. Sci. Eng. **17**, 713–727 (2001)
13. Yimit, A., Hagihara, Y., Miyoshi, T., Hagihara, Y.: 2-D direction histogram based entropic thresholding. Neurocomputing **120**(23), 287–297 (2013)
14. Xiao, Y., Cao, Z., Zhong, S.: New entropic thresholding approach using gray-level spatial correlation histogram. Opt. Eng. **49**(12), 127007 (2010)

15. Xiao, Y., Cao, Z., Yuan, J.: Entropic image thresholding based on GLGM histogram. Pattern Recogn. Lett. **40**, 47–55 (2014)
16. Akay, B.: A study on particle swarm optimization and artificial bee colony algorithms for multilevel thresholding. Appl. Soft Comput. **13**, 3066–3091 (2013)
17. Ali, M., Ahn, C.W., Pant, M.: Multi-level image thresholding by synergetic differential evolution. Appl. Soft Comput. **17**, 1–11 (2014)
18. Tao, W.B., Tian, J.W., Liu, J.: Image segmentation by three-level thresholding based on maximum fuzzy entropy and genetic algorithm. Pattern Recogn. Lett. **24**(16), 3069–3078 (2003)
19. Hammouche, K., Diaf, M., Siarry, P.: A multilevel automatic thresholding method based on a genetic algorithm for a fast image segmentation. Comput. Vis. Image Underst. **109**(2), 163–175 (2008)
20. Ghamisi, P., Couceiro, M.S., Benediktsson, J.A., Ferreira, N.M.: An efficient method for segmentation of images based on fractional calculus and natural selection. Expert Syst. Appl. **39**(16), 12407–12417 (2012)
21. Patra, S., Gautam, R., Singla, A.: A novel context sensitive multilevel thresholding for image segmentation. Appl. Soft Comput. **23**, 122–127 (2014)
22. Singla, A., Patra, S.: A fast automatic optimal threshold selection technique for image segmentation. SIViP **11**(2), 243–250 (2017)
23. Eberhart, R.C., Kennedy, J.: A new optimizer using particle swarm theory. In: Proceedings of the Sixth International Symposium on Micro Machine and Human Science, vol. 1, New York, pp. 39–43 (1995)
24. Kennedy, J., Eberhart, R.: Particle swarm optimization. In: IEEE International of First Conference on Neural Networks (1995)
25. Shi, Y., Eberhart, R.: A modified particle swarm optimizer. In: The 1998 IEEE International Conference on Evolutionary Computation Proceedings. IEEE World Congress on Computational Intelligence, pp. 69–73. IEEE (1998)
26. Davis, D.L., Bouldin, D.W.: A cluster separation measure. IEEE Trans. Pattern Anal. Mach. Intell. **PAMI-1**(2), 224–227 (1979)
27. Ghosh, S., Kothari, M., Halder, A., Ghosh, A.: Use of aggregation pheromone density for image segmentation. Pattern Recogn. Lett. (2009)

Extraction of Abnormal Portion of Brain Using Jaya Algorithm

Kanwarpreet Kaur[✉], Gurjot Kaur, and Jaspreet Kaur

Department of Electronics and Communication Engineering, Guru Nanak Dev Engineering College, Ludhiana, India
kpreet2392@gmail.com, gurjotwalia@yahoo.com,
jaspreetkaur513@yahoo.com

Abstract. Brain performs the task of homeostasis, control and coordination. Many times normal functioning of brain is hampered due to blockage, tumors etc. Neurologists generally recommend MRI technique for the detection of any type of abnormality in the brain. The optimization algorithms can be used for extraction of the abnormal portion of the brain. This paper discusses about extraction of abnormal part of brain using the Jaya algorithm after preprocessing of the MRI images.

Keywords: Magnetic Resonance Imaging (MRI) · Segmentation · Central Nervous System (CNS) · Optimization · Jaya algorithm

1 Introduction

Central Nervous system involves the most sensitive organs of the human body, brain and spinal cord. Brain performs the task of control and coordination of the body. It stores the responses of previous stimuli and reminiscence them later to guide the human in future. These responses are known as memory. It receives the instructions about any change that occur inside the body and then, coordinates the activities of those organs, so as to help in maintaining the balance in internal environment of the body. Many diseases related to the brain hampers the normal functioning of the brain such as blockage, tumors etc. The statistics from the American Society of Clinical Oncology (ANCO), in August 2015 showed that in United States about 22,850 people will be diagnosed of having the primary cancerous tumors of the brain and spinal cord. It is approximated that 15,320 people are going to die from this disease in the coming year. It is estimated that approximately 4,300 youngsters will be diagnosed with a CNS tumor in the following year [1, 2]. Brain tumor occurs because of the abnormal and uncontrolled division of the cells. Generally, the cells in the human body die with age and are replaced with the new cells. But, this cycle gets disturbed in tumors. Tumor can damage the normal cells of the brain directly as well as indirectly [3, 4]. The damage to brain is caused with the increase of pressure in the brain, shifting the brain or pushing across skull, damaging nerves. Brain tumor is classified into primary as well as metastatic tumors. Primary tumor commences in the brain while metastatic tumor is the one present

© Springer Nature Singapore Pte Ltd. 2017
K. Deep et al. (eds.), *Proceedings of Sixth International Conference on Soft Computing for Problem Solving*, Advances in Intelligent Systems and Computing 547, DOI 10.1007/978-981-10-3325-4_16

in other parts of body and then, spread to the brain via blood or through adjacent tissues. Primary tumor is further classified into benign and malignant tumor. Benign tumor is less harmful as compared to the malignant tumor, which grows faster [5].

The various techniques that can be used for detection of abnormalities in brain are Magnetic Resonance Imaging (MRI), Magnetic Resonance Spectroscopy (MRS), Computed Tomography (CT), Positron Emission Tomography (PET), Biopsy and Electroencephalogram (EEG). MRI is the standard imaging technique used for identifying the abnormalities in brain. It uses the magnetic field along with the computers to acquire brain images by capturing the signals emitted from normal and abnormal tissues present in the brain. MRS is also a type of MRI which estimates level of metabolite present in the body. The presence of amount of metabolites in abnormalities is different from that of the normal tissues. It helps in diagnosis by studying the pattern of activity. The various types of metabolites that are commonly used in MRS are choline, lactate, and N-Acetyl- Aspartate (NAA) etc. CT uses both x-ray and computer technology for the diagnosis of the abnormality. It has the ability to show a combination of soft tissues of the body as well as bones and blood vessels. They are able to decide about some types of tumors that may be present, as well as aid in identifying the swelling and bleeding. The contrast agent used in a CT scan is Iodine. PET is the auxiliary test that can be used for obtaining the further information after an MRI. It provides an image of the brain's activity, by calculating the rate at which the glucose is consumed by the tissues. This scan measures the activities occurring in the brain and uses this data to create a live image. After performing above mentioned scans, sometimes, biopsy is done to check out the type of the tumor. To access the tumor site and obtain sample tissue from the brain, either small hole in the brain is drilled or skull portion is removed. Then, the tissue is examined under the microscope by the pathologist. EEG records the electrical activity of the brain with the aid of electrodes placed on the head and also connected to the computer. It is used for diagnosis of unconsciousness, dementia, epilepsy etc. Sometimes, MRA is used to find the location of the blood vessels that lead to the tumor. It is generally used before performing surgery, so as to remove the tumor located in area having large number of the blood vessels [6].

The paper involves: Sect. 2 confers about the preprocessing. Section 3 explains segmentation process. Section 4 shows results while Sect. 5 presents conclusion.

2 Preprocessing

Preprocessing is done in order to enhance the MRI images. The images are resized to a specific size. The colored images are converted to gray scale images. The various noises such as Gaussian noise, shot noise, white noise, salt and pepper noise etc. get added to MRI images resulting in the distortion because of the noise added to it. Noise removal is done at the same stage in order to improve the quality of the images. The noise can be removed using median filter, high pass filter, adaptive filter etc. [3, 4]. The noise should be removed in such a way that the edges should be preserved. The original image shown in Fig. 1 is obtained from the BRATS [7].

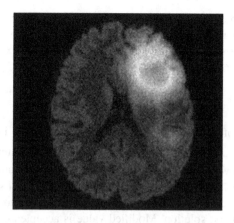

Fig. 1. Original MRI image (Source: MICCAI BRATS 2012 [7])

Here, Fig. 2 shows preprocessed image of the original MRI image shown in Fig. 1.

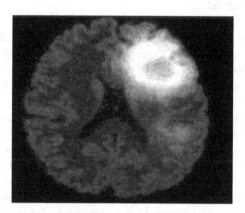

Fig. 2. MRI image after preprocessing

3 Segmentation

Image segmentation is generally used to analyze the image. In this, MRI images are divided into the multiple segments so as to extract the details about the each part of the brain. The various segmentation techniques normally used are sobel edge detection, adaptive thresholding, prewitt edge detection, canny edge detection, Fuzzy c-means, watershed thresholding etc. [8–11]. Segmentation can also be done using the optimization algorithms such as Ant Colony Optimization, Particle Swarm Optimization, Genetic Algorithm etc. [3, 4]. The Jaya algorithm proposed by Venkata Rao [12] is used for constrained as well as unconstrained problems. The result attained for the problem approaches to best solution while preventing the worst solution. These algorithms need an objective function for extraction of abnormal portion of the image. The objective function f(x) needs to be minimized or maximized. There can be m number of design variables and n number of contestant solutions for each i iterations. Best values of f(x) are obtained by the best contestant while the worst values of f(x) are obtained by the worst contestant. The equation that is used in this algorithm is:

$$Z'_{b,c,a} = Z_{b,c,a} + r_{1,b,a}\left(Z_{b,best,a} - |Z_{,b,c,a}|\right) - r_{2,b,a}\left(Z_{b,worst,a} - |Z_{,b,c,a}|\right) \tag{1}$$

where $Z_{b,c,a}$ is value of bth variable for c^{th} contestant during a^{th} iteration. $Z'_{b,c,a}$ is the modified value of $Z_{b,c,a}$. $Z_{b,best,a}$ is value of b variable for best contestant while $Z_{j,worst,i}$ is value of b variable for worst contestant. $r_{1,b,a}$ and $r_{2,b,a}$ are random numbers lying between 0 and 1 for b variable during a iteration. The second term in Eq. 1 shows the chances of solution to move nearer to best solution whereas third term shows the possibility of solution to prevent worst solution. Modified value is accepted only if the solution is better as compared to the previous one.

4   Proposed Method

The MRI images obtained from the BRATS-2012 are used for the extraction of the abnormal portion. The MRI images are converted to grayscale image after being resized to 256×256. The noise is removed using the median filter. The procedure of using Jaya algorithm for the segmentation as well as extraction of the abnormal part of brain is proposed in the following flowchart (Fig. 3):

Jaya algorithm is applied on preprocessed images. It tends to choose the best solution while preventing the worst solution. The equation is used to find the updated value of best solution. If the updated value is better as compared to previous value, the best solution is updated otherwise the earlier value is preserved. When the termination criteria are reached, the mean of the best solutions found is then used as a thresholding value to extract the abnormal part of the brain. The thresholding technique is applied to the MRI image. In this, a parameter θ is defined for the brightness threshold, which is found by the Jaya algorithm [8]. Here, the values above θ will appear as the abnormal portion of the brain.

5   Results

The algorithm is accomplished using the MATLAB software. The extracted abnormal part appears to be white. It can also appear black if the background is of white color. The results of the applied algorithm are presented in Fig. 4 which incorporates original MRI images, preprocessed images and the extracted images from the preprocessed images.

The performance of Jaya Algorithm is measured in terms of detection rate, positional accuracy as well as execution time. The Jaya algorithm is applied on the 25 images, and the abnormal portion is extracted in each image, thus having the detection rate of hundred percent in the available database whereas the positional accuracy of abnormal portion

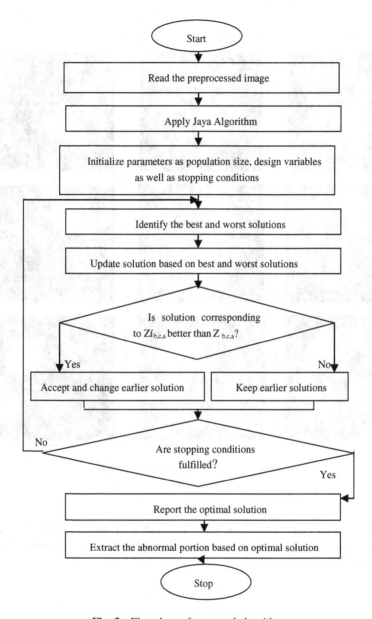

Fig. 3. Flowchart of proposed algorithm

varies. The position of the extracted abnormal portion is compared with the position of truth images. In this work, if there is overlap of 60% and above in extracted abnormal portion and truth image, then it is considered to be accurate. Earlier in the work done by Gopal and Karnan, 50% overlap is considered as accurate [3]. Table 1 shows the

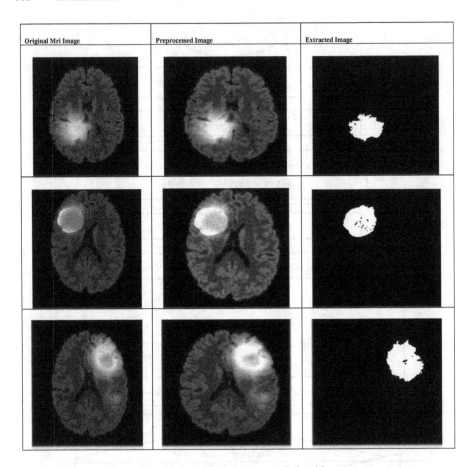

Fig. 4. Results of the proposed algorithm

performance parameters of the Jaya Algorithm. Table 2 compares the performance of Jaya Algorithm with the Particle Swarm Optimization (PSO) and Genetic Algorithm (GA) with FCM [3].

Table 1. Performance of Jaya algorithm

Sr. no.	Parameters	Values
1.	Detection Rate	100%
2.	Positional Accuracy	92%
3.	Execution Time	18.8 s

Table 2. Comparison of Jaya algorithm with other algorithms

Sr. no.	Optimization algorithm	Detection rate (%)	Positional accuracy (%)	Execution time (seconds)
1.	Genetic Algorithm with FCM	89.6	74.6	93.39
2.	Particle Swarm Optimization with FCM	98.87	92.3	100.03
3.	Jaya Algorithm	100	92.0	18.8

6 Conclusion

The proposed system helps in easy extraction of the abnormal portion present in the MRI images of brain. It proves to be accurate for available database. The location of the abnormal portion can also be estimated that further helps doctors in treatment. The execution time for Jaya algorithm is less as compared to others.

References

1. Cancer.Net. http://www.cancer.net/cancer-types/brain-tumor/statistics
2. Dhami, P.S., Chopra, G., Srivastava, H.N.: Neural control and coordination. In: A Textbook of Biology (Class XI), vol. II, Pradeep, India, pp. 314–320 (2011)
3. Gopal, N.N., Karnan, M.: Diagnose brain tumor through MRI using image processing clustering algorithms such as fuzzy c means along with intelligent optimization techniques. In: IEEE International Conference on Computational Intelligence and Computing Research, Coimbatore, India, pp. 1–4 (2010)
4. Karnan, M., Selvanayaki, K.: Improved implementation of brain MR image segmentation using meta heuristic algorithms. In: IEEE International Conference on Computational Intelligence and Computing Research, Coimbatore, India, pp. 1–4 (2010)
5. Sapra, P., Singh, R., Khurana, S.: Brain tumor detection using neural network. Int. J. Sci. Mod. Eng. **1**, 83–88 (2013)
6. National Brain Tumor Society. http://blog.braintumor.org
7. MICCAI BRATS 2012. http://www.imm.dtu.dk/projects/BRATS2012
8. Badran, E.F., Mahmoud, E.G., Hamdy, N.: An algorithm for detecting brain tumor in MRI images. In: IEEE International Conference on Computer Engineering and Systems, Cairo, Egypt, pp. 368–373 (2010)
9. Parveen, S.A.: detection of brain tumor in MRI images, using combination of fuzzy c-means and SVM. In: IEEE 2nd International Conference on Signal Processing and Integrated Networks (SPIN), Noida, India, pp. 98–102 (2015)
10. Roy, S., Bandyopadhyay, S.K.: Detection and quantification of brain tumor from MRI of brain and it's symmetric analysis. Int. J. Inf. Commun. Technol. Res. **2**, 477–483 (2012)
11. Narkhede, S.G., Khairnar, V.: Brain tumor detection based on symmetry information. Int. J. Eng. Res. Appl. **3**, 430–432 (2013)
12. Rao, R.: Jaya: a simple and new optimization algorithm for solving constrained and unconstrained optimization problems. Int. J. Ind. Eng. Comput. **7**, 19–34 (2016)

Script Identification from Offline Handwritten Characters Using Combination of Features

Akshi Bhardwaj[1(✉)] and Simpel Rani Jindal[2]

[1] Department of Computer Science and Engineering, Thapar University, Patiala, Punjab, India
bhardwajakshi26@gmail.com
[2] Department of Computer Science and Engineering, Yadwindra College of Engineering,
Talwandi Sabo, Punjab, India
simpel_jindal@rediffmail.com

Abstract. Script identification in multi-lingual text images will help in improving the efficiency of many real life applications, such as sorting, transcription of multilingual documents and OCR. In this paper, we have presented a technique for identification of three scripts, namely, Devanagari, Gurmukhi and Roman. We have identified the script of text based on statistical features, namely, zoning features; diagonal features; intersection and open end points based features; peak extent based features and combinations of these features. For classification, we have used multiple classification techniques, namely, Support Vector Machine (SVM), k-Nearest Neighbour (k-NN), and Convolutional Neural Network (CNN). The proposed strategy using CNN attains an average identification rate of 93.64%, with 5-fold cross-validation, for these three scripts when isolated offline handwritten characters of these scripts were considered.

Keywords: Feature extraction · Classification · Script identification · k-NN · SVM · LeNet-CNN · Cross-validation

1 Introduction

Optical Character Recognition (OCR) is a key area in the field of document analysis and recognition, which is generally defined as the process of text recognition of printed or handwritten document images and converting these into digital format. Nowadays, many techniques have been presented in the literature for an OCR of a particular script; however, such OCR will not work for multiscript text recognition. So, script identification system is essential to build successful multilingual text recognition systems. In this field, most of the available work on script identification of Indian scripts compacts with printed documents and a very few articles were found for handwritten script identification. For Indian scripts, paragraph level script identification work has been discussed in [1, 2] and line level work has been discussed in [1–3]. Spitz [4] examined the upward concavities of connected components. He proposed a technique where one could separate Asian and European languages. Similarly, texture analysis was used for automatic script and language identification from document images by Tan *et al.* [5]. They used multiple channel (Gabor), filters

© Springer Nature Singapore Pte Ltd. 2017
K. Deep et al. (eds.), *Proceedings of Sixth International Conference on Soft Computing for Problem Solving*, Advances in Intelligent Systems and Computing 547,
DOI 10.1007/978-981-10-3325-4_17

and Gray level co-occurrence matrices to study seven languages: Chinese, English, Greek, Korean, Malayalam, Persian and Russian. Hochberg *et al.* [6, 7] described a technique for automatic script identification of document images. They have also presented a technique for script and language identification using statistical feature extraction techniques. Wood *et al.* [8] provided a technique for identification of Roman, Russian, Arabic, Korean and Chinese characters using the projection profiles method. Pal *et al.* [9] proposed a technique for text line separation from 12 Indian scripts. Dhanya *et al.* [10] have used spatial spread features and Gabor filters for automatic script identification of Roman and Tamil words present in a bilingual document. Pal *et al.* [11] have presented conventional and water reservoir features for word-wise script identification from a document containing English, Devanagari, and Telugu text. Basavaraj *et al.* [12] have used neural network for script identification of Kannada, Devanagari and Roman scripts.

A technique for script identification of pre-segmented characters is proposed in this paper. The scripts considered in this paper are Devanagari, Gurmukhi and Roman. Devanagari character set of 36 basic characters (Fig. 1), Gurmukhi character set of 35 basic characters (Fig. 2) and Roman character set of 52 characters (26 upper case letters and 26 lower case letters) are considered in this work. In this paper, comparative study of four different types of features for script identification using different classifiers has been presented. The organization of the paper is as follows: the algorithms to extract features considered in this work are explained in Sect. 2. Classification techniques are discussed in Sect. 3. The details on dataset used are given in Sect. 4 and experimental results are depicted in Sect. 5. Section 6 concludes the paper with some future directions.

2 Feature Extraction

The relevant shape contained in a character is best described by its features. As such, employing the best suitable feature extraction technique becomes a very pertinent factor in any classification problem. The extracted features play an important role in the performance of the recognition/identification system. In this paper, we have considered zoning; diagonal; intersection and open end points; and peak extent based features for the identification of three scripts. We have also considered several combinations of these features.

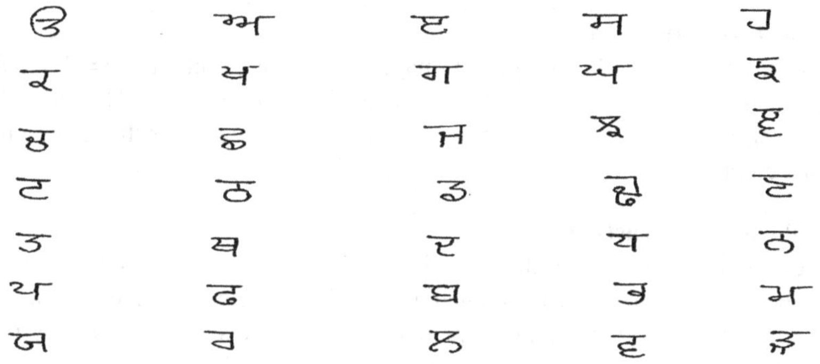

Fig. 1. Basic character set of Gurmukhi script

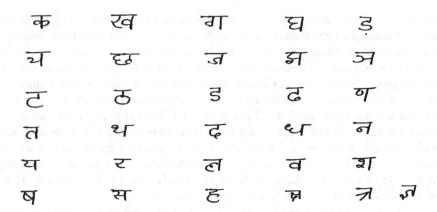

Fig. 2. Basic character set of Devanagari script

Z 1	Z 2	Z 3	Z 4	Z 5	Z 6	Z 7	Z 8	Z 9	Z 10
Z 11	Z 12	Z 13	Z 14	Z 15	Z 16	Z 17	Z 18	Z 19	Z 20
Z 21	Z 22	Z 23	Z 24	Z 25	Z 26	Z 27	Z 28	Z 29	Z 30
Z 31	Z 32	Z 33	Z 34	Z 35	Z 36	Z 37	Z 38	Z 39	Z 40
Z 41	Z 42	Z 43	Z 44	Z 45	Z 46	Z 47	Z 48	Z 49	Z 50
Z 51	Z 52	Z 53	Z 54	Z 55	Z 56	Z 57	Z 58	Z 59	Z 60
Z 61	Z 62	Z 63	Z 64	Z 65	Z 66	Z 67	Z 68	Z 69	Z 70
Z 71	Z 72	Z 73	Z 74	Z 75	Z 76	Z 77	Z 78	Z 79	Z 80
Z 81	Z 82	Z 83	Z 84	Z 85	Z 86	Z 87	Z 88	Z 89	Z 90
Z 91	Z 92	Z 93	Z 94	Z 95	Z 96	Z 97	Z 98	Z 99	Z 100

Fig. 3. Zones of an input character

Zoning feature extraction

In this technique, we divide the thinned image of a character into an n (= 100, as used in the present work) number of equal sized zones as shown in Fig. 3. The ratio of foreground pixels to background pixels is considered as a feature and we thus have a feature set with n elements.

Diagonal feature extraction

To achieve higher accuracy for the script identification system, diagonal features have found to be helpful. Here also, the thinned image of a character is divided into n (= 100) zones. Now, diagonal features are extracted from the pixels of each zone by moving along its diagonals.

The steps that have been used to extract these features are:

Step I: Thinned image of the character is divided into n (= 100) number of zones.

Step II: In order to get a single sub-feature, foreground pixels along each diagonal are summed up.

Step III: These sub-feature values are averaged to form a single value which is then placed in the relevant zone as its feature.

Step IV: Corresponding to the zones whose diagonals do not have a foreground pixel, the feature value is taken as zero.

These steps will give a feature set with n elements.

Intersection and open end point feature extraction

A pixel is called *intersection point* if it has more than one pixel in its neighborhood and *open end point* if it has only one pixel in its neighborhood.

The following steps have been implemented for extracting these features.

Step I: Thinned image of the character is divided into n (= 100) number of zones.

Step II: We calculated the number of intersections and open end points for each zone.

This will give $2n$ features for a character image.

Peak extent based feature extraction

In peak extent features, we consider the sum of the peak extents that fit successive black pixels horizontally in each row of a zone.

Following steps have been used to extract peak extent based features:

Step I: Divide the bitmap image into an n (= 100) number of zones, each of size 10×10 pixels.

Step II: Find the peak extent as the sum of successive foreground pixels in each row of a zone.

Step III: Replace the values of successive foreground pixels by peak extent value, in each row of a zone.

Step IV: Find the largest value of peak extent in each row. As such, each zone has 10 horizontal peak extent features.

Step V: Obtain the sum of these 10 peak extent sub-feature values for each zone and consider this as a feature for corresponding zone.

Step VI: For the zones that do not have a foreground pixel, take the feature value as zero.

Step VII: Normalize the feature vector.

These steps will give a feature set with n elements.

3   Classification

Classification phase is the phase of an identification system wherein one makes the decisions. It uses the features extracted in the feature extraction stage, for making class membership in the identification system. The preliminary aim of the classification phase of an identification system is to develop a constraint that can help to reduce the misclassification relevant to feature extraction. The effectiveness of any script identification system is highly dependent on the capability of identifying the unique features of a script and the capability of the classifier to relate features of a script to its class. In this work, we have considered Linear-SVM, k-NN and CNN classifiers for script identification work. SVM is a very useful technique for classification in the pattern recognition field. SVM classifier has been considered with linear kernel in the present work. For SVM classification, we have used C-SVC tool.

We have computed the Euclidean distances from candidate vector to a stored vector in the k-NN classifier. This distance is given by,

$$d = \sqrt{\sum_{k=1}^{N}(x_k - y_k)^2}$$

The total number of features in the feature set is denoted by N while x_k is the library stored feature vector and y_k is the candidate feature vector. The input character has been assigned the class of the library stored feature vector which produces the smallest Euclidean distance.

For classification, we have also considered multilayer Convolutional Neural Network (CNN). CNN is among the most suitable classifier in the field of pattern recognition. Recent CNN work is focused on pattern recognition and computer vision problems such as object recognition, natural images and traffic signs. In this paper, we have used LeNet (First successful application of Convolution Networks) of CNN for script classification (the parameters have been fixed as dropout rate = 0.2, patch size = 3×3, pool width and height = 2).

4   Data Set

This is worth mentioning here that we do not have a database for our problem that is available in public domain. As such, a dataset of 4,920 characters has been collected as described in Table 1. This dataset consists of 1,400 Gurmukhi script characters, 1,440 Devanagari script characters, and 2,080 Roman script characters (1,040 upper case and 1,040 lower case characters).

5   Experimental Results

As mentioned above, we have performed experiments using zoning features (F_1); diagonal features (F_2); intersection and open end point features (F_3); and peak extent based features (F_4) and combinations of these feature extraction techniques for script identification. Experiments are carried out with Linear-SVM, k-NN, and Convolutional Neural Network (CNN) classifiers. To evaluate the classifier performance and validate the effectiveness of the proposed techniques, whole dataset is randomly divided into five groups and a 5-fold cross validation was done to get optimum results. A maximum accuracy of 93.64% has been achieved for the classification of these three scripts using CNN classifier. The feature that participated in this accuracy is a combination of zoning; diagonal; and intersection and open end points based features, as depicted in Table 2. Script identification accuracies achieved with different features and classifiers are graphically depicted in Fig. 4.

Table 1. Data set details

Script	Number of characters	Number of writers involved	Total number of characters
Devanagari	36 (basic characters)	40	1,440
Gurmukhi	35 (basic characters	40	1,400
Roman	26 (upper case basic characters)	40	1,040
Roman	26 (lower case basic characters)	40	1,040

Table 2. Average identification accuracy using 5-fold cross validation

Feature extraction technique	Classification technique		
	L-SVM	k-NN	CNN
Zoning (F_1)	74.79%	85.17%	85.52%
Diagonal (F_2)	74.77%	85.42%	83.69%
Intersection (F_3)	80.79%	85.40%	85.00%
Peak extent (F_4)	73.25%	76.84%	79.91%
$F_1 + F_2$	78.21%	87.52%	87.98%
$F_1 + F_3$	84.53%	89.22%	91.09%
$F_1 + F_4$	79.18%	85.71%	87.58%
$F_2 + F_3$	82.88%	86.68%	89.43%
$F_2 + F_4$	77.62%	85.56%	86.72%
$F_3 + F_4$	84.14%	85.56%	89.59%
$F_1 + F_2 + F_3$	84.41%	88.86%	93.64%
$F_2 + F_3 + F_4$	85.18%	87.03%	92.74%
$F_1 + F_3 + F_4$	86.28%	89.22%	91.68%
$F_1 + F_2 + F_3 + F_4$	86.09%	89.04%	93.02%

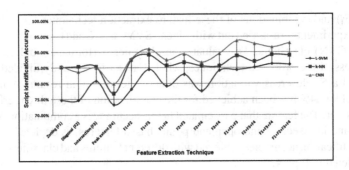

Fig. 4. Classifier wise average accuracy using 5-fold cross validation

6   Conclusion and Future Scope

We have seen that most of the existing techniques for script identification of Indian scripts are based on the whole document, block, line or word level. We focused our research on single character script identification and have got promising results. Promising results have been achieved by focusing the attention on single character script identification. In this paper, we have proposed a very simple method for offline handwritten script identification of three scripts: Devanagari, Gurmukhi and Roman. The primary focus of the paper is to help improving the efficiency of script based retrieval of offline handwritten document and also to facilitate the retrieval of multilingual offline handwritten OCR systems. By using 4,920 characters of three scripts (Devanagari, Gurmukhi and Roman), we have achieved maximum identification accuracy of 93.64% using CNN classifier and a combination of zoning; diagonal; and intersection and open end points based features. We have seen that LeNet-CNN performs better than the other classifiers for script identification task. This work can be extended for other Indian and non-Indian scripts.

References

1. Pal, U., Chaudhari, B.B.: Script line separation from Indian multi-script documents. In: Proceedings of International Conference on Documents Analysis and Recognition, pp. 406–409 (1999)
2. Pal, U., Chaudhuri, B.B.: Identification of different script lines from multi-script documents. Image Vis. Comput. **20**(13), 945–954 (2002)
3. Pal, U., Sinha, S., Chaudhri, B.B.: Multi-script line identification from Indian documents. In: Proceedings of International Conference on Documents Analysis and Recognition, pp. 880–884 (2003)
4. Spitz, A.L.: Determination of the script and language content of document images. IEEE Trans. Patt. Anal. Mach. Intell. **19**(3), 235–245 (1997)
5. Peake, G.S., Tan, T.N.: Script and language identification from document images. In: Proceedings of Third Asian Conference on Computer Vision Hong Kong, vol. 2, pp. 97–104 (1997)

6. Hochberg, J., Kelly, P., Thomas, T., Kerns, L.: Automatic script identification from document images using cluster-based templates. IEEE Trans. Patt. Anal. Mach. Intell. **19**(2), 176–181 (1997)
7. Hochberg, J., Bowers, K., Cannon, M., Kelly, P.: Script and language identification for handwritten document images. Int. J. Doc. Anal. Recogn. **2**(2), 45–52 (1999)
8. Wood, S.L., Yao, X., Krishnamurthi, K., Dang, L.: Language identification for printed text independent of segmentation. In: Proceedings of International Conference on Image Processing, vol. 3, pp. 428–431 (1995)
9. Pal, U., Chaudhuri, B.B.: Script line separation from Indian multi-script documents. In: Proceedings of 5th International Conference on Documents Analysis and Recognition, pp. 406–409 (1999)
10. Dhanya, D., Ramakrishnan, A.G.: Script identification in printed bilingual documents. In: Proceedings of 5th International Workshop on Document Analysis and System, pp. 13–24 (2002)
11. Pal, U., Sinha, S., Chaudhuri, B.B.: Word-wise script identification from a document containing English, Devnagari and Telugu Text. In: Proceedings of National Conference on Document Analysis and Recognition, pp. 213–220 (2003)
12. Patil, S.B., Subbareddy, N.V.: Neural network based system for script identification in Indian documents. SADHANA **27**(1), 83–97 (2002)

Method Noise Based Two Stage Nonlocal Means Filtering Approach for Gaussian Noise Reduction

Karamjeet Singh$^{(\boxtimes)}$, Sukhjeet Kaur Ranade, and Chandan Singh

Punjabi University, Patiala, India
karampup@gmail.com

Abstract. Method noise is the residual image containing significant structural information lost during the process of denoising and can be effectively used for image restoration. In this paper, we propose an efficient two stage filtering approach using nonlocal means and method noise for the reduction of Gaussian noise from the images. In first stage, the block-based NLM is applied to obtain the denoised image and in second stage, the nonlocal similarities present in the method noise and the denoised image are used for the computation of effective weights for weighted NLM denoising. Experimental results demonstrate significant improvements in the denoising performance of the proposed approach as compared to the classical and block-based NLM approaches.

Keywords: Nonlocal means (NLM) · Image denoising · Method noise · Gaussian noise

1 Introduction

Gaussian noise reduction has been considered highly desirable for various image processing applications including image analysis, pattern recognition and computer vision. Over the period, numerous approaches have been developed for attenuating the Gaussian noise while preserving the essential structural details of the image. Traditionally, linear and standard denoising filters like Gaussian filter [1], Wiener filter [2] and median filter [3] have been effectively used for the reduction of Gaussian noise. However, linear filtering results in the substantial loss of edge information and other details of the image. Therefore, the use of non-linear filters such as anisotropic diffusion (AD) [4], total variation (TV) [5], bilateral filter (BF) [6] and non-local means (NLM) filter [7, 8] have been proposed as they preserve edges and other finer details of the image using pixel of local neighbourhood. Among non-linear approaches, NLM filter given by Baudes et al. [7, 8] has set a new trend due to its simplicity and efficiency in utilizing pattern redundancy in the input image. The NLM approach locates a set of target patches that are similar to the patch of interest (or reference patch) in the given search window and the centre pixel of the reference patch is denoised by a weighted average of the centre pixels of each target patch. The weighted averaging of all selected patches is thus used to reduce the effect of noise. However, one of the major drawbacks of the approach is the computational load associated with search operation required for

© Springer Nature Singapore Pte Ltd. 2017

K. Deep et al. (eds.), *Proceedings of Sixth International Conference on Soft Computing for Problem Solving*, Advances in Intelligent Systems and Computing 547, DOI 10.1007/978-981-10-3325-4_18

locating the similar patch within the search window. In order to reduce this computational load and improve the speed, a number of fast implementations of *NLM* approach have been proposed by various authors [8–12]. Mahmoudi and Sapiro [9] proposed an algorithm that eliminates the unrelated neighbourhoods from the averaging process on the basis of local average gray values and gradients and hence lowering the complexity of *NLM* filter. In another approach proposed by Orchard and Ebrahmi et al. [10], fast Fourier transform is used to compute mean squared deviation (*MSD*) in order to enhance the speed of *NLM* while Taisden [11] proposed the use of principal neighborhood dictionary to accelerate the classical *NLM* filter. Only few researchers attempted to enhance the performance of *NLM* and reduce the loss of information occurred due to the averaging process. Method noise is the residual image obtained after denoising and contains significant structural information lost during the denoising process [12, 13]. Brunet et al. [12] analysed the usefulness of method noise in image denoising. Later, Zhong et al. [13] successfully enhanced the performance of *NLM* based denoising approach by first pre-filtering the image using classical *NLM* filter and then exploiting the nonlocal similarities of both the pre-denoised image and method noise to compute new weights for *NLM* filtering in the second stage. The proposed approach improves the performance of *NLM* effectively especially for large noise variance, but failed to exploit the residual image with weaker energy. The block wise nonlocal means *BNLM* [8] approach divides the image into overlapping blocks and apply nonlocal means restoration technique for image denoising. The approach enhances the performance of classical *NLM* and has reduced computational load.

In this paper, we propose a two stage filtering approach using block based *NLM* filter and method noise for the reduction of Gaussian noise. In the first stage, block-based *NLM* is applied to obtain the pre-filtered denoised image and in the second stage, the nonlocal similarities present in the method noise and the pre-filtered denoised image are used for the computation of effective weights for weighted *NLM* denoising. The rest of paper is organized as follows. Section 2 presents a brief introduction about the classical *NLM* and the block wise nonlocal means filter (BNLM). The proposed two stage denoising approach is explained in Sect. 3. Experimental analysis and conclusions are given in Sects. 4 and 5, respectively.

2   Nonlocal Means-Based Approaches

Non-local means (NLM)-based approaches have been widely used to denoise the Gaussian noise effectively. Thus, the following subsections represent the mathematical formulation of classical NLM and its blockwise implementation. Given the additive noise model for image signal z_i in a mathematical form as:

$$y_i = z_i + n_i, \; i = 1, 2, 3, \ldots, M \tag{1}$$

where, n_i is assumed to be as identically and independent distributed (*i.i.d*) random variables and M is the total number of pixels in the image.

2.1 Classical NLM

The *NLM* approach is provided by Baudes et al. [7] which are based upon the assumption that images contain repeated structure (the redundancy property of periodic images, textured images or natural images) and averaging them will reduce the random noise. Then, the *NLM* estimation of denoised signal $\hat{z}_i$ is defined as follows [7].

$$\hat{z}_i = \frac{\sum_{j \in S_i} w_{i,j} y_j}{\sum_{j \in S_i} w_{i,j}} \tag{2}$$

$$w_{i,j} = exp\left(-\frac{\|y_i - y_j\|_2^2}{h^2}\right) \tag{3}$$

where, i is the pixel being filtered, j represents each one of pixels in the search window S_i, $w(i,j)$ is the similarity between the pixel i and j, satisfying the conditions $0 \le w(i,j) \le 1$ and $\sum_{j \in S_i} w_{i,j} = 1$, S_i is search window around pixel i and h is the smoothing factor. Here, $\|.\|_2^2$ denotes square of L_2- norm to measure patch similarities.

2.2 Block Wise Nonlocal Means (*BNLM*) Filter

Let $\{i_1, i_2, \ldots, i_n\}$ be a subset of I, where I is noisy image. For each i_k, consider B_k as neighbourhood in y_k, such that $W_k = i_k + B_k$, where, $W_k \subset I$ is a neighbourhood centred at y_k. Here, each W_k is a connected subset of I, such that $I = W_1 \cup W_2 \cup \ldots \cup W_n$ and intersection between neighbourhood is non-empty. The size of NL(W_k) is a vector of the same size as W_k. To estimate the denoised pixel at location i, considering all W_k containing the pixel i, the block wise NLM filter [8] is defined as follows.

$$NL(W_k) = \frac{1}{C_k} \sum_{j \in S_i} z(j + B_k) \, exp\left(-\frac{\|d_{i_k j}\|_2^2}{h^2}\right) \tag{4}$$

$$C_k = \sum_{j \in S_i} exp\left(-\frac{\|d_{i_k j}\|_2^2}{h^2}\right) \tag{5}$$
$$\text{where, } d_{i_k j} = (z(i_k + B_k) - z(j + B_k))$$

where, h is a filtering parameter which is expressed in terms of noise variance σ^2 as $h^2 = 2\alpha\sigma^2|P|$ [13, 14] where P is the size of the patch, α is a parameter to be determined empirically. Normally, it is between **0.7** and **1.0**. The restored pixel at i, we considered all W_k containing the pixel i. Thus, restored pixel at i is defined as follows.

$$NL(i) = \frac{1}{|A_i|} \sum_{k \in A_i} NL(W_k(i)) \, and \, A_i = \{k | i \in W_k\} \tag{6}$$

3 Proposed Approach

Method noise is the difference between the noisy image and the corresponding denoised estimate and contains useful information lost due to some inadequacies of denoising process. In *NLM*-based approaches such inadequacies arise because of in appropriate computation of weights required for the averaging process. The proposed approach use the method noise for computing the reweights and enhance the performance of *BNLM*. The proposed approach consists of two stages. In the first stage, block wise *NLM* is applied to obtain the pre-filtered denoised image and the residual image. We propose the use of *BNLM* as in provide better performance with reduced computational load as compared to classical *NLM*. In the second stage, *NLM-reweight* [13] is applied with the new weights computed using the Euclidean distance to approximate the nonlocal similarity between the pre-denoised image and the residual image as follows.

$$w_{i,j} = exp\left(-\frac{\|z_i - z_j\|_2^2}{r^2}\right) \tag{7}$$

$$\|z_i - z_j\|_2^2 = \|\hat{z}_i - \hat{z}_j\|_2^2 + \|\Delta z_i - \Delta z_j\|_2^2 + 2\sum_k \left(\hat{z}_i^k - \hat{z}_j^k\right)\left(\Delta z_i^k - \Delta z_j^k\right) \tag{8}$$

where, $\Delta z_i = y_i - \hat{z}_i$ represent method noise after applying 3×3 mean filter to provide better smoothing to essential details left in the residual image and r denotes the smoothing parameter used at second stage. To demonstrate the advantage of the new weight scheme, larger weight values are assigned to those pixels which have similar edges or textural structures and hence improved denoising performance is obtained. The reasoning behind the improved performance of the approach is attributed to the fact that it exploits the joint nonlocal similarities between method noise and pre-denoised image. The similarities between noise free patches can be robustly measured with higher accuracy by utilizing the above weight scheme as mentioned by Eq. (7).

4 Experiment Results

In this section, we empirically investigate the performances of classical *NLM*, *BNLM*, *NLM-reweight* [13] and the proposed two stage denoising approaches using various quantitative and qualitative measures. All approaches are implemented using Visual C+ + 6.0 under Microsoft Windows environment on a PC with 2.13 GHz CPU and 4 GB RAM. All experiments are conducted on the set of nine different test images as shown in Fig. 1 with standard size of 512×512 pixels corrupted with white Gaussian noise with zero mean and standard deviation ($\sigma = 10, 20, 30, 40$, and 50) at five different noise levels.

Figure 2 shows the sample noised Boat image of with $\sigma = 20$ and the denoised images obtained after applying different approaches for a quick comparison. We chose Boat image as it contain high structural details. The size of search window and local

Fig. 1. Test images: Lena, Jetplane, Mandril, House, Boat, Lake, Barbara, Peppers and Pirate each of size 512 × 512 pixels.

windows is 15 × 15 and 7 × 7 pixels, respectively. The values of 'h' in Eq. (5) and 'r' in Eq. (7) are set empirically and are approximately 5σ and 4σ, respectively.

4.1 Comparative Performance Analysis Using Peak-Signal-to Noise Ratio (*PSNR*)

PSNR is used to measure the quality of the denoised image, $\hat{x}$ with respect to the original noise free image, x. It computed as function of *MSE* as follows:

$$PSNR = 10 \times log_{10}\left(\frac{255 \times 255}{MSE}\right) \tag{9}$$

$$\text{where, } MSE = \frac{1}{MN}\sum_{i=1}^{M}\sum_{J=1}^{N}\left[x_{ij} - \hat{x}_{ij}\right]^2 \tag{10}$$

We analyzed the denoising performance of classical *NLM*, *BNLM*, *NLM-reweight* and the proposed approaches using all the nine test images by computing *PSNR* under five different noise levels. The *PSNR* values obtained for each approach with different noise levels are presented in Table 1. As evident from the table, *BNLM* provides better

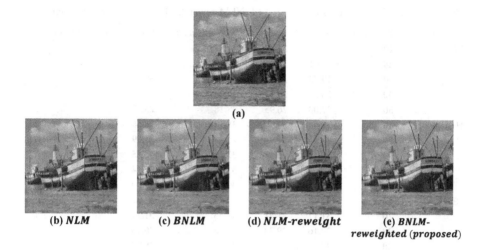

(a)

| (b) *NLM* | (c) *BNLM* | (d) *NLM-reweight* | (e) *BNLM-reweighted (proposed)* |

Fig. 2. (a) Boat image with noise level σ = 20 (b)–(e) denoised image after applying various approaches.

performance than the classical *NLM* and *NLM-reweight* approach at the lower noise levels and is comparable with the proposed approach. However, at higher noise the proposed approach outperforms all the other approaches with the gain of *PSNR* above + 1 dB. This significant improvement in the *PSNR* is attributed to the efficient use of pre-filtered and method noise image.

4.2 Comparative Performance Analysis Using Mean Structured Similarity Index Measurement (*MSSIM*)

MSSIM is used to measure the structural and perceptual closeness between the original (x) and denoised image ($\hat{x}$) [15] and is computed as follows.

$$MSSIM(x,\hat{x}) = \frac{1}{n1} \sum_{j=1}^{n1} SSIM(x,\hat{x})_j \qquad (11)$$

$$\text{where, } SSIM(x,\hat{x}) = \frac{(2\mu_x\mu_{\hat{x}} + C_1)(2\sigma_{x\hat{x}} + C_2)}{(\mu_x^2 + \mu_{\hat{x}}^2 + C_1)(\sigma_x^2 + \sigma_{\hat{x}}^2 + C_2)} \qquad (12)$$

where μ_x and $\mu_{\hat{x}}$ are the average gray values while σ_x^2 and $\sigma_{\hat{x}}^2$ are the variance of patches x and $\hat{x}$, respectively. $\sigma_{x\hat{x}}$ is the covariance of x and $\hat{x}$ and $C_1 = (K_1 * 255)^2$, $C_2 = (K_2 * 255)^2$ are two constants to stabilize the division with default values of $K_1 = 0.01$ and $K_2 = 0.03$ [11]. The value of MSSIM lies in the interval $[0, 1]$ where a larger value of MSSIM represents higher structural match between the input noise-free and denoised images. We analyzed the denoising performance of classical *NLM*, *BNLM*,

Table 1. PSNR comparison between NLM, NLM-reweight, BNLM and proposed scheme

	Noise (σ)	NLM	BNLM	NLM-reweight	BNLM-reweighted (proposed)
LEENA	10	34.76	**35.23**	34.87	34.86
	20	31.61	32.03	32.41	**32.51**
	30	29.73	29.82	30.64	**30.73**
	40	28.24	28.24	29.36	**29.43**
	50	27.03	26.95	28.18	**28.23**
HOUSE	10	37.65	37.96	38.32	**38.36**
	20	34.53	34.81	35.41	**35.69**
	30	32.46	32.56	33.46	**33.85**
	40	30.74	30.68	31.78	**32.14**
	50	29.13	28.93	30.14	**30.39**
JETPLANE	10	34.08	**34.72**	34.33	34.32
	20	31.17	31.51	31.65	**31.79**
	30	29.16	29.33	29.98	**30.15**
	40	27.58	27.65	28.71	**28.78**
	50	26.19	26.19	27.41	**27.43**
MANDRIL	10	30.71	**31.02**	30.91	30.94
	20	27.22	27.81	28.08	**28.09**
	30	25.02	25.56	26.35	**26.37**
	40	23.68	24.04	25.07	**25.13**
	50	22.78	23.01	24.09	**24.17**
PEPPER	10	33.30	**33.54**	33.12	33.16
	20	31.34	31.51	31.53	**31.64**
	30	29.76	29.76	30.23	**30.37**
	40	28.24	28.18	28.97	**29.09**
	50	27.04	26.92	27.91	**27.98**
BARBARA	10	33.18	**33.59**	33.07	33.18
	20	30.16	**30.57**	30.36	30.53
	30	28.16	28.34	28.56	**28.66**
	40	26.46	28.64	27.14	**27.18**
	50	24.92	25.02	25.62	**25.64**
LAKE	10	31.25	**31.73**	31.36	31.28
	20	28.68	29.23	29.26	**29.34**
	30	27.03	27.29	27.82	**27.88**
	40	25.59	25.81	26.69	**26.71**
	50	24.36	24.53	25.62	**25.68**
PIRATE	10	30.31	31.02	**32.56**	32.94
	20	28.45	28.81	30.01	**30.08**
	30	26.56	26.64	28.29	**28.35**
	40	24.03	24.35	27.17	**27.23**
	50	23.01	23.37	26.31	**26.37**
BOAT	10	32.44	**32.84**	32.51	32.48
	20	29.53	29.86	30.04	**30.07**
	30	27.53	27.75	28.39	**28.42**
	40	26.06	26.22	27.33	**27.42**
	50	24.85	24.93	26.05	**26.06**

Table 2. MSSIM comparison between NLM, NLM-reweight, BNLM and proposed scheme

Image	NLM	BNLM	NLM-reweight	Proposed method
LEENA	0.809	0.843	0.854	0.865
HOUSE	0.893	0.899	0.912	0.918
JETPLANE	0.850	0.864	0.879	0.883
MANDRIL	0.799	0.856	0.877	0.885
PEPPER	0.791	0.824	0.856	0.877
BARBARA	0.852	0.872	0.867	0.879
LAKE	0.796	0.799	0.805	0.813
PIRATE	0.783	0.798	0.812	0.824
BOAT	0.778	0.785	0.791	0.798

NLM-reweight and the proposed approaches using all the nine test images by computing *PSNR* under five different noise levels. The *MSSIM* values obtained for each approach with different chosen noise level $\sigma = 20$ for Boat and House images are presented in Table 2. As evident from the table, the proposed approach outperforms at noise levels indicating better preservance of the structural details during denoising.

4.3 Comparative Performance Analysis Using Method Noise

Method noise is used to evaluate the loss of structural or edge information after denoising and is defined as the difference between noisy observation and the denoised image [7, 8] as follows.

$$m_{noise} = y - \hat{x}$$

where, m_{noise} is method noise, y is noisy image and $\hat{x}$ is denoised image. A denoising algorithm must produce the difference or residual image that looks like a noisy image which should contain minimum structural information. Figure 3 shows the method noise obtained after applying various approaches for the sample image Boat with noise level $\sigma = 20$. As seen from the figure the proposed two-stage approach based on *BNLM* and method noise, does not present any noticeable structures in the method noise image.

(a) *NLM*   (b) *BNLM*   (c) *NLM-reweight*   (d) *BNLM-reweighted (proposed)*

Fig. 3. Method noise image of Boat at $\sigma = 20$ after applying various approaches

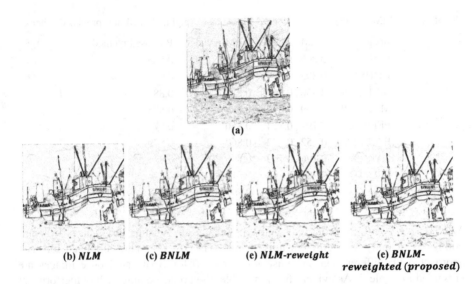

(a)

(b) *NLM*   (c) *BNLM*   (e) *NLM-reweight*   (e) *BNLM-reweighted (proposed)*

Fig. 4. Edge map of Boat image for (a) noise free (b)–(e) denoised images obtained after applying various approaches

4.4   Comparative Performance Analysis Using Edge Map

We also performed the qualitative comparative analysis using edge map for various approaches under investigation. Figure 4 shows the images obtained after edge restoration of the noise free sample Boat image and that of the denoised image after applying the classical *NLM*, *BNLM*, *NLM-reweight* and the proposed approach. As observed from the figure the edge restoration by the proposed approach is similar to that of the original noise free image and far better than the classical *LM*, *BNLM* and *NLM-reweight* apporaches. Thus, the proposed approach more efficiently preserves the essential structural or edge details in the denoised image than all the other approaches.

5   Conclusion and Future Scope

Based on the comparative analysis we can conclude that the proposed two stage approach outperforms all the existing approaches under investigation. These significant improvements in the proposed approach are attributed to the efficient use of pre-filtered and method noise image for the computation of new weights during the second stage of denoising. In future, speed optimization can be obtained by developing a single stage denoising algorithm that include edge map for the computation of weights for the better preservance of the finer details.

Acknowledgments. One of the authors (Karamjeet Singh) is grateful to University Grant Commission (UGC) and the Ministry of Minority Affairs (MOMA), Govt. of India, for providing Maulana Azad National Fellowship (F1-17.1/2012-13/MANF-2012-13-SIK-PUN-13364) for carrying out the research work.

References

1. Aurich, V., Weule, J.: Non-linear Gaussian filters performing edge preserving diffusion. In: Sagerer G., Posch S., Kummert F. (eds.) Mustererkennung 1995. Informatik aktuell, pp. 538–545. Springer, Heidelberg (1995)
2. Lee, J.S.: Digital image enhancement and noise filtering by use of local statistics. IEEE Trans. Pattern Anal. Mach. Intell. **2**, 165–168 (1980)
3. Huang, T.S., Yang, G.J., Tang, G.Y.: A fast two-dimensional median filtering algorithm. IEEE Trans. Acoust. Speech Signal Process. **27**, 13–18 (1979)
4. Perona, P., Malik, J.: Scale-space and edge detection using anisotropic diffusion. IEEE Trans. Pattern Anal. Mach. Intell. **12**, 629–639 (1990)
5. Rudin, L.I., Osher, S., Fatemi, E.: Non-linear total variation based noise removal algorithms. Physica D **60**, 259–268 (1992)
6. Tomasi, C., Manduchi, R.: Bilateral filtering for gray and color images. In: 1998 Proceedings of the Sixth International Conference on Computer Vision, pp. 839–846 (1998)
7. Buades, A., Coll, B., Morel, J.: A nonlocal algorithm for image denoising. In: 2005 Proceedings of CVPR, vol. 2, no. 7, pp. 60–65. IEEE Computer Society Conference on Computer Vision and Pattern Recognition (2005)
8. Buades, A., Coll, B., Morel, J.: A review of image denoising algorithms, with a new one. Soc. Ind. Appl. Math. (SIAM) **4**, 490–530 (2005)
9. Mahmoudi, M., Sapiro, G.: Fast image and video denoising via nonlocal means of similar neighborhoods. IEEE Sig. Process. Lett. **12**, 839–842 (2005)
10. Orchard, J., Ebrahimi, M., Wong, A.: Faster nonlocal means image denoising using the FFT. IEEE Trans. Image Process., 1–6 (2007)
11. Tasdizen, T.: Principal neighborhood dictionaries for nonlocal mean image denoising. IEEE Trans. Image Process. **18**, 2649–2660 (2009)
12. Brunet, D., Vrscay, E.R., Wang, Z.: The use of residuals in image denoising. In: Kamel, M., Campilho, A. (eds.) ICIAR 2009. LNCS, vol. 5627, pp. 1–12. Springer, Heidelberg (2009). doi:10.1007/978-3-642-02611-9_1
13. Zhong, H., Yang, C., Zhang, X.: A new weight for nonlocal means denoising using method noise. IEEE Sig. Process. Lett. **19**, 535–538 (2012)
14. Coupé, P., Yger, P., Prima, S., Hellier, P., Kervrann, C., Barillot, C.: An optimized blockwise non local means denoising filter for 3D magnetic resonance images. IEEE Trans. Med. Imaging **27**, 425–441 (2008)
15. Wang, Z., Bovik, A.C., Skeikh, H.R., Simoncelli, E.P.: Image quality assessment: from error visibility to structural similarity. IEEE Trans. Image Process. **13**, 600–612 (2008)
16. Herrera, J.V.M., Baudes, A.: NLmeans filter. http://www.mi.parisdescartes.fr\~baudes\recerca.html
17. Gonzalez, R.C., Woods, R.E.: Digital image processing, 3rd edn. Pearson Education, India (2008)

Solving Multi-objective Two Dimensional Rectangle Packing Problem

Amandeep Kaur Virk[1(✉)] and Kawaljeet Singh[2]

[1] Sri Guru Granth Sahib World University, Fatehgarh Sahib, India
anu_virk10@yahoo.co.in
[2] University Computer Centre, Punjabi University, Patiala, India

Abstract. The work presented here solves rectangular packing problem in which rectangular items are packed on a rectangular stock sheet. Multiple objectives have been considered which are optimized using rectangle packing algorithm with different heuristics. A mathematical formulation has been presented to solve the problem. Computational experiments have been conducted to find the best packing layout for the problem.

Keywords: Nesting problem · Multi-objective optimization · Non-guillotine cutting

1 Introduction

Nesting problems are two-dimensional cutting and packing problems which select the best possible arrangement for two-dimensional regular or irregular shapes on a larger stock sheet such that minimum stock sheet material is wasted. These problems have been a classic subject in computer science research. Nesting problems find wide applicability in many industries ranging from small scale industries like leather[1], paper [2, 3], glass, wood, sheet metal cutting to large scale industries related to ship building, automobiles and VLSI design. The two-dimensional nesting problems belong to the class of NP-complete problems [4–6] where finding exhaustive solutions becomes difficult as the size of the problem increases. Thus, heuristic techniques are used to find optimum arrangement for pieces on a stock sheet.

In this paper, a special variation of nesting problem, two-dimensional non-guillotine rectangular stock cutting problem with multiple objectives is considered. The rectangle packing algorithm used by Singh and Jain in 2009 [7] has been redrafted to generate 60 feasible patterns using heuristics.

The remainder of the paper is organized as follows: Sect. 2 discusses the problem definition, Sect. 3 covers the multiobjective nature of the problem, Sect. 4 describes the mathematical formulation, Sect. 5 discusses the results and conclusion and future scope is covered in Sect. 6.

© Springer Nature Singapore Pte Ltd. 2017
K. Deep et al. (eds.), *Proceedings of Sixth International Conference on Soft Computing for Problem Solving*, Advances in Intelligent Systems and Computing 547,
DOI 10.1007/978-981-10-3325-4_19

2 Problem Definition

The two-dimensional non-guillotine rectangle packing problem consists of a large rectangular stock sheet of given length L and width W and an order list of small rectangular items i of specified length l_i and width w_i, i = 1,2,3...n, to be cut from stock sheet such that no two items overlap each other. The items are allowed to rotate by 90°. The cuts are non-guillotine which means that the cuts may not go from one end of the stock sheet to another end [8] (Fig. 1). The primary objective of this problem is to find a layout of items on stock sheet which maximizes the utilization of stock sheet. Wascher et al. in 2007 [9] gave a typology of cutting and packing problems. They have classified two-dimensional rectangle packing problem as two-dimensional rectangular single large object placement problem (2D-SLOPP).

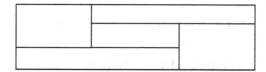

Fig. 1. Non-guillotine cuts

The heuristic used for placing rectangular pieces on the stock sheets has been given by Singh and Jain in 2009 [7]. Brief description of the heuristic is as follows:

- The rectangles are first sorted in decreasing order according to the aspects related to rectangle. The sequencing done in this paper is on the basis of length, area, aspect ratio, profit and maximum allowed lateness of rectangles.
- The item first picked is placed horizontally and then vertically on the left bottom corner of the stock sheet. The item placed gives rise to two pivot points-top left and bottom right corners of the item placed. Pivot points are probable positions where the next item can be placed. The next item to be placed is placed both length-wise and breadth-wise at the two pivot points. The position which gives minimum wastage is selected. New pivot points are defined and used ones are deleted. The pivot points are sequenced according to:
  - minimum radial distance (PPD)
  - minimum x-distance, in case of tie, point with minimum y-distance is taken as next pivot point (PPL)
  - minimum y-distance, in case of tie, point with minimum x-distance is taken as next pivot point (PPB)
- The process is repeated until all items are placed.
- The stock sheet is also oriented in both directions-horizontally and vertically.

Thus the total patterns possible for a problem are 60 which is the product of probable orientation of stock sheet, probable orientation of first item placed on the stock sheet, probable sequencing patterns, different basis for sequencing of items considered, different sequencings of pivot points (along) considered = $2 \times 2 \times 1 \times 5 \times 3 = 60$.

Thus OL-IL-D-SL-PPD stands for a pattern obtained, when object is oriented length-wise; first-item placed on the reference point is also oriented length-wise; items are sorted in decreasing order and are sequenced on the basis of length; pivot points are arranged in increasing order along the diagonal of the object (Fig. 2).

Possible Orientation of Object	Length-wise			Breadth- wise	
	OL			OB	
Possible Orientation of First item placed on Object	Length-wise			Breadth- wise	
	IL			IB	
Sequencing Pattern	Decreasing				
	D				
Different Basis Sequencing of Items Considered	Length	Area	Aspect Ratio	Value	Due Date
	SL	SA	SAR	VA	DD
Sequencing of Pivot Points Considered	Length	Diagonal/ Radial		Breadth	
	PPL	PPD		PPB	

Fig. 2. Heuristics for rectangle packing

3   Multi-objective Nature of the Problem

The principal aim of the rectangle packing problem is to attain maximum utilization of the stock sheet. Due to many real world applications of the cutting stock problem, multiple, often conflicting objectives arise naturally. Researchers are of the view that cutting stock problem with single objective does not consider all the aspects of the production process. This study takes into account number of cuts, maximum lateness of rectangles and total profit as additional objectives along with utilization factor maximization.

1. Utilization factor

Utilization factor is defined as the amount of stock sheet space used to place rectangular items. Our aim is to place rectangles in an optimum arrangement so that maximum space is utilized. The items are placed according to all the possible 60 orientations stated above and the values are noted.

2. Number of cuts

Optimization or efficient usage of cutting equipment can be achieved by minimizing the number of independent cuts required by a packing arrangement. Cuts are simply defined as the number of distinct edges within a packing arrangement. We take into account non-guillotine cuts.

The required number of cuts is counted using the following procedure [10]. The total number of cuts required by a single rectangle is four as it has four edges. For each rectangle, all edges of the rectangle are checked whether they lie on the edge of mother sheet or not. If an edge lies on mother sheet edge, then the number of cuts for each such occurrence is reduced by one as sheet edge does not require a cut. The figure below shows a rectangle with two edges lying on the boundary of the sheet. So, the number of cuts required to cut this rectangle is reduced to two (Fig. 3).

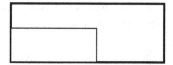

Fig. 3. Rectangle with edges on the stock sheet boundary

Next we check the alignment of each rectangle with all other rectangles placed on the stock sheet. A rectangle can share a common edge with other rectangles horizontally or vertically. There are two possible cases:

Case1: When rectangles are aligned sidewise (fully or partially).

Case 2: When rectangles are aligned one above the other (fully or partially).

For both cases, if rectangles touch each other, then the number of cuts is decreased by one (Fig. 4).

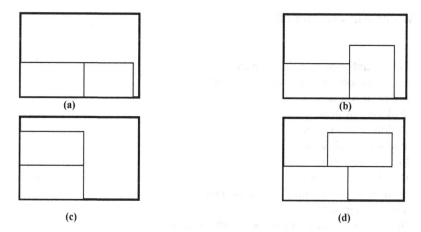

Fig. 4. (a) Fully sidewise aligned rectangles (b) Partially sidewise aligned rectangles (c) Fully top wise aligned rectangles (d) partially top wise aligned rectangles

3. Maximum lateness of rectangles

This objective deals with time by which a rectangle should be available to the customer. Each rectangle has an associated due date d_i, for $i = 1,2,\ldots.n$, that specifies the time by which it should ideally be cut and available to the customer. Each stock sheet has a fixed processing time/cutting time T. The goal is to minimize the maximum lateness of the rectangles with respect to their due dates (DD). The due date of a stock sheet δ_j will the minimum due date of all the rectangles packed on that stock sheet [11]. Therefore, ideally a stock sheet should be cut before the due date of the sheet. Thus

$$\delta_j = \min_{i \in j} d_i \quad 0 \le i \le n, \, 0 \le j \le N \qquad (1)$$

Further the completion time of a stock sheet is T which is kept constant. Thus, maximum lateness of rectangles packed is given by:

$$\sum_{j=1}^{N} T - \delta_j \quad 0 \le j \le N \tag{2}$$

4. Total profit

Each rectangle is assigned a value and the objective is to maximize the total profit by cutting the pieces with higher value first.

$$\text{Maximize} \sum_{j=1}^{N} \sum_{i=1}^{n} v_i X_j \quad 0 \le i \le n, 0 \le j \le N \tag{3}$$

4 Mathematical Modelling

Let
n   Total number of items to be packed
N   Total Number of stock sheets available in reserve
l_i Length of ith item
w_i Width of ith item
L   Length of stock sheet
W   Width of stock sheet
q_i Requirement of ith shape item
d_i Due date of ith item
c_i Number of cuts required to cut ith item
v_i Value assigned to ith item
P_{ij} Number of pieces of ith shape on jth stock sheet
X_j Number of times jth stock sheet enters into an optimal solution
S_j Cost of jth stock sheet
T   Processing time of single stock sheet
δj Minimum due date of jth stock sheet

Objective function is defined as:

1. Utilization Factor

$$\text{Maximize} \sum_{j=1}^{N} \sum_{i=1}^{n} \frac{l_i w_i}{N(L_j W_j)} \quad 0 \le i \le n, 0 \le j \le N \tag{4}$$

2. Number of Cuts

$$\text{Minimize} \sum_{j=1}^{N} \sum_{i=1}^{n} C_i P_{ij} \qquad 0 \le i \le n, \, 0 \le j \le N \tag{5}$$

3. Maximum Lateness of Rectangles

$$\text{Minimize} \sum_{j=1}^{N} T - \delta_j \qquad 0 \le j \le N \tag{6}$$

4. Total Profit

$$\text{Maximize} \sum_{j=1}^{N} \sum_{i=1}^{n} v_i X_j \qquad 0 \le i \le n, \, 0 \le j \le N \tag{7}$$

Subject to following constraints:

5. Requirement of each object in the order list

$$\sum_{j=1}^{N} \sum_{i=1}^{n} P_{ij} X_j \ge q_i \qquad 0 \le i \le n, \, 0 \le j \le N \tag{8}$$

6. Reserve of stock sheets

$$\sum_{j=1}^{N} X_j \ge 1 \qquad 0 \le j \le N \tag{9}$$

5 Experimental Setup and Results

The algorithm was implemented on MATLAB 7.3. A test data set, consisting of the objects of different sizes and the items required to be placed on these objects, was considered as given in Table 1 below.

For the given data set, a total of 60 feasible patterns were generated using the given heuristics. In total 30 different layouts were observed and they have been tabulated in the Table 2 given below along with values for total number of pieces placed (TPP), utilization factor (UF), total number of cuts, total value/profit and maximum lateness.

Table 1. Data Set

Stock sheets: 2
Length = 70 Width = 40 Quantity: 5
Length = 40 Width = 40 Quantity: 4

Requirement										
S.No	1	2	3	4	5	6	7	8	9	10
Length	22	31	35	24	30	13	14	14	12	13
Width	21	13	9	9	7	11	10	8	8	7
Qty	7	9	4	9	12	4	11	3	12	14

Table 2. Different Patterns Generated

Layout	Qty	22×21	31×13	35×9	24×9	30×7	13×11	14×10	14×8	12×8	13×7	TPP	UF	Cuts	Value	Lateness
X1		7	9	4	9	12	4	11	3	12	14	8	88.07	12	2963.1	51.09
X2			3	3	1						3	9	96.1	16	3112.2	44.84
X3			6						3	11	7	22	86.85	45	3693.4	45.7
X4		1	1			1	1					6	79.17	11	2379.8	45.70
X5		1	4	4		2	1		1			7	76.2	14	2251.1	45.7
X6			3	1								8	88.7	11	2630.8	51.09
X7		1	3	3			1		1			7	80.03	13	2329.3	51.09
X8			4			7	5		1			8	95.3	13	3070	44.8
X9				2			12				3	12	91.5	21	5046.2	36.8
X10					4	5						12	90	22	2803.2	41.64
X11				5		4	2					10	77.1	10	2763.3	51.09
X12				2			7				3	11	85.6	19	2325.6	56.1
X13						1	1		1			7	86.9	12	2676.6	56.1
X14				5		8			3		1	11	94	18	2731.2	56.1
X15				6							5	13	87.3	19	4774.1	51.09
X16				2	4		1					10	88.2	16	2896	44.8
X17					4		1			1		7	93.85	12	2942	56.1
X18						3	1					8	85.2	10	2493.1	51.09
X19							11					8	75.6	13	2059.2	7.39
X20						9			1		2	11	82.5	18	2569.6	41.04
X21				4	2	2						12	80.9	20	4558.6	56.1
X22			2	4		2						8	95.5	13	3114.4	51.09
X23			2			7		3				8	88.7	15	2630.8	51.09
X24		3			1			3			1	10	89	19	2577.8	56.1
X25		3	1		1							7	67.75	11	2269.6	51.09
X26				2	7	1						5	71.6	9	1831.8	4.07
X27			2		6							10	84	16	4306.4	51.09
X28			1			8					4	12	88.07	23	3170.8	51.09
X29			1		8	2						9	74.3	18	2000.6	56.1
X30					8							10	76.7	17	4289.2	51.09

Mathematical formulation for the result obtained above for the given dataset is stated below. x1 represents pattern 1, x2 represents pattern 2 and so on.

- Maximize (Utilization factor)
  88.07 x1 + 96.1 x2 + 86.85 x3 + 79.17 x4 + 76.2 x5 + 88.7 x6 + 80.03 x7 + 91.3 x8 + 95.5 x9 + 90 x10 + 77.1 x11 + 85.6 x12 + 86.9 x13 + 94 x14 + 87.3 x15 + 88.2 x16 + 93.85 x17 + 85.2 x18 + 75.6 x19 + 82.5 x20 + 80.9 x21 + 95.5 x22 + 88.7 x23 + 89 x24 + 67.75 x25 + 71.6 x26 + 84 x27 + 88.07 x28 + 74.3 x29 + 76.7 x30

- Minimize (number of cuts)
  12 x1 + 16 x2 + 45 x3 + 11 x4 + 14 x5 +11 x6 + 13 x7 + 13 x8 +21 x9 + 22 x10 + 10 x11 + 19 x12 + 12 x13 +18 x14 + 19 x15 + 16 x16 + 12 x17 + 10 x18 + 13 x19 + 18 x20 + 20 x21 + 13 x22 + 15 x23 + 19 x24 + 11 x25 + 9 x26 + 16 x27 + 23 x28 + 18 x29 + 17 x30

- Maximize (Profit)
  2963.1   x1 + 3112.2   x2 + 3693.4   x3 + 2379.8   x4 + 2251.1   x5 + 2630.8 x6 + 2329.3 x7 + 3070 x8 + 5046.2 x9 + 2803.2 x10 + 2763.3 x11 + 2325.6 x12 + 2676.6 x13 + 2731.2 x14 + 4774.1 x15 + 2896 x16 + 2942 x17 + 2493.1 x18 + 2059.2   x19 + 2569.6   x20 + 4558.6   x21 + 3114.4   x22 + 2630.8 x23 + 2577.8   x24 + 2269.6   x25 + 1831.8   x26 + 4306.4   x27 + 3170.8 x28 + 2000.6 x29 + 4289.2 x30

- Minimize (Maximum Lateness)
  51.09 x1 + 44.84 x2 + 45.7 x3 + 45.7 x4 + 45.7 x5 + 51.09 x6 + 51.09 x7 + 44.8 x8 + 36.8 x9 + 41.64 x10 + 51.09 x11 + 56.1 x12 + 56.1 x13 + 56.1 x14 + 51.09 x15 + 44.8   x16 + 56.1   x17 + 51.09   x18 + 7.39   x19 + 41.04   x20 + 56.1 x21 + 51.09   x22 + 51.09   x23 + 56.1   x24 + 51.09   x25 + 4.07   x26 + 51.09 x27 + 51.09 x28 + 56.1 x29 + 51.09 x30

The solution is selected by exhaustively searching for the layout pattern that best satisfies all the objectives. The result is tabulated below in Table 3.

Table 3. Results obtained

Sheet type	UF	Cuts	Value	Lateness	TPP	Heuristic
70 × 40	96.1	16	3112.2	44.84	9	OL_IL_D_SA_PPL
70 × 40	95.3	13	3070	44.8	8	OB_IB_D_SA_PPL
70 × 40	91.5	21	5046.2	36.8	12	OL_IB_D_VA_PPL
70 × 40	86.85	45	3693.4	45.7	22	OL_IL_D_SAR_PPD
70 × 40	88.07	12	2963.1	51.09	8	OL_IL_D_SL_PPL
40 × 40	95.5	13	3114.4	51.09	8	OL_IB_D_DD_PPL
40 × 40	89	19	2577.8	56.1	10	OL_IL_D_SA_PPL
40 × 40	88.07	23	3170.8	51.09	12	OL_IL_D_VA_PPL
40 × 40	93.85	12	2942	56.1	7	OL_IL_D_SA_PPL

6   Conclusion and Future Scope

This paper investigates the two-dimensional rectangle packing problem where rectangular items are to be arranged on a stock sheet to maximize the profit. The single objective problem has been converted into a multi-objective problem to meet the real world scenarios. The new multi-objective rectangle packing problem has been solved using various heuristics and the results obtained have been exhaustively searched to look for the best solution. More objectives and heuristics can be considered to satisfy the real world industrial requirements.

References

1. Yang, H.H., Lin, C.L.: On genetic algorithms for shoe making nesting. A Taiwan Case Expert Syst. Appl. **36**(2), 1134–1141 (2009)
2. Selow, R., Junior, F.N., Heitor, S., Lopes, H.S.: Genetic algorithms for the nesting problem in the packing industry. In: The International Multi Conference of Engineers and Computer Scientists (IMECS), pp. 1–6 (2007)
3. Yaodong, C., Yiping, L.: Heuristic algorithm for a cutting stock problem in the steel bridge construction. Comput. Oper. Res. **36**, 612–622 (2009)
4. Garey, M., Johnson, D.: Computers and Intractability, A Guide to the Theory of NP-Completeness. W.H. Freeman and Company, San Francisco (1979)
5. Hopper, E., Turton, B.C.H.: A review of the application of metaheuristic algorithms to 2D strip packing problems. Artif. Intell. Rev. **16**, 257–300 (2001)
6. Liu, H.Y., He, Y.J.: Algorithm for 2D irregular-shaped nesting problem based on the NFP algorithm and lowest-gravity-center principle. J. Zhejiang Univ.–Sci. **7**(4), 570–576 (2006)
7. Singh, K., Jain, L.: Optimal solution for 2-D Rectangle Packing Problem. Int. J. Appl. Eng. Res. **4**, 2203–2222 (2009)
8. Leung, T.W., Yung, C.H., Troutt, M.D.: Applications of genetic search and simulated annealing to the two-dimensional non-guillotine cutting stock problem. Comput. Ind. Eng. **40**, 201–214 (2001)
9. Wascher, G., Hausner, H., Schumann, H.: An improved typology of cutting and packing problems. Eur. J. Oper. Res. **183**, 1109–1130 (2007)
10. Uday, A., Goodman, E.D., Debnath, A.A.: Nesting of irregular shapes using feature matching and parallel genetic algorithms. In: Genetic and Evolutionary Computation Conference Late Breaking Papers, San Francisco, California, USA, pp. 429–434 (2001)
11. Bennell, J.A., Lee. L.S., Potts, C.N.: A genetic algorithm for two-dimensional bin packing with due dates. Int. J. Prod. Econ. **145**(2), 547–560 (2013)

Multi-parameter Retrieval in a Porous Fin
Using Binary-Coded Genetic Algorithm

Rohit K. Singla[✉] and Ranjan Das

Department of Mechanical Engineering, IIT Ropar, Punjab, India
rohit.singla@iitrpr.ac.in, ranjandas81@gmail.com

Abstract. In this paper, the implementation of the binary-coded Genetic Algorithm (GA) for multi-parameter retrieval through inverse analysis is demonstrated. A porous rectangular fin with constant thermo-physical parameters is
investigated. The porous fin involves Fourier law of heat conduction along with
natural convection and surface radiation phenomena. Due highly nonlinear
phenomenon owing to the radiative effect and because of the associated
complexity in the gradient evaluation, gradient-free method based on the GA has
been used for unknown parameter retrieval. The analysis is done for satisfying a
given temperature distribution on the fin surface generated using a well-validated
forward solver based on the Runge-Kutta method. It is observed from the simulated experiments that for simultaneous multi-parameter retrieval, the GA yields
multiple combinations of unknown parameters satisfying a particular distribution
of temperature.

Keywords: Porous fin · Inverse problem · Optimization · Parameter retrieval

1 Introduction

Fins are one of the passive methods of enhancing the rate of heat transfer from a primarily
surface [1, 2]. The rate of heat transfer through fins primarily depends on the temperature
gradient, the available surface area, the surface heat transfer coefficient along with other
thermal parameters [3]. Fins are widely used in various applications, such as air conditioning and refrigeration, electronic equipments, space radiators and many more [4–6].
Among different types of fins, porous fins increase the effective surface area of heat
convection to the surrounding fluid medium. For the heat transfer analysis of porous
fins, the local temperature is considered to distinctly change along the longitudinal
direction, whereas, the fluid-solid interaction is taken to obey the Darcy's law and the
surface convection effect is normally ignored [7, 8]. In order to analyze a system's
response to a known set of input and operating conditions, a forward/direct analysis is
made [9]. Alternatively, when some of the pertinent conditions are required to be either
predicted or retrieved, then an inverse analysis is required to be accomplished [10]. Many
studies have been undertaken addressing the application of porous fins which include
both the forward and the inverse analyses. For example, Bhanja and Kundu [11]
performed a closed form forward analysis on a T-shaped porous fin using the Adomain

© Springer Nature Singapore Pte Ltd. 2017
K. Deep et al. (eds.), *Proceedings of Sixth International Conference on Soft Computing
for Problem Solving*, Advances in Intelligent Systems and Computing 547,
DOI 10.1007/978-981-10-3325-4_20

decomposition method. Gorla and Bakier [12] implemented the fourth order accurate Runge-Kutta method to obtain a forward numerical solution for local temperature distribution in a rectangular porous fin problem. Das and Ooi demonstrated an inverse solution methodology based on the simulated annealing algorithm for simultaneously retrieving five unknown parameters satisfying a given temperature distribution [13]. Hatami and Ganji [14] studied the heat transfer phenomenon of ceramic porous fins. They solved the forward problem using the least squares method along with the Runge-Kutta method. Moradi et al. [15] adopted a forward solver based on the differential transformation method to investigate a triangular porous fin involving variable heat diffusion parameter. Das [16] demonstrated the application of the Runge-Kutta method as a forward algorithm along with the hybrid combination of nonlinear programming and the Differential Evolution (DE) algorithm-based inverse procedure for parameter retrieval. Das and Prasad [17] reported an inverse algorithm incorporating the DE in a porous fin to simultaneously predict the porosity and thermal diffusivity.

The available literature on porous fins in particular reveals that very few studies deal with the application of the binary-coded GA for multi-parameter retrievals using inverse analysis [18]. Furthermore, it is well-known that the evolutionary methods of optimization are preferable for performing an intelligent searching and they are suitable where gradients are not easily computed due to associated difficulties such as nonlinearity, discontinuity, non-convexity, etc. [19, 20]. Due to this reason, the motivation in the present paper is to apply the binary-coded GA for solving an inverse heat transfer problem in a rectangular porous fin.

2 Formulation and Solution Methodology

A rectangular porous fin geometry operating under environment temperature, T_a as shown in Fig. 1 has been studied. The fin involves porosity, ϕ along with the thermal conductivity of solid and fluid as, k_s and k_f, respectively. The fin is attached to the base involving temperature, T_b with its tip being adiabatic. The local temperature, T changes along the axial (i.e., x) direction, the fluid-solid interaction follows Darcy's law [21] and the surface convection effect has been neglected [7, 8]. Performing an energy balance between any two locations (x and $x + \Delta x$) along the fin length, L, the relevant heat transfer rate, Q under steady-state can be stated as,

$$Q(x) - Q(x + \Delta x) = \dot{m} c_p \left(T - T_a \right) \tag{1}$$

where, c_p is the specific heat capacity of the fluid medium and the mass flow rate, $\dot{m}$ can be computed in the following way [8],

$$\dot{m} = \rho \, v_w \Delta x W \tag{2}$$

where, ρ is the fluid density and W is the fin width. The fluid velocity, v_w can be obtained as [8],

$$v_w = \frac{gK_p\beta_{vol}}{v}\left(T - T_a\right) \tag{3}$$

where, v, K_p and β_{vol} are the kinematic viscosity of fluid, the fin permeability and the coefficient of volumetric thermal expansion, respectively.

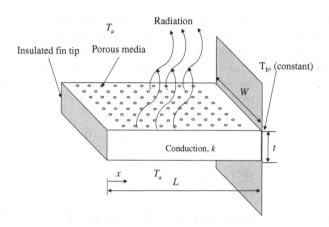

Fig. 1. Geometry of the porous rectangular fin.

In the present work, the heat dissipation occurs due to inside conduction obeying Fourier law, surface radiation and natural convection [12]. Realizing the effective fin thermal conductivity as $k_{eff} = \phi k_f + (1 - \phi)k_s$, cross section area, $A = Wt$ and fin perimeter, $P = 2(W + t)$ and considering the width, W to be considerably larger than the fin thickness, t (i.e., $W + t \approx W$), the governing energy equation can be represented as [12, 17],

$$k_{eff}\frac{d^2T}{dx^2} - \frac{2\sigma\varepsilon}{t}\left(T^4 - T_a^4\right) - \frac{\rho_f c_p gK\beta_{vol}\left(T - T_a\right)^2}{t\,v} = 0 \tag{4}$$

where, $\sigma = 5.67 \times 10^{-8}\,\mathrm{W/(m^2.K^4)}$ denotes the Stefan-Boltzmann constant. The governing equation, Eq. (4) has been solved using the following boundary conditions,

$$\frac{dT}{dx}\bigg|_{x=0} = 0;\ T(x)|_{x=L} = T_b \tag{5}$$

For generalization, the above-mentioned, Eq. (4) is non-dimensionalized using the following parameters [12],

$$\theta = \frac{(T - T_a)}{(T_b - T_a)}; X = \frac{x}{L}; Da = \frac{K_p}{t^2}; Ra = \frac{\beta_{vol}g(T_b - T_a)t^3}{v\,\alpha}; K_r = \frac{k_{eff}}{k_f};$$

$$C_a = \frac{T_a}{(T_b - T_a)}; S_h = \frac{Ra.Da.\left(\frac{L}{t}\right)^2}{K_r}; N_r = \frac{2\sigma\varepsilon}{k_{eff}t}L^2(T_b - T_a)^3 \tag{6}$$

where, C_a represents the temperature ratio, θ represent dimensionless temperature, X denotes dimensionless distance, K_r indicates the conductivity ratio, S_h signifies porous parameter, N_r denotes the radiation parameter and α pertains to the thermal diffusivity of the fluid. Additionally, Da and Ra characterize Darcy and Rayleigh numbers, respectively. Using Eq. (6), the dimensionless version of Eqs. (4) and (5) can be characterized by Eqs. (7) and (8), respectively, i.e.,

$$\frac{d^2\theta}{dX^2} - S_h\theta^2 - N_r\left[(\theta + C_a)^4 - C_a^4\right] = 0 \tag{7}$$

$$\frac{d\theta}{dX}\bigg|_{X=0} = 0;\ \theta(X)|_{X=1} = 1 \tag{8}$$

In the present work, the forward problem has been solved using the Matlab-based solver (bvp4c) implanting a fourth order accurate implicit Runge-Kutta method. Since this method is discussed elsewhere [16], therefore its details are not provided here. Furthermore, the code validation study is also accomplished in many previously-published literature [17, 21], consequently this discussion is also not elaborated here. Next, we proceed for the inverse analysis, which is the main objective of the present study.

Some of the important parameters relatable to a porous fin are the porosity, ϕ, the permeability, K_p, thermal conductivity of the solid, k_s and thermal conductivity of the fluid, k_f. These parameters have been estimated to meet a given dimensionless temperature profile, θ considering other parameters to be known. The simulated temperature measurement data are considered at three axial locations, two at the fin base and tip, while another at the middle location of the fin. For concurrently predicting the parameters, the objective function, J is described as,

$$\text{Minimize: } J = (\tilde{\theta}_1 - \theta_1)^2 + \left(\tilde{\theta}_{(i+1)} - \theta_{(i+1)}\right)^2 + (\tilde{\theta}_i - \theta_i)^2 \tag{9}$$

where, i (an odd number, 61 for the current case) is the index for the simulated temperature measurement location, whereas, the superscript ($\sim$) represent the exact simulated temperature data. The temperature involving guessed value of unknowns (i.e., ϕ, K_p, k_s and k_f) which is iteratively updated using the GA is represented without superscript. The optimization algorithm is explained next.

3   Optimization Algorithm

In the current study, evolutionary optimization algorithm based on the binary-coded GA has been tested for multi-parameter retrieval. The termination condition is considered upto 100 iterations. The population size in the GA is considered 20. The configuration of the computing system is Intel(R) Core i7-3770, 8 GB RAM with clock speed of 3.40 GHz. In the present binary-coded GA, depending on the number of unknowns, say n_p, $10 \times n_p$ binary bits are randomly created to signify one string (individual) of the GA. First 10 bits are allocated to denote the first parameter and so on. Binary bits are converted into decimal system to evaluate the objective function, J for each string. The set of solutions in any population is called the generation of the GA, which is iteratively updated. For converting a binary into a decimal number, randomly-generated 10 binary bits (either 0 or 1) are used, which generate $2^{10} = 1024$ distinct numbers ranging from 0 and 1023 (i.e., $2^{10}-1$). To characterize a particular parameter, y in decimal system satisfying a given upper and lower limits (i.e., LL, UL), the following concept is used [22],

$$y = LL + \frac{(UL - LL)}{(2^{10} - 1)} \times (\text{decimal correspondent of the binary number}) \qquad (10)$$

where, $LL = [0.10, 1 \times 10^{-20}$ m^2, 10 W/(m.K), 0.01 W/(m.K)] and $UL = [0.90, 1 \times 10^{-8}$ m^2, 400 W/(m.K), 0.50 W/(m.K)] for porosity, ϕ, permeability, K_p, solid thermal conductivity, k_s and fluid thermal conductivity, k_f, respectively. If the termination condition is not satisfied, then the reproduction is performed through crossover and mutation through predefined probabilities. In the present work, the crossover and the mutation probabilities have been considered as 0.80 and 0.03, respectively. For four parameters, i.e., $n_p = 4$, the crossover method is exemplified below,

$$
\begin{aligned}
&\text{Parents (before crossover)}\\
&\text{'01101-10101-}\textit{00101-11'101}\text{-11001-10110-}\underline{10001\text{-}10100}\\
&\text{'11011-10010-}\textit{11010-01'110}\text{-10001-11100-}\underline{00001\text{-}10010}
\end{aligned}
\qquad (11a)
$$

$$
\begin{aligned}
&\text{Children (after crossover)}\\
&\text{'11011 -10010-}\textit{11010-01'101}\text{-11001-10110-}\underline{10001\text{-}10100}\\
&\text{'01101 -10101-}\textit{00101-11'110}\text{-10001-11100-}\underline{00001\text{-}10010}
\end{aligned}
\qquad (11b)
$$

The mutation (occurring within an individual child) is denoted below,

$$
\begin{aligned}
&\text{Child-1 (Before mutation)}\\
&\text{11011-10010-}\textit{11010-011'0'1}\text{-11001-10110-}\underline{10001\text{-}10100}\\
&\text{Child-1 (After mutation)}\\
&\text{11011-10010-}\textit{11010-011'1'1}\text{-11001-10110-}\underline{10001\text{-}10100}
\end{aligned}
\qquad (12)
$$

The iterative process is executed until the termination condition attainment.

4 Results and Analysis

As mentioned earlier that the forward solver based on the Matlab function (bvp4c) implementing the Runge-Kutta method of fourth order accuracy has been validated against the literature results [12]. Since its details are available elsewhere in many previously published literature (for instance, [17, 21]), consequently its details are not repeated in the current study. The objective here is to implement the GA for simultaneous prediction of four parameters (ϕ, K_p, k_s and k_f) meeting a given simulated temperature profile. This is accomplished by the minimization of the objective function, J defined in Eq. (9). The fin length, L and the fin thickness, t of the porous surface have been considered as 0.10 m and 0.03 m, respectively. Other parameters considered for creating the exact simulated temperature distribution, $\tilde{\theta}$ through the forward solver are, $S_h = 1.4471$, $N_r = 0.0535$ and $C_a = 1.0105$. Table 1 presents the parameters along with their relevant objective function and computational time inversely-retrieved using the binary-coded GA. Five distinct runs of the GA have been executed in this study with results as presented in Table 1. It is found that the final combination of retrieved parameters for each run differ from one another. So, it is clear that the inverse optimization technique provide multiple combinations of the parameters satisfying a given condition. The average values of objective function and computational time are also mentioned in the table. It is worthy to point here that each set of estimated parameters (ϕ, K_p, k_s and k_f) obtained in Table 1 may yield a distinct set of dimensionless parameters (S_h, G, C_t) that is different from the one considered in the forward method.

Table 1. Retrieved porous fin parameters obtained using the binary-coded GA; forward method: [$S_h = 1.4471$, $G = 0.0535$ and $C_t = 1.0105$].

Sl. No.	Retrieved parameters				Objective function	CPU time
	ϕ	K_p, m^2	k_s, W/(m.K)	k_f, W/(m.K)	J	seconds
1.	0.2056	5.06×10^{-9}	40.12	0.3999	2.304×10^{-7}	96.18
2.	0.1438	9.85×10^{-9}	26.39	0.1422	4.043×10^{-6}	104.70
3.	0.2322	9.77×10^{-9}	10.00	0.0378	7.418×10^{-9}	96.77
4.	0.8922	5.18×10^{-10}	45.46	0.3980	6.611×10^{-6}	101.92
5.	0.1063	4.99×10^{-9}	10.00	0.0847	2.017×10^{-4}	97.18
				Average	4.251×10^{-5}	99.35

Next, it is important to check the accuracy of the retrieved parameters, which can be confirmed from the reconstructed temperature profiles shown in Fig. 2. It is found that all distributions evaluated using the retrieved parameters are very well in agreement with the exact distribution. From the GA-reconstructed profiles, the maximum residual is found to be approximately 1.97%. The iterative variation of the objective function, J for all runs of Table 1 is presented in Fig. 3. The value of the objective function, J regularly decreases with iterations as can be observed from Fig. 3. As the termination condition was considered either when the objective function, J attains zero or number of generations (i.e., iterations) reaches 100. Thus, in this analysis, all runs have been terminated at the 100th generation.

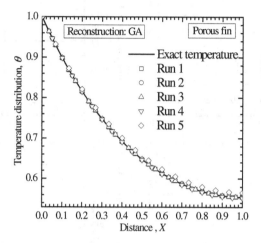

Fig. 2. Comparison of the reconstructed and the exact temperature distributions for different runs.

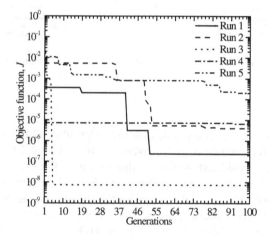

Fig. 3. Variation of the objective function, J with generations of GA.

The iterative variations of all the retrieved parameters are illustrated in Fig. 4. It is found that multiplicity of solutions exists satisfying a given criterion (i.e., yielding a given temperature profile). This is because from the objective function trends (Fig. 3), for all runs, the value of the objective function, J reaches O ($\leq 10^{-3}$) by at most 50 generations (iterations), however, in Fig. 4, many concerned parameters (ϕ, K_p, k_s and k_f) estimated from the inverse approach differ considerably from one run to another. Additionally, for a given run (for instance, Run 5 in Fig. 4a and d), the values of the parameters significantly change even beyond 50 generations.

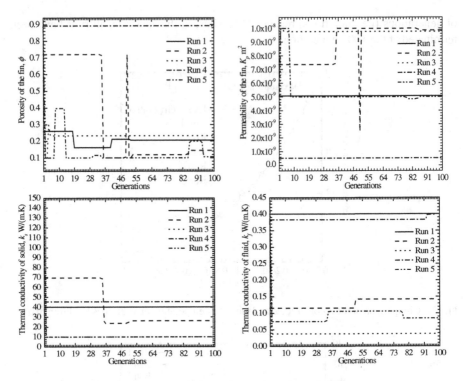

Fig. 4. Iterative variation of inversely estimated parameters

5   Conclusions

The implementation of the binary-coded Genetic Algorithm (GA) for multi-parameter estimation in porous rectangular fin by inverse analysis is established. The governing energy equation is considered nonlinear, due to which the GA has been used for obtaining the solution of the inverse problem. The forward problem is solved using the well-validated fourth order accurate Runge-Kutta method that was also used to synthesize the simulated temperature distribution. Four critical parameters have been successfully predicted using the binary GA. The present work also reveals that for the simultaneous retrieval of existing parameters, multiplicity of solutions exists which simultaneously satisfy a particular thermal objective. This consequently serves as the selection strategy to decide the appropriate operating parameters from the available alternatives.

References

1. Kern, D.Q., Kraus, A.D.: Extended Surface Heat Transfer. McGraw-Hill, New York (1972)
2. Lienhard, J.H.: A Heat Transfer Textbook, 4th edn. Dover Publicatons, New York (2011)
3. Lane, H.J., Heggs, P.J.: Extended surface heat transfer—the dovetail fin. Appl. Therm. Eng. **25**, 2555–2565 (2005)

4. Peng, C.S.P., Howell, J.R.: Analysis and design of efficient absorbers for low-temperature desiccant air conditioners. J. Sol. Energy Eng. **103**, 67–74 (1981)
5. Alawadhi, E.M., Amon, C.H.: PCM thermal control unit for portable electronic devices: experimental and numerical studies. IEEE Trans. Compon. Packag. Technol. **26**, 116–125 (2003)
6. Singla, R.K., Das, R.: Adomain decomposition method for a stepped fin space radiator with internal heat generation. In: 2nd IEEE International Conference on Recent Advances in Engineering and Computational Sciences, Chandigarh, pp. 1–6 (2015)
7. Kiwan, S.: Effect of radiative losses on the heat transfer from porous fins. Int. J. Therm. Sci. **46**, 1046–1055 (2007)
8. Kiwan, S.: Thermal analysis of natural convection porous fins. Transp. Porous Media **67**, 17–29 (2007)
9. Delprat-Jannaud, F., Lailly, P.: Ill-posed and well-posed formulations of the reflection travel time tomography problem. J. Geophys. Res. Solid Earth **98**, 6589–6605 (1993)
10. Girault, M., Petit, D., Penot, F.: Identification of velocity distribution in a turbulent flow inside parallel-plate ducts from wall temperature measurements. Inverse Probl. Sci. Eng. **12**, 247–262 (2004)
11. Bhanja, D., Kundu, B.: Thermal analysis of a constructal T-shaped porous fin with radiation effects. Int. J. Refrig. **34**, 1483–1496 (2011)
12. Gorla, R.S.R., Bakier, A.Y.: Thermal analysis of natural convection and radiation in porous fins. Int. Commun. Heat Mass Transf. **38**, 638–645 (2011)
13. Das, R., Ooi, K.T.: Predicting multiple combination of parameters for designing a porous fin subjected to a given temperature requirement. Energy Convers. Manag. **66**, 211–219 (2013)
14. Hatami, M., Ganji, D.D.: Thermal behavior of longitudinal convective-radiative porous fins with different section shapes and ceramic materials (SiC and Si_3N_4). Ceram. Int. **40**, 6765–6775 (2014)
15. Moradi, A., Hayat, T., Alsaedi, A.: Convection-radiation thermal analysis of triangular porous fins with temperature-dependent thermal conductivity by DTM. Energy Convers. Manag. **77**, 70–77 (2014)
16. Das, R.: Forward and inverse solutions of a conductive, convective and radiative cylindrical porous fin. Energy Convers. Manag. **87**, 96–106 (2014)
17. Das, R., Prasad, D.K.: Prediction of porosity and thermal diffusivity in a porous fin using differential evolution algorithm. Swarm Evol. Comput. **23**, 27–39 (2015)
18. Panda, S., Das, R.: Inverse analysis of a radial porous fin using genetic algorithm. In: Eighth IEEE International Conference on Contemporary Computing (IC3), Noida, pp. 167–170 (2015)
19. Parthasarathy, S., Balaji, C.: Estimation of parameters in multi-mode heat transfer problems using bayesian inference - effect of noise and a priori. Int. J. Heat Mass Transf. **51**, 2313–2334 (2008)
20. Zhang, S., Liu, S.: A novel artificial bee colony algorithm for function optimization. Math. Probl. Eng. **2015**, 1–10 (2015)
21. Das, R., Singla, R.K.: Inverse heat transfer study of a nonlinear straight porous fin using hybrid optimization. In: ASME 2014 Gas Turbine India Conference, pp. V001T04A001–V001T04A010 (2014)
22. Deb, K.: Optimization for Engineering Design: Algorithms and Examples. PHI Learning Pvt. Ltd., New Delhi (2012)

Effectiveness of Constrained Laplacian Biogeography Based Optimization for Solving Structural Engineering Design Problems

Vanita Garg$^{(\boxtimes)}$ and Kusum Deep

Department of Mathematics, Indian Institute of Technology, Roorkee, India
vanitagarg16@gmail.com, kusumdeep@gmail.com

Abstract. There is number of engineering design problems which are modeled as non-linear optimization problems. These problems are considered as benchmark problems for testing newly design optimization techniques. The objective of this paper is to demonstrate the effectiveness of a newly proposed constrained Laplacian Biogeography Based Optimization algorithm on three engineering design problems. The results are compared with existing efforts reported in literature. It is shown that in a majority of the cases constrained Laplacian Biogeography Based Optimization algorithm provides superior results.

Keywords: Constrained laplacian biogeography based optimization (C-LX-BBO) · Welded beam design problem · Pressure design problem · Tension string problem

1 Introduction

Many researchers and engineers consider structural engineering optimization problems as the benchmark problems to test the performance of an algorithm which is proposed for solving constrained optimization problems. These problems are very basic yet complex and important in the field of Structural Engineering design optimization. Engineering design optimization is the process of finding the optimal parameters which are responsible for the design of the any machine. The welded beam design problem, Pressure vessel design problem and tension string design problems are well known engineering design problems. The problems differ in their complexity, non-linearity of the constraints, number of decision variables (Arora 1989).

Many researchers have attempted their respective method to solve the engineering design problems. Sandgren (1988) has used integer and discrete programming for solving these engineering design problems. Zhang and Wang (1993) have used mixed discrete optimization with simulated annealing. Later Zhang et al. (2008) have used Differential evolution for solving these constrained optimization problems. Ray and Liew (2003) have proposed an algorithm based on social behavior to solve engineering design problems. Tsai (2005) has put some light on solving non-linear fractional programming problems by using structural engineering design optimization problems as benchmark functions. Hedar and Fukushima (2006) have given the derivative free filter simulated annealing method for constrained optimization problems. Dimopoulos

© Springer Nature Singapore Pte Ltd. 2017
K. Deep et al. (eds.), *Proceedings of Sixth International Conference on Soft Computing for Problem Solving*, Advances in Intelligent Systems and Computing 547,
DOI 10.1007/978-981-10-3325-4_21

(2007) have also made an attempt to solve mixed variable engineering optimization problems. In Cagnina et al. (2008) Constrained Particle Swarm Optimization is tested on four standard constrained optimization problems. In Coelho (2010), a new approach of PSO for solving constrained optimization problem is proposed. Kaveh and Talatahari (2010) have solved engineering optimization problems using hybrid Particle Swarm Optimization and Artificial Bee colony. Mehta and Dasgupta (2012) have used a new constrained optimization technique based on simplex search method. This new approach is called Gaussian quantum-behaved particle swarm optimization approaches for constrained engineering design problems. In Gandomi et al. (2011) and Gandomi et al. (2013), Firefly Algorithm and Cuckoo search is used respectively to solve the mixed variable structural optimization problems. In Garg (2016), a hybrid technique called PSO-GA has been used.

Due to their limitations, many traditional methods fail to solve these engineering design problems. On the other hand, heuristic methods can be used for solving any complex and highly non-linear optimization problems. Biogeography Based Optimization is proposed by Simon (2008) for solving unconstrained optimization problems. For solving constrained optimization problems, BBO has been generalized for solving constrained optimization problems. Recently proposed Constrained Laplacian BBO (Garg and Deep 2016) has been tested and proved to be a good constrained optimization algorithm.

In this paper, Constrained Laplacian BBO has been applied to solve the engineering design problems. The paper has been designed as: Sect. 2 gives the description of the engineering design problem i.e. welded beam design Problem, Pressure Design Problem, Tension string problem. Section 3 gives the Constrained Laplacian BBO which is used for solving engineering design problems. Section 4 presents the numerical results obtained after solving these engineering optimization problems by Constrained Laplacian Biogeography Based Optimization. In the last Sect. 5, conclusion and future aspects of this study is provided.

2 Structural Engineering Design Problems

In this Section, three standard structural engineering design problems are considered i.e. Pressure vessel design Problem, Welded Beam Design Problem and Tension String Design Problem. The mathematical models and a brief introduction to all three design problems are as follows:

2.1 Pressure Vessel Design Problem

A cylindrical vessel capped at both ends by hemi spherical heads is into two parts. The compressed air storage tank with working pressure of 2000 psi and a maximum volume of 750 ft^3 is considered. The two parts are joined by welding two longitudinal welds to form a cylinder. The objective is to minimize the cost which includes cost of material and forming and welding. This optimization problem contains four design variables namely thickness of the pressure vessel (T_s), thickness of the head (T_h), inner radius of the vessel (R) and the length of the vessel without heads (L) (Fig. 1).

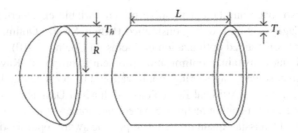

Fig. 1. Pressure vessel design

To model this design optimization problem, we are considering, $T_s = x_1$, $T_h = x_2$, $R = x_3$ and $L = x_4$ are four variables. Mathematical model of the problems is given as follows:

$$\min f(X) = 0.6224x_1x_3x_4 + 1.7781x_2x_3^2 + 3.1661x_1^2x_4 + 19.8x_1^2x_3$$
$$s.t. g_1(X) = -x_1 + 0.0193x_3 \leq 0$$
$$g_2(X) = -x_2 + 0.009543x_3 \leq 0$$
$$g_3(X) = -\pi x_3^2 x_4 - \frac{4}{3}\pi x_3^3 + 1296000 \leq 0$$
$$g_4(X) = x_4 - 240 \leq 0$$

The variable region is defined as:
Region I:

$$1 * 0.0625 \leq x_1, x_2 \leq 99 * 0.0625;$$
$$10 \leq x_3, x_4 \leq 200$$

A different version of pressure vessel design Problem is given in following manner:
Region II: In Region I, the upper bound of fourth decision variable(x_4) is 200.

Thus, the fourth constrain (g_4) is automatically satisfied. Thus, for more extensive study, the upper limit of fourth decision variable is set equal to 240. The range of the decision variable is taken as follows:

$$1 * 0.0625 \leq x_1, x_2 \leq 99 * 0.0625;$$
$$10 \leq x_3 \leq 200; 10 \leq x_4 \leq 240$$

In Madhvi et al. (2007), the problem of Region II is solved.

2.2   Welded Beam Design Problem

Rasgedelle and Philips (1976) have given the detailed introduction to welded beam design problem. The objective of the optimal design problem is to minimize the cost of fabrication. The decision variables are thickness of the weld (h), length of the welded

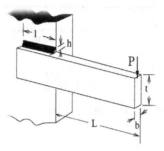

Fig. 2.   welded beam engineering design

joint (l), width of the beam (t) and thickness of the beam (b). Deb (1991) describes the welded beam engineering design problem in detail (Fig. 2).

The decision variables are given as vector $X = (x_1, x_2, x_3, x_4) = (h, l, t, b)$

The mathematical model of the problem is as follows:

$$\min f(X) = 1.10471x_1^2x_2 + 0.04811x_3x_4(14 + x_2)$$
$$s.t. g_1(X) = \tau(X) - \tau_{max} \leq 0$$
$$g_2(X) = \sigma(X) - \sigma_{max} \leq 0$$
$$g_3(X) = x_1 - x_4 \leq 0$$
$$g_4(X) = 0.125 - x_1 \leq 0$$
$$g_5(X) = \delta(X) - 0.25 \leq 0$$
$$g_6(X) = P - P_c(X) \leq 0$$
$$0.1 \leq x_1 \leq 2; 0.1 \leq x_2 \leq 10; 0.1 \leq x_3 \leq 10;$$
$$0.1 \leq x_4 \leq 2$$

Where τ is the shear stress in the weld, $\tau_{max}(= 13600\,psi)$ is the maximum allowable shear stress in the weld. σ is the normal stress in the beam, σ_{max} is the allowable normal stress for the beam material (=30000 psi), P_c the bar buckling load, P the load (=6000 lb), and δ the beam end deflection (Garg 2014).

$$\tau(X) = \sqrt{\tau_1^2 + 2\tau_1\tau_2\left(\frac{x_2}{2R}\right) + \tau_2^2}; \tau_1 = \frac{P}{\sqrt{2}x_1x_2}; \tau_2 = \frac{MR}{J}.$$

Where τ_1 and τ_2 are called primary and secondary stress.

$$M = P\left(L + \frac{x_2}{2}\right); J(X) = 2\left\{\sqrt{2}x_1x_2\left[\frac{x_2^4}{4} + \left(\frac{x_1 + x_2}{2}\right)^2\right]\right\}$$

Are known as moments and polar moment of inertia respectively while the other terms of the model are as follows:

$$R = \sqrt{\frac{x_2^2}{4} + \left(\frac{x_1 + x_3}{2}\right)^2}; \sigma(X) = \frac{6PL}{x_4 x_3^2}; \delta(X) = \frac{6PL^3}{Ex_3^3 x_4}$$

$$P_c(X) = \frac{4.013E\sqrt{\frac{x_3^2 x_4^6}{36}}}{L^2}\left(1 - \frac{x_3}{2L}\sqrt{\frac{E}{4G}}\right)$$

$$G = 12 \times 10^6 \text{psi}, E = 30 \times 10^6 \text{psi}, P = 6000lb, L = 14in$$

2.3   Tension String Design Problem

The objective function of the tension string design problem is to minimize the weight of string with respect to minimum shear stress, deflection etc. The three decision variables are: coil diameter (x_1), the wire diameter (x_2) and the number of active coil (x_3) (Fig. 3).

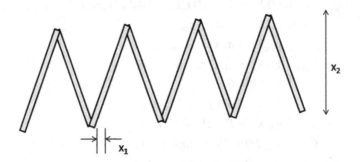

Fig. 3.   Tension String Design

The formulation of the mathematical problems is as follows:

$$\min f(X) = x_1^2 x_2 (2 + x_3)$$

$$s.t. g_1(X) = 1 - \frac{x_2^3 x_3}{71785 x_1^4} \leq 0$$

$$g_2(X) = \frac{4x_2^2 - x_1 x_2}{12566(x_2 x_1^3 - x_1^4)} + \frac{1}{5108 x_1^2} - 1 \leq 0$$

$$g_3(X) = 1 - \frac{140.45 x_1}{x_2^2 x_3} \leq 0$$

$$g_4(X) = \frac{x_1 + x_2}{1.5} - 1 \leq 0$$

$$0.05 \leq x_1 \leq 2; 0.25 \leq x_2 \leq 1.3; 2 \leq x_3 \leq 15$$

3 Constrained Laplacian Biogeography Based Optimization

Biogeography Based optimization is based on the study of extinction and rising of species. On the basis of the suitability index variables, species migrate from one habitat to another. Migration is the key operator which is done probabilistically using immigration and emigration rates.

$$\lambda_i = I(1 - \frac{k(i)}{n})$$

$$\mu_i = E(\frac{k(i)}{n})$$

μ_i: emigration rate for i^{th} species
λ_i: immigration rate for i^{th} species
n: population size.
I, E: maximum possible immigration and emigration rate
k(i): fitness rank of the ith species.

After initializing the random set of solutions, two main operators' migration and mutation are performed until some pre defined criteria is achieved.

Laplacian BBO (LX-BBO) (Garg and Deep 2015) is an improved version of BBO. LX-BBO was further extended for solving constrained optimization problems in Garg and Deep (2016). Deb's feasibility rules are used to handle the constrains (Deb 2000) for Constrained LX-BBO. Constrained LX-BBO is tested on CEC 2006 benchmark functions and five different constrained optimization problems having varying complexity. Two operators are designed in constrained LX-BBO as follows:

3.1 Migration Operator of Constrained LX-BBO

Laplace crossover of real coded Genetic Algorithm is incorporated in the migration operator of Constrained LX-BBO. Two habitats x_1 and x_2 are selected based upon immigration and emigration rates.

These habitats are used to generate two new habitats y_1 and y_2 which are given by the following equation:

$$y_1^i = x_1^i + \beta(x_1^i - x_2^i)$$

$$y_2^i = x_2^i + \beta(x_1^i - x_2^i)$$

Where random number β which follows Laplace distribution is generated given by the equation:

$$\beta = \begin{cases} a - b * log(u), u \leq 1/2 \\ a + b * log(u), u > 1/2 \end{cases}$$

$u_i \in [0,1]$ is a uniform random number. $a \in R$ is called location parameter and $b > 0$ is called scale parameter.

These two new habitats further give rise to a new habitat z which is used in further generations.

$$z = \gamma y_1^i + (1 - \gamma) y_2^i$$

Where $\gamma = \gamma_{min} + (\gamma_{max} - \gamma_{min})^{k^t}$

Where γ_{min} and γ_{max} are minimum and maximum values of γ. These two parameters lies in [0, 1] and t is the generation count number. $k < 1$ is a user defined parameter. The new solution replaces the worse solution.

3.2 Mutation Operator of Constrained LX-BBO

To improve the exploring ability of the algorithm, random mutation is used. Mutation operator is based on mutation probability. Mutation probability is the probability of selecting an individual to go through mutation process. The mutation rate (0.005) is set same as for the unconstrained Laplacian BBO. On the basis of mutation probability, the solution is replaced with a randomly generated solution z'.

$$z' = x_{min} + rand * (x_{max} - x_{min})$$

x_{min} is the minimum bound of decision variable x.
x_{max} is the maximum bound of decision variable x.
rand is the uniformly generated random number in [0,1].

3.3 Deb's Constrained Handling Rules

In Constrained Laplacain Biogeography Based Optimization, Deb's constraint handling rules are applied to check the constraints of the optimization problem. In Deb (2000), a fitness function based on Deb's feasibility rules is defined as follows:

$$fitness(x_i) = \begin{cases} f(x_i), & \text{if } x_i \text{ is feasible} \\ f_{worst} + \sum_{j=1}^{p} |\emptyset_j(x_i)|, & \text{if} x_i \text{ is not feasible} \end{cases}$$

Where fitness (x_i) is the fitness value of the individual x_i. $f(x_i)$ is the objective function value of the solution x_i. f_{worst} is the worst value of all the functional value obtained from all feasible solutions. $\emptyset_j(x_i)$ is the value of left hand side of inequality constraint (equality constraint is also converted into inequality constraint) of solution x_i. Total number of constraints is p.

According to Deb's technique one solution is preferred over the other one based on these three conditions:

(1) A feasible solution is preferred over infeasible solution.
(2) If two solutions are feasible, then the solution with better objective function value is preferred.
(3) If two solutions are infeasible, then the solution having less constraint violation will be preferred.

The use of constraint violation in Deb's constrained feasibility rules allows pushing infeasible solution into feasible solution. These rules are implemented to check which of the solution will go in the future generation after migration and mutation operator.

4 Numerical Results

All three structural engineering design problems are solved by Constrained Laplacian BBO. The population size is set equal to 30, as all the problems are of three or four variable problems.30 independent runs are performed to check the robustness of the algorithm. Random population is generated using the clock time seed.

Followed by the criteria set by Garg (2014), the best solution among all the 30 runs is reported. Tables 1, 2 and 3 record the comparitive results of all the three problems solved by constrained LX-BBO and all the other algorithms reported in Garg (2014).

In Table 1, the best solution of pressure vessel design problem out of all the 30 independent runs is given. Sandgren (1988); Zhang and Wang (1993); Kannan and Kramer (1994), Deb and Gene (1997), Coello (2000); Coello and Montes (2002); Hu et al. (2003); Gandomi et al. (2013); He et al. (2004); Lee and Geem (2005); He and Wang (2007); Montes et al. (2007); Montes and Coello (2008); Cagnina et al. (2008); Kaveh and Talatahari (2009); Kaveh and Talatahari (2010); Coelho (2010) and Akay and Karaboga (2012) are some of the well known attempts for solving pressure vessel design problems (for Region I). pressure vessel design problems (for Region II) are solved in Hedar and Fukushima (2006); Dimopoulos (2007); Mahdavi et al. (2007) and Gandomi et al. (2011). In Garg (2014), all these results are taken to compare the results obtained by ABC. In the present study, we are comparing with all the above results. Although Constrained LX-BBO has obtained the minimum objective function value for all the other reported results of other heuristic methods. But, ABC has produced better minimum value for this problem only. However the difference between the results of two algorithms is very small.

In Table 2, the best solution of welded beam design problem out all the 30 runs is presented. This optimization problem is solved by many researchers: Ragsdell and Phillips (1976); Deb (1991), Rao (1996); Deb (2000); Ray and Liew (2003); Lee and Geem (2005); Hwang and He (2006) and Mehta and Dasgupta (2012). In Garg (2014)

Table 1. Comparison of results of pressure vessel design problem obtained by Constrained LX-BBO and other algorithms

Region	S. no.	Method	Design variables				f(X)
I	1	Sandgren (1988)	1.1250	0.6250	47.7000	117.7010	8129.1036
	2	Zhang and Wang (1993)	1.1250	0.6250	58.2900	43.6930	7197.7000
	3	Kannan and Kramer (1994)	1.1250	0.6250	58.2910	43.6900	7198.0428
	4	Deb and Gene (1997)	0.9375	0.5000	48.3290	112.6790	6410.3811
	5	Coello (2000)	0.8125	0.4375	40.3239	200.0000	6288.7445
	6	Coello and Montes (2002)	0.8125	0.4375	42.0974	176.6541	6059.9460
	7	Hu et al. (2003)	0.8125	0.4375	42.0985	176.6366	6059.1313
	8	Gandomi et al. (2003)	0.8125	0.4375	42.0984	176.6366	6059.7143
	9	He et al. (2004)	0.8125	0.4375	42.0984	176.6366	6059.7143
	10	Lee and Geem (2005)	1.1250	0.6250	58.2789	43.7549	7198.4330
	11	He and Wang (2007)	0.8125	0.4375	42.0913	176.7465	6061.0777
	12	Montes et al. (2007)	0.8125	0.4375	42.0984	176.6360	6059.7017
	13	Montes and Coello (2008)	0.8125	0.4375	42.0981	176.6405	6059.7456
	14	Cagnina et al. (2008)	0.8125	0.4375	42.0984	176.6366	6059.7143
	15	Kaveh and Talatahari (2009)	0.8125	0.4375	42.1036	176.5732	6059.0925
	16	Kaveh and Talatahari (2010)	0.8125	0.4375	42.0984	176.6378	6059.7258
	17	Coelho (2010)	0.8125	0.4375	42.0984	176.6372	6059.7208
	18	Akay and Karaboga (2012)	0.8125	0.4375	42.0984	176.6366	6059.7143
	19	Garg (2014)	0.7782	0.3847	40.3211	199.9802	5885.4033
	20	**Present study (C-LX-BBO)**	**0.7940**	**0.3925**	**41.12337**	**189.1067**	**5915.4130**
II	1	Dimopoulos (2007)	0.7500	0.3750	38.8601	221.3655	5850.3831
	2	Mahdavi et al. (2007)	0.7500	0.3750	38.8601	221.3655	5849.7617
	3	Hedar and Fukushima (2006)	0.7683	0.3798	39.8096	207.2256	5868.7648
	4	Gandomi et al. (2011)	0.7500	0.3750	38.8601	221.3655	5850.3831
	5	Garg (2014)	0.7276	0.3597	37.6991	239.9998	5804.4487
	6	**Present study (C-LX-BBO)**	**0.7451**	**0.3689**	**38.5727**	**225.8359**	**5837.1528**

all these results produced are compared. Garg (2014) have outperformed all the mentioned results. However, Constrained LX-BBO has minimum objective function value than Garg (2014).

In Table 3, the results of tension string design problem is given. Belegundu (1982); Arora (1989); Coello (2000); Ray and Saini (2001); Coello and Montes (2002);

Table 2. Comparison of results of welded beam design problem obtained by Constrained LX-BBO and other algorithms

	Method	Design variables				f(X)
1	Ragsdell and Phillips (1976)	0.2455	6.1960	8.2730	0.2455	2.3859
2	Rao (Rao 1996)	0.2455	6.1960	8.2730	0.2455	2.3860
3	Deb (1991)	0.2489	6.1730	8.1789	0.2533	2.4331
4	Deb (2000)	NA	NA	NA	NA	2.3812
5	Ray and Liew (2003)	0.2444	6.2380	8.2886	0.2446	2.3854
6	Lee and Geem (2005)	0.2442	6.2231	8.2915	0. 2443	2.3800
7	Hwang and He (2006)	0.2231	1.5815	12.8468	0.2245	2.2500
8	Mehta and Dasgupta (2012)	0.2444	6.2186	8.2915	0.2444	2.3811
9	Garg (2014)	0.2444	6.2177	8.2916	0.2444	2.3810
10	**Present study (C-LX-BBO)**	**0.2411**	**3.0829**	**8.3181**	**0.24419**	**1.8673**

Table 3. Comparison of results of tension string design problem obtained by Constrained LX-BBO and other algorithms

	Method	Design variables			f(X)
1	Belegundu (1982)	0.0500	0.3159	14.250	0.0128
2	Arora (1989)	0.0534	0.3992	9.1854	0.0127
3	Coello (2000)	0.0515	0.3517	11.6322	0.0127
4	Ray and Saini (2001)	0.0504	0.3215	13.9799	0.0131
5	Coello and Montes (2002	0.0520	0.3640	10.8905	0.0127
6	Ray and Liew (2003)	0.0522	0.3682	10.6484	0.0127
7	Hu et al. (2003)	0.0515	0.3514	11.6087	0.0127
8	He et al. (2004)	0.0517	0.3567	11.2871	0.0127
9	Hedar and Fukushima (2006)	0.0517	0.3580	11.2139	0.0127
10	Raj et al. (2005)	0.0539	0.4113	8.6844	0.0127
11	Tsai (Tsai 2005)	0.0517	0.3567	11.2890	0.0127
12	Mahdavi et al. (2007)	0.0512	0.3499	12.0764	0.0127
13	Montes et al. (2007)	0.0517	0.3567	11.2905	0.0127
14	He and Wang (2007)	0.0517	0.3576	11.2445	0.0127
15	Cagnina et al. (2008)	0.0516	0.3542	11.4387	0.0127
16	Zhang et al. (2008)	0.0517	0.3567	11.2890	0.0127
17	Montes and Coello (2008)	0.0516	0.3554	11.3979	0.0127
18	Omran and Salman (2009)	0.0517	0.3566	11.2965	0.0127
19	Keveh and Talatahari (2010)	0.0519	0.3615	11.0000	0.0126
20	Coelho (2010)	0.0515	0.3525	11.5389	0.0127
21	Akay and Karaboga (2012)	0.0517	0.3582	11.2038	0.0127
22	Garg (2014)	0.0517	0.3567	11.2888	0.0127
23	**Present Study (C-LX-BBO)**	**0.05505**	**0.4774**	**6.0586**	**0.0117**

Table 4. Best, worst, std dev, average and median of objective function value obtained from 30 runs using Constrained LX-BBO

Problem	Best	Worst	Std Dev	Average	Median
Pressure vessel design problem (region 1)	5915.4130	7580.6050	449.3805	6641.2334	6559.5687
Pressure vessel design problem (region 2)	5837.1528	7358.3346	406.6185	6536.9169	6508.7070
Welded beam design problem	1.8673	3.2298	0.3715	2.2352	2.1080
Tension string design problem	0.0117	0.01795	0.0018	0.0147	0.0151

Ray and Liew (2003); Hu et al. (2003); He et al. (2004); Hedar and Fukushima (2006); Raj et al. (2005); Tsai (2005); Mahdavi et al. (2007); Montes et al. (2007); He and Wang (2007); Cagnina et al. (2008); Zhang et al. (2008); Montes and Coello (2008); Omran and Salman (2009); Keveh and Talatahari (2010); Coelho (2010) and Akay and Karaboga (2012) are some honorable mentions used to solve tension string design problem. Garg (2014) has the best result produced so far. ($f(X) = 0.0127$). Constrained LX-BBO has obtained $f(X) = 0.0117$. Thus, constrained LX-BBO has emerged as a winner for solving tension string design problems.

In Table 4, the statistical results of all the three design problems solved by Constrained LX-BBO are given. Best value represents the minimum function value obtained by Constrained LX-BBO out of 30 runs. Worst value is the maximum value obtained from 30 runs. Std is the standard deviation of all the functional values obtained in 30 runs. Average and Mean value of 30 functional values are also given in Table 4.

To check the efficiency of the Constrained LX-BBO on design problems, convergence graph of all the problems are drawn. Horizontal axis gives the number of iterations and vertical axis gives the corresponding function value obtained by using Constrained LX-BBO. Convergence graphs of all the design problems are given in Fig. 4.

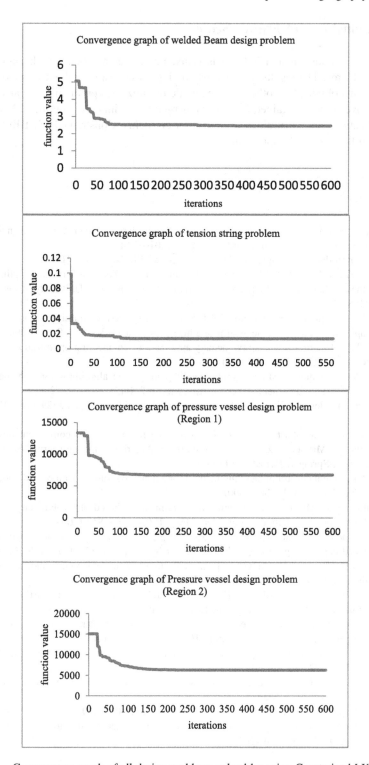

Fig. 4. Convergence graph of all design problems solved by using Constrained LX-BBO.

5 Conclusion and Future Scope

In this paper, constrained LX-BBO is tested upon standard benchmark engineering design problems. The results obtained by solving these design problems are compared by the results obtained by other meta-heuristic optimization problems. The analysis of results proves that constrained LX-BBO emerged as a winner among them. Many real life applications are constrained optimization problems. Constrained LX-BBO can be applied to those real life applications.

References

Akay, B., Karaboga, D.: Artificial bee colony algorithm for large-scale problems and engineering design optimization. J. Intell. Manufact. **23**(4), 1001–1014 (2012)

Arora, J.S.: Introduction to Optimum Design. McGraw-Hill, New York (1989)

Belegundu, A.D.: A Study of Mathematical Programming Methods for Structural Optimization, Ph.D. thesis, Department of Civil and Environmental Engineering, University of Iowa, Iowa, USA (1982)

Cagnina, L.C., Esquivel, S.C., Coello, C.A.C.: Solving engineering optimization problems with the simple constrained particle swarm optimizer. Informatica **32**(3), 319–326 (2008)

Coello, C.A.C.: Use of a self-adaptive penalty approach for engineering optimization problems. Comput. Ind. **41**(2), 113–127 (2000)

Coello, C.A.C., Montes, E.M.: Constraint-handling in genetic algorithms through the use of dominance-based tournament selection. Adv. Eng. Inf. **16**(3), 193–203 (2002)

Deb, K.: Optimal design of a welded beam via genetic algorithms. AIAA J. **29**(11), 2013–2015 (1991)

Deb, K.: GeneAS: a robust optimal design technique for mechanical component design. In: Dasgupta, D., Michalewicz, Z. (eds.) Evolutionary Algorithms in Engineering Applications, pp. 497–514. Springer, Heidelberg (1997)

Deb, K.: An efficient constraint handling method for genetic algorithms. Comput. Methods Appl. Mech. Eng. **186**(2), 311–338 (2000)

Dimopoulos, G.G.: Mixed-variable engineering optimization based on evolutionary and social metaphors. Comput. Methods Appl. Mech. Eng. **196**(4), 803–817 (2007)

dos Santos Coelho, L.: Gaussian quantum-behaved particle swarm optimization approaches for constrained engineering design problems. Expert Syst. Appl. **37**(2), 1676–1683 (2010)

Gandomi, A.H., Yang, X.S., Alavi, A.H.: Mixed variable structural optimization using firefly algorithm. Comput. Struct. **89**(23), 2325–2336 (2011)

Gandomi, A.H., Yang, X.S., Alavi, A.H.: Cuckoo search algorithm: a metaheuristic approach to solve structural optimization problems. Eng. Comput. **29**(1), 17–35 (2013)

Garg, H.: Solving structural engineering design optimization problems using an artificial bee colony algorithm. J. Ind. Manag. Optim. **10**(3), 777–794 (2014)

Garg, H.: A hybrid PSO-GA algorithm for constrained optimization problems. Appl. Math. Comput. **274**, 292–305 (2016)

Garg, V., Deep, K.: Constrained laplacian biogeography based optimization. Communicated in Int. J. Syst. Assur. Eng. Manag.

Garg, V., Deep, K.: Performance of Laplacian Biogeography-Based Optimization Algorithm on CEC 2014 continuous optimization benchmarks and camera calibration problem. Swarm Evol. Comput. **27**, 132–144 (2015)

He, Q., Wang, L.: An effective co-evolutionary particle swarm optimization for constrained engineering design problems. Eng. Appl. Artif. Intell. **20**(1), 89–99 (2007)

He, S., Prempain, E., Wu, Q.H.: An improved particle swarm optimizer for mechanical design optimization problems. Eng. Optim. **36**(5), 585–605 (2004)

Hedar, A.R., Fukushima, M.: Derivative-free filter simulated annealing method for constrained continuous global optimization. J. Global Optim. **35**(4), 521–549 (2006)

Hu, X., Eberhart, R.C., Shi, Y.: Engineering optimization with particle swarm. In: Proceedings of the 2003 IEEE Swarm Intelligence Symposium, SIS 2003, pp. 53–57. IEEE, April 2003

Hwang, S.F., He, R.S.: A hybrid real-parameter genetic algorithm for function optimization. Adv. Eng. Inf. **20**(1), 7–21 (2006)

Kannan, B.K., Kramer, S.N.: An augmented Lagrange multiplier based method for mixed integer discrete continuous optimization and its applications to mechanical design. J. Mech. Des. **116** (2), 405–411 (1994)

Kaveh, A., Talatahari, S.: Engineering optimization with hybrid particle swarm and ant colony optimization. Asian, J. Civil Eng. **10**(6), 611–628 (2009)

Kaveh, A., Talatahari, S.: An improved ant colony optimization for constrained engineering design problems. Eng. Comput. **27**(1), 155–182 (2010)

Lee, K.S., Geem, Z.W.: A new meta-heuristic algorithm for continuous engineering optimization: harmony search theory and practice. Comput. Methods Appl. Mech. Eng. **194**(36), 3902–3933 (2005)

Mahdavi, M., Fesanghary, M., Damangir, E.: An improved harmony search algorithm for solving optimization problems. Appl. Math. Comput. **188**, 1567–1579 (2007)

Mehta, V.K., Dasgupta, B.: A constrained optimization algorithm based on the simplex search method. Eng. Optim. **44**, 537–550 (2012)

Montes, E.M., Coello, C.A.C.: An empirical study about the usefulness of evolution strategies to solve constrained optimization problems. Int. J. General Syst. **37**, 443–473 (2008)

Montes, E.M., Coello, C.A.C., Reyes, J.V., Davila, L.M.: Multiple trial vectors in differential evolution for engineering design. Eng. Optim. **39**, 567–589 (2007)

Omran, M.G.H., Salman, A.: Constrained optimization using CODEQ. Chaos, Solitons Fractals **42**(2009), 662–668 (2009)

Ragsdell, K.M., Phillips, D.T.: Optimal design of a class of welded structures using geometric programming. ASME J. Eng. Ind. **98**, 1021–1025 (1976)

Raj, K.H., Sharma, R.S., Mishra, G.S., Dua, A., Patvardhan, C.: An evolutionary computational technique for constrained optimisation in engineering design. J. Inst. Eng. India Part Me Mech. Eng. Div. **86**, 121–128 (2005)

Rao, S.S.: Engineering Optimization: Theory and Practice, 3rd edn. Wiley (1996)

Ray, T., Saini, P.: Engineering design optimization using a swarm with an intelligent information sharing among individuals. Eng. Optim. **33**, 735–748 (2001)

Ray, T., Liew, K.M.: Society and civilization: An optimization algorithm based on the simulation of social behavior. IEEE Trans. Evol. Comput. **7**, 386–396 (2003)

Sandgren, E.: Integer and discrete programming in mechanical design. In: Proceedings of the ASME Design Technology Conference, F.L. Kissimine, pp. 95–105 (1988)

Simon, D.: Biogeography-based optimization. IEEE Trans. Evol.Comput. **12**(6), 702–713 (2008)

Tsai, J.F.: Global optimization of nonlinear fractional programming problems in engineering design. Eng. Optim. **37**(4), 399–409 (2005)

Zhang, C., Wang, H.P.: Mixed-discrete nonlinear optimization with simulated annealing. Eng. Optim. **21**(4), 277–291 (1993)

Zhang, M., Luo, W., Wang, X.: Differential evolution with dynamic stochastic selection for constrained optimization. Inf. Sci. **178**(15), 3043–3074 (2008)

Soft Computing Based Software Testing – A Concise Travelogue

Deepak Sharma[✉] and Pravin Chandra

University School of Information and Communication Technology,
Guru Gobind Singh Indraprastha University, Dwarka 110078, Delhi, India
deepakdixit151@gmail.com

Abstract. Soft computing is an accumulation of procedures, which intend to adventure resistance for the defect, deception, ambiguity and incomplete truth to accomplish tractability, strength, and low arrangement cost. In this paper, a comprehensive overview of software testing based on soft computing is presented. In this survey, we try to elaborate some problems of software engineering specifically software testing and their solutions, which are based on soft computing approaches. The paper presents an overview of the usage of soft computing techniques including Neural Networks, Fuzzy Logic, Ant Colony Optimization, and Particle Swarm Optimization and Genetic algorithm in software testing.

Keywords: Soft computing · Software testing · Neural networks · Fuzzy logic · Genetic programming · Ant colony optimization · Particle swarm optimization

1 Introduction

Soft computing approaches (SCA) techniques are powerful methods that have been shown useful in solution of difficult problem. In the mid of 1990 s, SCA comes into a precise region of research practice in Computer Science [1]. Previous methodologies of computation could design and evaluate absolutely for basic and easy system frameworks. More unpredictable frameworks emerging in science, pharmaceutical, humanities, administration sciences, science related to management and comparative fields regularly stayed unmanageable to routine numerical and scientific techniques [2, 19]. n any case, it ought to be known that straight forwardness and unpredictability of system frameworks are relative, and numerous customary scientific models have been both testing and gainful. Processing by SCA manages imprecision, vulnerability, fractional truth, and estimate to accomplish practicability, power, and low evaluation cost. Accordingly, it frames the premise of many techniques related to machine learning. Late patterns have a tendency to include developmental and insight based calculations or algorithms and bio-inspired computation as suggested by [2, 3].

There are primary contrasts between SCA and probability. Probability is utilized when we do not have enough data to determine on an issue yet processing by SCA is utilized when we do not have enough data about the issue itself. These types of issues

© Springer Nature Singapore Pte Ltd. 2017
K. Deep et al. (eds.), *Proceedings of Sixth International Conference on Soft Computing for Problem Solving*, Advances in Intelligent Systems and Computing 547,
DOI 10.1007/978-981-10-3325-4_22

begin in the human brain with each of its questions, individuality and feelings; an illustration can be deciding a suitable temperature for space in a room to make individuals feel relaxed. In this paper, we try to explain application areas of SC approaches for solving different issues about software testing (ST). Moreover, research directions in which efficient and effective use of SCA is made in ST. Adjoining Fig. 1 for ST illustrates that software requirements specifications will be tested with design and code independently then it make error free software.

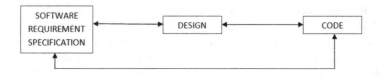

Fig. 1. Software testing

1.1   Conventional Testing Techniques

In software engineering, SC is the utilization of answers for computationally hard undertakings, for example, the result of NP-complete issues, for which there is an unknown sequence of steps that can figure approximately correct result in a polynomial period. SCA is a contraposition and supplementary to typical estimation in that, dissimilar to typical estimation it is receptive of deception, instability, incomplete accuracy, and estimation. In actuality, a good example for SC is the human brain. Gray Box (Functionality with code or vice versa) testing is a way to analyze the application with restricted learning about the inner more workings of an application. Structural testing is the deep analysis of the inner working of logic and design of the code. Structural testing is also called code testing, glass testing or white box or open box testing [40] The process of testing the functionality of software or an application is functional testing.

1.2   Testing Levels

Testing levels are essential to analyze the areas, which are missing, and to prevent the overlapping and recurrence between the phases of system development life cycle. In the model of software development life cycle models define phases from requirement gathering to testing and deployment. Each phase has different experiences of testing. Hence, there are different levels of testing. The different levels of testing are: first level is unit testing which means, of tests that confirms the usefulness and correctness of a particular area of programming code, normally at the activity level. In an O-O (object oriented) situation, this is commonly at the class level, and the minimal unit tests incorporate testing of the constructors and destructors [4]. Second is Integration testing which attempts to uncover areas in the interfaces and connection between incorporated segments (modules). According to [5], a bigger collection of ST segments, comparable to components of the constructive design layout has integrated and tested until the software activity works as a system. Third is Component interface testing in which the demonstration of portion interface testing can be used to check the correctness of passed data between distinctive units, or subsystem portions, prior to complete joining evaluation between those units [6]. The information being run can be considered as "message bundle packets" and the dimension or data sorts can be tested, for information delivered from one unit, and strive for authenticity. One decision for this testing is to keep an alternate chunk record file of information things being run, consistently with a period stamp chunked to allow examination of a large number of instances of information run between units for very long time. Tests look at the treatment of some convincing data qualities although other interface valuable variables are passed as orderly quality values. Unusual statistical values in an interface support sudden execution in the adjacent nearing unit. Segment (part) interface testing is an assortment of dark-box testing, [6] with the consideration on the information values passed, essentially the related exercises of a subsystem part testing. Fourth is System testing in which a complete software application environment of structure testing in a circumstance that impersonates genuine use, for example, associating with a database, utilizing system interchanges, or interfacing with hardware equipment's, applications, or frameworks if suitable. Quality analyst groups perform it. End-to-end testing, tests a completely organized system to watch that it meets its necessities or not (Fig. 2).

Fig. 2. Word cloud generated using the responses to our informal survey

2 Soft Computing Techniques for Software Testing

In this, we will discuss some of the basic Soft computing approaches and their current uses for testing's. This section will also include many other related techniques for the better efficiency and effectiveness. The main idea behind using Soft Computing in current software testing is to increase the quality of software testing and to make these testing automated. Before discussing research directions, it is important to know that what we resolve in our research or what we want to "optimize." Optimization means the best possible solution to a particular problem. According to Harman [7] a search issue is an issue where about perfect or almost perfect results looked for in a space of exploration of adversary results, managed by a fitness function that recognizes better and more terrible results. When we talk about ST Firstly, we should know that testing software is not just to test the application or a system, but we should test everything of that particular software or web application. It implies that testing phase, level, stage, module from information gathering and enterprise engineering to support and reengineering (Table 1).

Table 1. Comparison between conventional testing techniques

Sr. no.	Criteria	Black box testing	White box testing	Gray box testing
1	Level of inner knowledge required for testing	No	Yes	Yes
2	Known as	Functional, data driven and closed box	Clear box, structural or code based	Translucent/Glass testing
3	Users	End user's, tester's and developer's	Software test professional's and developer's	Limited internal know-how for End user's, tester's and developer's
4	Level of internal knowledge required for design test cases	Depends on outer desires - Internal conduct of the application is unknown	The software tester can plan test information in like manner	High level diagrams of database and DFD's
5	Performance	Less time taken and exhaustive	Mostly exhaustive and more time taken	Partly time taken and exhaustive
6	Suitability of algorithm testing	No	Yes	No
7	Level of testing	Only done by trial and error method	Data field/area/block and Internal boundaries can be better analyzed	Data field/area/block and Internal boundaries can be tested, if known
8	Granularity	Low	High	Medium
9	Testing method suitability	Functional or business domain	All methods are suitable in it.	Functional/business domain testing bit in depth.

2.1 Genetic Programming and Algorithm Based Software Testing

Genetic Programming (GP) is a developmental optimization technique in artificial intelligence, which is an algorithm-based approach activated by biological development to search programs of computers that execute a user-defined task. GP is a set of steps and a fitness function to measure how well a computer has performed a task. This technique has been applied successfully to a large number of fatigue problems such as automatic design, pattern recognition and test data generation in ST. Genetic algorithm helps in generation of basis test paths automatically [8], including problems like test sequence generation [9], Test case generation and optimization [10] and test data generation [11, 39]. In recent studies we found that the regeneration genetic algorithm (GA) which is effective and lightweight for coverage-oriented software test data generation is proposed [36].

2.2   Neural Network Based Software Testing

The combination of soft computing approach like a neural network with Software Testing gives many solutions for different types of issues like software reusability etc. In 2011, an NN miniature was proposed by Yogesh Singh and others for software reusability [38]. This form of model miniature can be utilized to model any arbitrarymapping of input and output and are fit for approximating any quantifiable capacity function. Subsequently, NN ought to have the capacity to model the usefulness of the system reusability moreover. In a wide sense, the neural system itself is a miniature in light of the fact that the topology and exchange elements of the hubs are generally detailed to coordinate the present issue [12, 13]

The Neural Network miniature is further approved on the two acceptance sets produced by the fuzzy miniature. For this type work, backpropagation algorithm is utilized for training prominently.

The Backpropagation algorithm is the leading preparation/training, method for multilayer systems networks [14–17]. The miniature model is prepared with four inputs in particular: Changeability, Interface Complexity, Understandability of Software system and Documentation Quality with one output i.e. software reusability. Baysian regularization based training algorithm has been recommended for training backpropagation networks. The Fig. 3, illustrates the research directions of software testing with applying the neural network [31, 38].

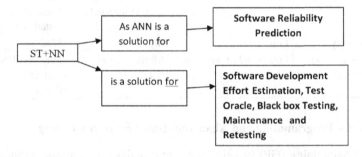

Fig. 3. Shows Solutions of Software testing with neural network

2.3   Ant Colony Based Software Testing

Ant colony optimization is best for test suite size and test suitecoverage probleminstead of a greedy algorithm. Herein, the attempt is to make best possible solutions for automation in test cases and test sequence generation problem [9, 18]. Ant colony optimization in software testing has been utilized test sequence generation, test data generation in structural testing and other relevant problems [18, 19, 36].

2.4 Particle Swarm Optimization Based Software Testing

Particle swarm optimization (PSO) is an approach, which is used to solve problems in software testing like testlet-based test generation system. This can be performed by discrete PSO. The intention behind for using this technique is to solve the problem in a dynamic way of computation. With the help of simplified Swarm Optimization (SSO) technique we can generate combinatorial test cases for automated GUI black box testing. [25, 32–34]. Swarm optimization is important in some testing approaches for cloud applications [37] and verification of intent test cases [35].

3 Recent Trends in Fuzzy Approach Based Software Testing

In a recent survey on Fuzzy approach based software testing (FL+ST) it was found that software quality estimation is an important topic in software engineering [27, 28]. To fulfill the exploration and research in the quality field is an important issue. Here the quality of software means the direction to attributes like the utility, extensibility, and complexity.

Table 2. Summary of SCA proposals in software engineering

Existing work	Problems in software testing areas	Solutions proposed
Schumann and Nelson [20]	Verification and Validation	NN
Aggarwal et al. [21]	Resource estimation	NN and training algorithms
Gokce et al. [22]	Test case prioritization based on coverage criteria	NN clustering
	Regression testing	
Engel and Last [23]	Verification and Validation	FL
	Testing risks and cost	
Lokasyuk et al. [24]	Software retesting estimation	NN: Artificial NN
Tsai et al. [25]	Test generation	Discrete PSO and Partial ontology
		Dynamic testlet-based computerized approach
Kumar and Singh [26]	Time estimation	ANN, FL, Neuro-Fuzzy, SVM
Pizzi, N.J. [27]	Software quality estimation	FL: Fuzzy classifier approach (FCA)
	Software metrics	
Tyagi and Sharma [28]	Reliability estimation	FL, Neuro-Fuzzy
Bhasin and Khanna [29]	Functional or black box testing	NN
	Test case design	
	Test case prioritization	
Ferrer et al. [9]	Black box testing or Functional testing:	GA and ACO
	Test sequence generation	Genetic test sequence generator
		Memory operator
		ACO test sequence
Yang et al. [36]	Test data generation	GA

On the other hand, the general nature of object, which is an object of software may show itself in ways that the basic understanding of measurements neglects to distinguish.

A superior methodology is to decide the best, conceivably non-direct, a subset of numerous software measurements for precisely assessing the quality of software. This technique is primarily utilized for classification that is, to decide a mapping from a group of measurements of software to a group of class names and labels to depicting the quality of software [30]. FL approach is used in Software development life cycle (SDLC) phase wise predicting defects of software. For further references on ST+FL: - [23, 26, 28, 33, 35, 38] The Fig. 4, illustrates the research directions of software testing with applying the fuzzy logic (Table 2).

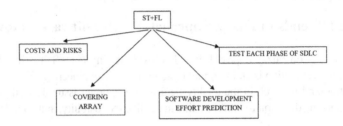

Fig. 4. Problems solved by fuzzy logic in software testing

4   Conclusion and Future Research Directions

This survey article presents a general overview of SC approaches. Research issues have been highlighted. The paper fulfills several goals of ST. It includes a conceptual discussion of all such approaches/trends/techniques/strategies/methodologies, looking at different criteria of classification and earlier efforts to develop categories for effective and efficient ST for building models of error, fault and failure-proneness. This survey may become the basis to develop a proposal for a new anatomy, which is a helpful as a conceptual tool to both understand and organize the existing work, and to identify possible areas for future research.

The work also includes a review of the literature in the area, starting from the pioneering works in soft computing with software testing. The reviewed papers are organized according to the new anatomy, and the main characteristics of the techniques engaged, as well as the application problems, future directions and results obtained, are presented. The important discussion of each approach concludes with the utilization of main features and a year-wise list of issues and proposed solutions about the efficacy of the corresponding methods presented.

References

1. Zadeh, L.A.: Fuzzy logic, neural networks, and soft computing. Commun. ACM **37**(3), 77–85 (1994)
2. Yang, X.S., Cui, Z., Xiao, R., Gandomi, A.H., Karamanoglu, M. (eds.): Swarm Intelligence and Bio-Inspired Computation: Theory and Applications. Newnes, Oxford (2013)
3. Chaturvedi, D.K.: Soft Computing: Techniques and its Applications in Electrical Engineering. SCI, vol. 103. Springer, Heidelberg (2008)

4. Binder, R.V.: Testing Object-Oriented Systems: Objects, Patterns, and Tools (1999)
5. Beizer, B.: Software Testing Techniques (1990)
6. Clapp, J.A.: Software Quality Control, Error Analysis, and Testing. William Andrew (1995)
7. Harman, M., Mansouri, S.A., Zhang, Y.: Search-based software engineering: Trends, techniques and applications. ACM Comput. Surv. (CSUR) **45**(1), 11 (2012)
8. Ghiduk, A.S.: Automatic generation of basis test paths using variable length genetic algorithm. Inf. Process. Lett. **114**(6), 304–316 (2014)
9. Ferrer, J., Kruse, P.M., Chicano, F., Alba, E.: Search based algorithms for test sequence generation in functional testing. Inf. Softw. Technol. **58**, 419–432 (2015)
10. Khurana, N., Chillar, R.S.: Test Case Generation and Optimization using UML Models and Genetic Algorithm. Procedia Comput. Sci. **57**, 996–1004 (2015)
11. Varshney, S., Mehrotra, M.: Search based software test data generation for structural testing: a perspective. ACM SIGSOFT Softw. Eng. Not. **38**(4), 1–6 (2013)
12. Fausett, L.: Fundamentals of Neural Networks: Architectures, Algorithms, and Applications. Prentice-Hall Inc., Upper Saddle River (1994)
13. Haykin, S.: Neural Network, A Comprehensive Foundation. Prentice Hall India, New Delhi (2003)
14. Aggarwal, K.K., Singh, Y., Kaur, A., Malhotra, R.: Application of artificial neural network for predicting maintainability using object-oriented metrics'. Trans. Eng. Comput. Technol. **15**, 285–289 (2006)
15. Aggarwal, K.K., Singh, Y., Kaur, A., Sangwan, O.P.: A neural net based approach to test oracle. ACM SIGSOFT Softw. Eng. Not. **29**(3), 1–6 (2004)
16. Singh, Y., Bhatia, P.K., Kaur, A., Sangwan, O.: Application of neural networks in software engineering: a review. In: Prasad, S.K., Routray, S., Khurana, R., Sahni, S. (eds.) ICISTM 2009. CCIS, vol. 31, pp. 128–137. Springer, Heidelberg (2009). doi: 10.1007/978-3-642-00405-6_17
17. Singh, Y., Bhatia, P.K., Sangwan, O.: ANN model for predicting software function point metric. ACM SIGSOFT Softw. Eng. Not. **34**(1), 1–4 (2009)
18. Pedemonte, M., Nesmachnow, S., Cancela, H.: A survey on parallel ant colony optimization. Appl. Soft Comput. **11**(8), 5181–5197 (2011)
19. Mao, C., Xiao, L., Yu, X., Chen, J.: Adapting ant colony optimization to generate test data for software structural testing. Swarm Evol. Comput. **20**, 23–36 (2015)
20. Schumann, J., Nelson, S.: Toward V&V of neural network based controllers. In: Proceedings of the First Workshop on Self-Healing Systems, pp. 67–72 ACM (2002)
21. Aggarwal, K.K., Singh, Y., Chandra, P., Puri, M.: Evaluation of various training algorithms in a neural network model for software engineering applications. ACM SIGSOFT Softw. Eng. Not. **30**(4), 1–4 (2005)
22. Gökçe, N., Eminov, M., Belli, F.: Coverage-based, prioritized testing using neural network clustering. In: Levi, A., Savaş, E., Yenigün, H., Balcısoy, S., Saygın, Y. (eds.) ISCIS 2006. LNCS, vol. 4263, pp. 1060–1071. Springer, Heidelberg (2006). doi:10.1007/11902140_110
23. Engel, A., Last, M.: Modeling software testing costs and risks using fuzzy logic paradigm. J. Syst. Softw. **80**(6), 817–835 (2007)
24. Lokasyuk, V.M., Pomorova, O.V., Govorushchenko, T.O.: Neural nets method for estimation of the software retesting necessity. In: Proceedings of the 2008 International Workshop on Software Engineering in East and South Europe, pp. 9–14. ACM (2008)
25. Tsai, K.H., Wang, T.I., Hsieh, T.C., Chiu, T.K., Lee, M.C.: Dynamic computerized testlet-based test generation system by discrete PSO with partial course ontology. Expert Syst. Appl. **37**(1), 774–786 (2010)

26. Kumar, P., Singh, Y.: Assessment of software testing time using soft computing techniques. ACM SIGSOFT Softw. Eng. Not. **37**(1), 1–6 (2012)
27. Pizzi, N.J.: A fuzzy classifier approach to estimating software quality. Inf. Sci. **241**, 1–11 (2013)
28. Tyagi, K., Sharma, A.: An adaptive neuro fuzzy model for estimating the reliability of component-based software systems. Appl. Comput. Inf. **10**(1), 38–51 (2014)
29. Bhasin, H., Khanna, E.: Neural network based black box testing. ACM SIGSOFT Softw. Eng. Not. **39**(2), 1–6 (2014)
30. Wang, J., Lin, Y.I.: A fuzzy multicriteria group decision making approach to select configuration items for software development. Fuzzy Sets Syst. **134**(3), 343–363 (2003)
31. Fenton, N.E., Ohlsson, N.: Quantitative analysis of faults and failures in a complex software system. IEEE Trans. Softw. Eng. **26**(8), 797–814 (2000)
32. Ahmed, B.S., Sahib, M.A., Potrus, M.Y.: Generating combinatorial test cases using Simplified Swarm Optimization (SSO) algorithm for automated GUI functional testing. Eng. Sci. Technol. Int. J. **17**(4), 218–226 (2014)
33. Mahmoud, T., Ahmed, B.S.: An efficient strategy for covering array construction with fuzzy logic-based adaptive swarm optimization for software testing use. Expert Syst. Appl. **42**(22), 8753–8765 (2015)
34. Darwish, S. M.: Software test quality rating: A paradigm shift in swarm computing for software certification. Knowl.-Based Systems (2016)
35. Masri, W., Zaraket, F.A.: Coverage-Based Software Testing: Beyond Basic Test Requirements. Advances in Computers (2016)
36. Yang, S., Man, T., Xu, J., Zeng, F., Li, K.: RGA: A lightweight and effective regeneration genetic algorithm for coverage-oriented software test data generation. Inf. Softw. Technol. **76**, 19–30 (2016)
37. Siddiqui, T., & Ahmad, R.: A review on software testing approaches for cloud applications. Perspect. Sci. **8**, 689–691 (2016)
38. Singh, Y., Bhatia, P.K., Sangwan, O.: Software reusability assessment using soft computing techniques. ACM SIGSOFT Softw. Eng. Not. **36**(1), 1–7 (2011)
39. Srivastava, P.R., Kim, T.H.: Application of genetic algorithm in software testing. Int. J. Softw. Eng. Appl. **3**(4), 87–96 (2009)
40. Saglietti, F., Oster, N., Pinte, F.: White and grey-box verification and validation approaches for safety-and security-critical software systems. Inf. Sec. Tech. Rep. **13**(1), 10–16 (2008)

Re-visiting the Impact of the Euro on Trade Flows: New Evidence Using Gravity Equation with Poisson Count-Data Technique

Mohd Hussain Kunroo[1]([✉]), Irfan A. Sofi[2], Mansi Khurana[3], and Sandeep K. Mogha[4]

[1] Department of Economics, Central University of Rajasthan, Bandarsindri, Ajmer 305817, India
mhkunroo@gmail.com
[2] Indian Institute of Technology, Guwahati 781039, Assam, India
Sofiirfan.irfan@gmail.com
[3] School of Management, The NorthCap University, Gurgaon 122017, Haryana, India
mansi2784@gmail.com
[4] Department of Mathematics, SGT University, Gurgaon 122505, India
moghadma@gmail.com

Abstract. This paper quantifies the most likely trade effects of the euro introduction using a panel data set of 29 European economies extended over the period 1994 to 2011. For this purpose, a gravity model of international trade is used. Following the recommendations of Santos Silva and Tenreyro (2006) paper [The log of gravity. Rev Econ Stat. 88 (4), 641–658], the gravity equation is estimated using Poisson pseudo-maximum-likelihood (PPML) technique. The main finding of this study is that the introduction of the euro has a small but statistically significant effect on export flows of European economies. The PPML estimates report this effect (euro effect) to be around 7%. This effect, although small, matches with the findings of the recent studies.

Keywords: Bilateral exports · Euro · Gravity equation · PPML

1 Introduction

Robert Mundell, in his seminal paper "A Theory of Optimum Currency Areas, (1961)" initiated the debate that it is possible for a group of countries to share a common currency. This set of countries which use a single currency is known as optimum currency area (OCA) or optimum currency region (OCR). In standard economic theory, optimum currency area or optimum currency region is defined as a geographic region in which it would maximize economic efficiency to have the entire region share a common currency. This optimal currency area is generally larger than a country, but it could also be smaller than a country. The only criterion is that the adoption of a single currency should result in maximum economic efficiency. Thus, it describes the optimal characteristics which should be possessed by the countries under question, for the merger of currencies or the creation of a new currency. The various criteria developed

© Springer Nature Singapore Pte Ltd. 2017
K. Deep et al. (eds.), *Proceedings of Sixth International Conference on Soft Computing for Problem Solving*, Advances in Intelligent Systems and Computing 547,
DOI 10.1007/978-981-10-3325-4_23

230 M.H. Kunroo et al.

under the theory are often used to argue whether or not a certain region is ready to become a currency union, one of the final stages in economic integration. A detailed review about the various criteria of OCA theory can be found in Kunroo (2015).

The theory found its success by the formulation of the 'euro' in 1999. Part of the rationale behind the creation of the euro was that European countries do not form an optimal currency region individually, but Europe as a whole does form an optimal currency region. The creation of the euro is often cited because it provides the most modern and largest-scale case study of an attempt to engineer an optimum currency area, and provides a comparative pre-and-post model by which to test the principles of the theory. One such principle of the theory (of OCA) is: *"in contrast to different monies of different countries, single currency adoption eliminates the transaction costs arising from the need to operate with multiple currencies; it thus, helps to boost the flow of trade in a currency area."* Many studies have made an attempt to find out the impact of common currency adoption on trade flows. The earlier studies suggested that single currency has a very large impact on trade flows (Rose 2000). However, recent studies prove that using a single currency has very small effect on trade (De Sousa 2012; Camarero et al. 2014; Kunroo et al. 2016). Though, all these studies use gravity model of international trade to find out the impact of common currency adoption on trade flows, the reason of these divergent results lies with the methodological improvements that took place over the period of time. For instance, Rose (2000) estimated the traditional gravity equation using simple Least Square technique. Also, the data set was seen very flawed (the dataset contained both large and small countries and colonies of countries). It is generally believed that least square estimation of a log-linearized gravity equation produces biased estimates of true elasticities when there is the presence of heterogeneity across panels, heteroskedasticity in trade flows and zero trade flows (Santos Silva and Tenreyro 2006). This paper makes an attempt to quantify the trade effect of using the euro as a common currency by taking care of above mentioned issues. Thus, the paper makes a valuable contribution towards the existing literature in two ways:

First, in order to avoid the estimation bias faced by Rose (2000), the study uses a homogenous data set of 29 European countries for the period 1994 to 2011. The selection of this homogenous country-set will reduce the influence of any dominant or very large country on the currency union estimates. Also, since its inception, the euro has celebrated 17 long years of its success. This is a substantial amount of time to trace out the actual effect of the euro introduction on bilateral trade flows. By now, the euro effect will be completely discernible.

Second, Santos Silva and Tenreyro (2006) suggest that estimation of a log-linearized gravity equation using ordinary least square (OLS) technique may produce misleading and biased results in the presence of heteroskedasticity. This generally happens because the expected value of the logarithm of a random variable, say Y is not equal to the logarithm of the expected value of the random variable i.e., $E(\ln Y) \neq \ln E(Y)$. This is the general implication of Jensen's inequality. Also, log-linearized model fails to deliver if there exist zero trade flows for a pair of countries because logarithm of zero does not exist. The inconsistency of OLS estimates becomes worse if this heteroskedasticity is associated with heterogeneity across panels. To avoid these estimation problems, this paper uses Poisson pseudo-maximum-likelihood

(PPML) technique as suggested by Santos Silva and Tenreyro (2006). However, OLS estimates are also shown for comparison purpose.

The rest of the paper is structured as follows. Section 2 highlights the model framework and the econometric technique used in this study. Section 3 provides a preliminary analysis of the data set and its descriptive statistics. Section 4 reports the estimator's estimates and their analysis. Section 5 concludes.

2 Methodology and the Model

Traditional gravity equation includes gross domestic product (GDP) of the reporting country, GDP of the partner country, the geographic distance between the two trading partners and various other trade supporting and trade impeding determinants such as common border sharing, common language, colonial ties, landlocked status, preferential trading agreements, and currency union dummy. In its simplest form, the gravity model of international trade suggests that exports from reporting country i to partner country j, denoted by EXP_{ij}, is directly proportional to the product of the two trading partners' GDPs, denoted by Y_i and Y_j, and inversely proportional to the geographic distance, denoted by $Dist_{ij}$, between them. Mathematically,

$$EXP_{ij} = \alpha \frac{Y_i^\beta Y_j^\gamma}{Dist_{ij}^\delta} \qquad (1)$$

Where, α, β, γ and δ are unknown parameters.

Besides these variables, the gravity model can incorporate other trade supporting and trade impending variables such as common border dummy, common language, colonial ties, free trade agreements, landlocked status, and currency union dummy. In this study, we extend the gravity equation to control for these bilateral effects as given below:

$$EXP_{ijt} = \alpha + \beta_1 Y_{it} + \beta_2 Y_{jt} + \beta_3 Dist_{ijt} + \beta_4 Cntgty_{ijt} + \beta_5 Comlango_{ijt} + \beta_6 Comlange_{ijt}$$
$$+ \beta_7 Landlkd_{it} + \beta_8 Landlkd_{jt} + \beta_9 CU_{it} + \beta_{10} CU_{jt} + \beta_{11} CU_{ijt} + \beta_{12} MinCU_{ijt} + \varepsilon_{ijt} \qquad (2)$$

Equation (2) represents the gravity model used in this study. The complete description of the variables used in the gravity Eq. (2) -along with their sources of data- is given in Table 1 below.

The above gravity model represented by Eq. (2) is estimated using: (i) OLS technique run on both full sample and positive export flows, and (ii) poisson non-linear count-data model. If Y_{it} represents the scalar dependent variable with X_{it} as the regressors, then a fully parametric model, with the conditional density, may be specified as

$$f(Y_{it} | \alpha_i, X_{it}) = f(Y_{it}, \alpha_i + X_{it}'\beta, \gamma); \ t = 1, \ldots, T_i, \ i = 1, \ldots, N \qquad (3)$$

Where i denotes the individual, t denotes time, γ denotes additional model parameters such as variance parameters and α_i represents individual effects.

A conditional mean model may be specified with additional effects or with multiplicative effects, respectively, as

$$E(Y_{it}|\,\alpha_i, X_{it}) = \alpha_i + h(X'_{it}\beta); \quad \text{for the specified function } h(\cdot) \qquad (4a)$$

$$E(Y_{it}|\,\alpha_i, X_{it}) = \alpha_i \times h(X'_{it}\beta); \quad \text{for the specified function } h(\cdot) \qquad (4b)$$

In the absence of individual effects, a poisson estimator assumes Y_{it} to follow Poisson distribution with a mean of

$$E(Y_{it}|\,X_{it}) = \exp(X'_{it}\beta) \qquad (5)$$

In the presence of individual-effects, Eq. (5) takes the form

$$E(Y_{it}|\,\alpha_i, X_{it}) = \exp(\gamma_i + X'_{it}\beta) = \alpha_i \exp(X'_{it}\beta) \qquad (6)$$

Where $\gamma_i = \ln \alpha_i$, and X_{it} includes an intercept.

For OLS estimation, it is necessary that $E(\ln \alpha_i | X_{it})$ does not depend on X_{it}. This condition gets violated in the presence of heteroskedasticity. Therefore, in such a situation, regressing $\ln Y_{it}$ on X_{it} by OLS estimation technique will lead to biased and inconsistent estimates of β. This makes, in general, the PPML estimates more reliable.

3 Data

The study is wholly restricted to European countries. The sample covers 29 countries, including the 19 Eurozone economies. A list of these sample countries is given in Table A1. The period of estimation is from 1994 to 2011. Following earlier studies, 1994 was chosen as the starting year. The end-time period of 2011 was chosen in order to have a balanced panel. The permutations of 29 countries into country pairs yield $29 \times 28 = 812$ bilateral cross-sectional units. The total number of observations on the full sample is 14560. This count reduces to 14288 when zero trade flows are left out. Table 2 gives the summary statistics of the sample panel data set.

In the beginning, the sample data set is checked for heteroskedasticity, serial correlation and heterogeneity across panels. The tests suggest the sample dataset suffers from heteroskedasticity, serial correlation and heterogeneity in cross-sectional groups. For instance, modified Wald test for groupwise heteroskedasticity generates Chi^2 (812) value equal to 1.6e+07, with Prob > Chi^2 = 0.00, thereby rejecting the null hypothesis of homoskedasticity. The Pesaran CD test of cross-sectional dependence provides a value of 540.59 with Pr of 0.00, again rejecting the null of no serial correlation. Same is the case of Wooldridge test for autocorrelation in panel data. In this test, $F(1, 811) = 193.39$ with Prob > F = 0.00. F Test also supports the presence of heterogeneity in groups [F test that all u_i = 0: $F(811, 13743) = 13.26$ and Prob > F = 0.00]. This clearly suggests the inconsistency of OLS estimates for the dataset under analysis. Therefore, our analysis will mainly focus on Poisson estimates. The OLS and PPML estimates are reported in Table 3.

Table 1. Description of variables used in the gravity equation and their data sources

Variable	Description	Data source
EXP_{ijt}	Log of volume of exports from country i (reporting country) to country j (partner country) in year t measured in current US dollars at current exchange rates	International Monetary Fund (IMF) Direction of Trade Statistics (DOTS) CD_ROM (Mar. 2012)
Y_{it}	Log of reporting country's GDP in year t measured in current US dollars at current exchange rates	World Development Indicators (2012) database of the World Bank
Y_{jt}	Log of partner country's GDP in year t measured in current US dollars at current exchange rates	World Development Indicators (2012) database of the World Bank
$Dist_{ijt}$	Log of distance between the reporting country i and the partner country j (in kilometers) in year t, measured according to the *Great Circle* formula (Head and Mayer 2002)	Centre d'Etudes Prospectives et d'Informations Internationales (CEPII)
$Cntgty_{ijt}$	dummy variable whose value is 1 if countries i and j share common border in year t; otherwise 0	*Centre d'Etudes Prospectives et d'Informations Internationales (CEPII)*
$Comlango_{ijt}$	dummy variable whose value is 1 if countries i and j have common official primary language in year t; otherwise 0. Table A2 provides details of the sample countries sharing the same official language	*Centre d'Etudes Prospectives et d'Informations Internationales (CEPII)*
$Comlange_{ijt}$	dummy variable whose value is 1 if at least 9% of the population in both the countries speak common language in year t; otherwise 0. The information about the sample countries where either 9% or more than 9% of the population speak same language is classified in Table A3	*Centre d'Etudes Prospectives et d'Informations Internationales (CEPII)*
$Landlkd_{it}$	dummy variable whose value is 1 if reporting country i is a landlocked country and partner country j is not in year t; otherwise 0	*Centre d'Etudes Prospectives et d'Informations Internationales (CEPII)*
$Landlkd_{jt}$	dummy variable whose value is 1 if country i is not a landlocked country but country j is a landlocked country in year t; otherwise 0	*Centre d'Etudes Prospectives et d'Informations Internationales (CEPII)*
CU_{it}	dummy variable whose value is 1 if country i is a member of the Eurozone and country j is not a member of the Eurozone in year t; otherwise 0	Authors' creation

(*continued*)

Table 1. (*continued*)

Variable	Description	Data source
CU_{jt}	dummy variable whose value is 1 if country j is a member of the Eurozone and country i is not a member of the Eurozone in year t; otherwise 0	Authors' creation
CU_{ijt}	dummy variable whose value is 1 when both the countries belong to the Eurozone in year t; otherwise 0. A year-wise list of countries joining the euro currency union is given in Table A4	Authors' creation
$MinCU_{ijt}$	variable that takes the value of minimum number of years the two trading partners have been using the Euro as their currency	Authors' creation

4 Results and Discussions

This section presents the estimated results of Eq. (2), on the complete dataset of 29 EMU economies described above, using OLS and PPML techniques. Column 1 and Column 4 of Table 3 report the OLS estimates and PPML estimates, respectively, based on the positive exports flows (Exports > 0). Column 2 of Table 3 presents the OLS estimates after compensating for the zero trade flows by the addition of 1 to the exports in levels. This addition of 1 to the exports avoids the problem of zero trade flows as all the export values are now greater than zero, thereby facilitating the logarithm of these export values. Column 3 (Table 3) is our main focus of analysis. It reports the PPML results of gravity Eq. (2) on the full data set.

In column 3 of Table 3, the expected signs for the estimators associated with the variables are based on the traditional gravity arguments. As suggested by the theory, the PPML estimates show a positive and significant impact of the reporting and partner countries' GDPs on exports. Bilateral distance has a negative and significant impact on export flows. This means that the economic size of a country, represented by its level of GDP, increases trade between two countries while geographic distance- a proxy of transport costs- reduces the volume of trade. The other estimators also show the same relationship of GDP and distance with exports flows. Regarding other variables, we expect a positive impact of contiguity, common language, currency union agreement, and the number of years in the currency union on trade flows; but a negative effect of landlockedness. The PPML estimates (column 3, Table 3) report a positive and significant effect of contiguity and common spoken language and a significantly negative impact of landlockedness on export flows. This was expected because two countries which share a common border or of which a good percentage of inhabitants speak the same language will have higher probability to trade among themselves. Besides, lack of

Table 2. Summary Statistics

Variable	Full sample				Exports > 0			
	Mean	Std. dev.	Min	Max	Mean	Std. dev.	Min	Max
ID	–	–	1	812	–	–	1	812
Year	–	–	1994	2011	–	–	1994	2011
Exports (EXPij)	2.90E+09	9.50E+09	0	1.70E+11	2.96E+09	9.58E+09	8.235	1.70E+11
Log of exports (Ln EXP$_{ij}$)	–	–	–	–	19.051	2.831	2.108	25.861
Log of reporter country's GDP (Y_i)	25.425	1.780	21.821	28.918	25.446	1.773	21.821	28.918
Log of partner country's GDP (Y_j)	25.425	1.776	21.821	28.919	25.453	1.772	21.821	28.919
Log of distance (Dist$_{ij}$)	7.163	0.618	5.081	9.641	7.164	0.617	5.081	9.641
Contiguity dummy (Cntgty$_{ij}$)	0.089	0.284	0	1	0.089	0.284	0	1
Common official language dummy (Comlango$_{ij}$)	0.034	0.182	0	1	0.033	0.179	0	1
Common ethnic language dummy (Comlange$_{ij}$)	0.032	0.176	0	1	0.033	0.178	0	1
Landlocked-reporter country dummy (Landlkd$_i$)	0.172	0.378	0	1	0.168	0.374	0	1
Landlocked-partner country dummy (Landlkd$_j$)	0.169	0.376	0	1	0.163	0.370	0	1
Euro-reporter currency union dummy (CU$_i$)	0.195	0.397	0	1	0.199	0.399	0	1
Euro-partner currency union dummy (CU$_j$)	0.195	0.396	0	1	0.198	0.399	0	1
Euro currency union dummy (CU$_{ij}$)	0.153	0.360	0	1	0.155	0.362	0	1
Minimum years of using the Euro-both reporter and partner country (MinCU$_{ij}$)	2.759	4.939	0	14	2.720	4.904	0	14

Table 3. The Gravity equation estimates

Estimator	OLS	OLS	PPML	PPML
Dependent variable:	Ln(EXPij) > 0	Ln(1+EXPij)	Ln(EXPij)	Ln(EXPij) > 0
Y_i	0.996*** (0.006)	1.072*** (0.013)	0.057*** (0.002)	0.053*** (0.001)
Y_j	0.795*** (0.006)	0.924*** (0.013)	0.049*** (0.002)	0.042*** (0.001)
Dist	− 1.501*** (0.018)	− 1.483*** (0.043)	− 0.087*** (0.006)	− 0.082*** (0.004)
Cntgty	0.382*** (0.038)	0.695*** (0.091)	0.024* (0.013)	0.007 (0.008)
Comlango	− 0.257*** (0.068)	− 2.194*** (0.154)	− 0.119*** (0.022)	− 0.016 (0.015)
Comlange	0.526*** (0.065)	2.234*** (0.151)	0.117*** (0.022)	0.023 (0.014)
Landlkd$_i$	0.088*** (0.024)	− 0.410*** (0.055)	− 0.020** (0.008)	0.009* (0.005)
Landlkd$_j$	− 0.620*** (0.024)	− 1.387*** (0.055)	− 0.077*** (.008)	− 0.030*** (0.005)
CU$_i$	0.012 (0.025)	0.426*** (0.059)	0.037*** (0.007)	0.002 (0.006)
CU$_j$	− 0.028 (0.025)	0.249*** (0.059)	0.030*** (0.007)	0.001 (0.006)
CU	− 0.234*** (0.038)	1.264*** (0.088)	0.071*** (0.008)	− 0.012 (0.008)
MinCU	0.025*** (0.003)	− 0.093*** (0.007)	− 0.005*** (0.001)	0.001* (0.001)
Observations	14288	14560	14560	14288

Notes: ***, **, & * show significance at 1%, 5%, and 10% level, respectively. Constants are not reported. Values in parenthesis report standard errors. Attempt was made to estimate the standard gravity model using PPML regression with fixed effects. However, the estimation confronted some problems since the maximization algorithm could not generate results. All the estimations are carried out using STATA Version 12 econometric software.

access to sea will reduce the volume of trade between two trading partners. What comes to our surprise is the sign of common official language dummy. Like common spoken language dummy, it should imply a significantly positive impact on trade flows. However, it shows a negative BUT significant impact in three out of four estimators. The number of years since two Euro-zone economies has been using the euro as their currency does not produce any conclusive result.

The main focus of this paper is to estimate the 'euro effect' on trade. The PPML estimates of the full data set show a positive and significant impact of the euro currency adoption on export flows. The magnitude of the respective variable i.e., CU_{ijt} is 0.071. This shows that two countries which are using the euro as their currency increase their trade (exports) by 7.36%.[1] Though not a surprising finding, this effect is smaller and supports the findings of recent studies (Kunroo 2016). For instance, using a panel dataset of 203 countries for the period 1948–2009, De Sousa (2012) finds that the currency union effect on trade flows is decreasing over time. Like this study, his OLS estimates also somewhat contrast with the PPML estimates. Camarero et al. (2014) adopt the second generation panel co-integration techniques to estimate the euro effect on trade. In their finding, the euro dummy has a coefficient of 0.16 which equally is very small compared to earlier findings. Kunroo et al. (2016), using panel fixed effects and random effects techniques, estimate that the effect of the euro on bilateral trade flows is around 14%. The other variables (CU_{it} and CU_{jt}) result in smaller effect compared to CU variable. These two variables generate a currency union effect of 3.76% and 3.04%, respectively. This suggests that it will be beneficial for two countries to boost their trade if they both merge their independent currencies. However, the PPML estimates on positive exports (Exports > 0 panel data set) show an insignificant effect of the euro currency adoption on export flows. The conclusion can be drawn that the euro effect is very small and may even reduce to almost insignificance if homogeneous dataset and zero trade flows are also taken into consideration.

5 Conclusion

The main purpose of this study was to investigate the trade effects of the euro for 29 European economies using gravity model of international trade. The gravity equation is estimated using PPML panel count-data estimation technique. The period of analysis is 1994 to 2011. The study finds that the euro introduction has a small but statistically significant effect on trade among EMU members. On the basis of this analysis (dataset), the exact euro effect is 7.36% on export flows.

---

[1] Since the coefficient of CU dummy is 0.071, the increase in bilateral exports induced by the adoption of the euro is $\left[\left(\frac{e^{0.071 \times 1}}{e^{0.071 \times 0}} - 1\right) \times 100\right] = \left[\left(\frac{1.07358}{1.00} - 1\right) \times 100\right] = 0.07358 \times 100 = 7.36\%$.

Appendix

Table A1. List of Countries

Austria	Denmark	Hungary	Luxembourg	Slovakia
Belgium	Estonia	Iceland	Malta	Slovenia
Bulgaria	Finland	Ireland	Netherlands	Sweden
Croatia	France	Italy	Poland	Spain
Cyprus	Germany	Latvia	Portugal	United Kingdom
Czech Republic	Greece	Lithuania	Romania	

Table A2. Common official language

English	French	Swedish	German	Dutch
Malta Ireland United Kingdom	France Luxembourg	Finland Sweden	Austria Germany	Belgium and Luxembourg Netherlands
Bulgarian	Greek	Portuguese	Danish	Latvian
Bulgaria	Cyprus Greece	Portugal	Denmark	Latvia
Czech	Spanish	Romanian	Estonian	Lithuanian
Czech Republic	Spain	Romania	Estonia	Lithuania
Polish	Slovenian	Serbo-Croatian	Hungarian	Icelandic
Poland	Slovenia	Croatia	Hungary	Iceland
Italian	Slovak			
Italy	Slovakia			

Table A3. Common Ethnic language

German	English	Hungarian	Greek	Dutch
Austria Germany Belgium and Luxembourg	Ireland United Kingdom	Hungary Romania Slovakia	Cyprus Greece	Belgium and Luxembourg Netherlands

Table A4. Euro currency adoption

Year	Country
1999	Euro was officially introduced with Portugal, the Netherlands, Luxembourg, Italy, Ireland, Germany, France, Finland, Belgium, Austria, and Spain as member countries
2001	Greece adopted the Euro as its currency (12th member)
2002	Euro as currency was physically introduced
2007	Slovenia adopted the Euro as its currency (13th member)
2008	Cyprus and Malta adopted the Euro as their currency (14th and 15th member)
2009	Slovakia adopted the Euro as its currency (16th member)
2011	Estonia adopted the Euro as its currency (17th member)
2014	Latvia adopted the Euro as its currency (18th member)
2015	Lithuania adopted the Euro as its currency (19th member)

References

Camarero, M., Gomez, E., Tamarit, C.: Is the 'euro effect' on trade so small after all? New evidence using gravity equations with panel cointegration techniques. Econ. Lett. (2014). doi:10.1016/j.econlet.2014.04.033

De Sousa, J.: The currency union effect on trade is decreasing over time. Econ. Lett. (2012). doi:10.1016/j.econlet.2012.07.009

Head, K., Mayer, T.: Illusory border effects: distance mismeasurement inflates estimates of home bias in trade. CEPII Working Paper No. 2002–01, Paris (2002). http://www.cepii.fr/PDF_PUB/wp/2002/wp2002-01.pdf

Kunroo, M.H.: Theory of optimum currency areas: a literature survey. Rev. Market Integr. (2015). doi:10.1177/0974929216631381

Kunroo, M.H., Sofi, I.A., Azad, N.A.: Trade implications of the Euro in EMU countries: a panel gravity analysis. Empirica: J. Euro. Econ. (2016). doi:10.1007/s10663-016-9334-6

Mundell, R.A.: A theory of optimum currency areas. Am. Econ. Rev. **51**, 657–665 (1961)

Rose, A.K.: One money, one market: the effect of common currencies on trade. Econ. Policy. **15**, 7–46 (2000)

Santos Silva, J.M.C., Tenreyro, S.: The log of gravity. Rev. Econ. Stat. **88**, 641–658 (2006)

Kunroo, M. H.: The effect of the euro on bilateral trade and exports of EMU economies. In: Chakraborty, D., Mukherjee, J. (eds.) Trade, Investment and Economic Development in Asia: Empirical and Policy Issues 2016, pp. 136–157. Routledge (2016)

Detection and Mitigation of Spoofing Attacks by Using SDN in LAN

Amandeep Kaur[✉] and Abhinav Bhandari

Department of Computer Engineering, Punjabi University, Patiala, India
Deepaman3093@gmail.com, bhandarinit@gmail.com

Abstract. Software Defined Networking (SDN) is defined as a solution for security, reliability and flexibility problems in traditional networks. It decouples the data plane and control plane. SDN is a programmable approach of networking. In this technology network can be controlled by a program. The key characteristic of Software Defined Networking is that it proposes the separation of the data plane and the control plane, this control is given to a centralize controller and Open data devices are used for data forwarding. These devices can act either as switch, hub or as any other network device it depends upon the application. Those applications also can be attack mitigation applications. In our proposed work we develop an IP spoofing attack mitigation application. That application can detect and mitigate the IP spoofing attack in SDN in LAN. Here, we will use POX controller so have to develop program in python language. The emulation tool Mininet is used for experiment.

Keywords: Software Defined Networking · Data plane · Control plane · Controller · Mininet · Emulation · Spoofing

1 Introduction

Today the use of internet (network of networks) is escalated day by day and number of users also increasing so due to this traffic computer networks become more complex and difficult to manage. By a "programmable approach" [2] managing of network become easy and also give the option to the network operator to control the network according to the requirements. Many network equipments are used in such a network like switch, router, firewall, load balancer etc. In traditional network these equipments are strongly coupled means control and data plane are bind in device e.g. traditional router and network operator can only configure the device. But in SDN [1] data plane and control plane are separated. In this, OpenFlow data device is used as data plane and a centralize controller [4] is used as a control plane. Different applications programs are connected to the controller through the northbound interface (NBI). In this technology, different OpenFlow controllers are used (1) NOX [8] that is based on C ++ and python languages (2) POX [8] that is based on python language.

In this technology different OpenFlow controllers are used (1) NOX [7] that is based on C ++ and python languages (2) POX [8] that is based on python language. Users can

© Springer Nature Singapore Pte Ltd. 2017
K. Deep et al. (eds.), *Proceedings of Sixth International Conference on Soft Computing for Problem Solving*, Advances in Intelligent Systems and Computing 547, DOI 10.1007/978-981-10-3325-4_24

also develop their own controller. Mininet [3] tool is used for programming and development of SDN. It is an emulation tool (Fig. 1).

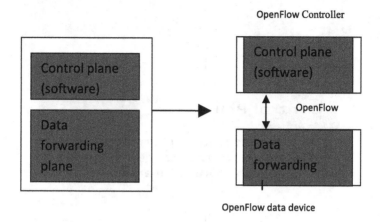

Fig. 1. SDN (separation of control and data plane)

1.1 SDN Architecture

This SDN architecture [2] includes following elements

Applications. These are applications that decide the network behavior according to the coding and communicate with controller via northbound interface. Further controller gives Instructions to the data devices about the handling of data packets and about the exception handling also, these application's main responsibility is to perform a function for device deal with the packet, either forwards it to Controller or to the destination (if there are entry in the flow table) or drop the packet. These applications are programs that are coded into programming languages (C ++, python, and java).

Controller. It is the operating system of the SDN implemented network. The controller maintains the network, manage the flow and control all the OpenFlow data devices. The main functions of the controller are (1) to discover end user (personal computer, laptop, server) (2) to discover network devices (router, switch, load balancer) (3) to manage the network flow and (4) to manage the network topology (how the devices are connected with each other). It communicates with devices via southbound interface (OpenFlow) and communicates with applications via northbound interface. Controller also has its own predefined modules such as hub and learning switch, it also has different application interface i.e. java API, python API or any other API.

SDN Data Devices. These are the data forwarding devices means data plane of the network. These are not static hardware like traditional devices, these are dynamic means an OpenFlow device can act as hub, switch or any other device according to the software implementation. SDN data devices have flow tables which have different matching entries according to these entries, these devices deal with the packets (Fig. 2).

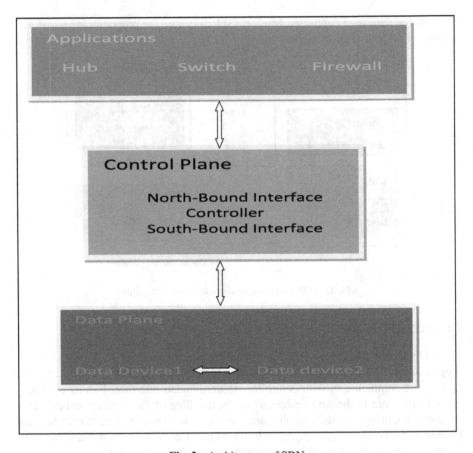

Fig. 2. Architecture of SDN

SDN Interfaces. These are the interfaces through which SDN data devices, controller and applications are communicated with each other. SBI is southbound interface and NBI is northbound interface. There are different application interface i.e. java API, python API or any other API.

2   IP Spoofing

IP spoofing is making of IP packets (fake packets) using some other host's IP address. IP header contains 32 bit long source IP address and destination IP address. So in fake packet attacker can steal the source IP of any other host. In IP spoofing, attacker's main goal is to gain unauthorized access and sensitive data of target host. In any network attacker do not have permission to reply or abuse trust link hosts so using spoofing attacker can perform man-in the middle attack. After the attack, attacker: can-

- sniff the traffic on network.
- intercept, block and delay the traffic.
- modify the traffic.

In IP spoofing in SDN, attacker create IP packet with duplicate IP address of any host in the network. So attacker manipulates the controller. Here, we can develop an application that can detect and mitigate the IP spoofing in SDN. That application runs on the controller in SDN.

3 Related Work

Security [5] is the major issue in SDN. It is a programmable approach towards networking. It is modern technology and such an evolution in networking. There are many applications that are developed in SDN i.e. hub, switch, router, load balancer etc. But security related applications are not much developed in SDN. Here we discuss the existing application related SDN.

3.1 Firewall

It is control the traffic so data can receive by only the right receiver. Any other host cannot access the data. In this mechanism traffic from a particular host can be blocked or traffic coming from particular host can be forward and other is blocked. There are two approaches presented.

White–list approach. In this approach a list is of that hosts is created which's traffic is allow to forward. So, this firewall application forward the traffic that coming from these hosts and block the traffic if coming from any other host.

Black-list approach. In this approach a list of that hosts is created which's traffic is disallow to forward. It is opposite to white–list approach. So controller blocks data from these hosts and allow traffic from all other hosts.

3.2 DoS Attack Mitigation

A denial-of-service(DoS) attack [6] in OpenFlow SDN networks involves overwhelming computing or networking resource such that a switch is unable to forward packets as expected. A successful attack involves sending a large number of packets. In this article two type of DoS attacks presented.

A. The control plane–In this a large number of new packets are send to the data device so data device forward the packets to controller because data device have no entry in table related that packet. With large number of packets heavy consumption of control plane bandwidth takes.

B. Attacking the switch's flow table–Data devices have limited memory to store flow rules. So this is utilized for Dos attacks. In this, attacker adds a large number of rules in the flow table of data device. So data device can't add more rules and unavailable for service.

4   IP Spoofing Detection and Mitigation

IP spoofing is recognized as a serious threat in SDN. IP spoofing can be detected by an application. That application runs on SDN controller and it can detect and mitigate the IP spoofing.

In this application in packet_in() function packet is parsed. If packet is incomplete then OpenFlow data device ignore the packet. If incoming packet is a complete then it further checks IP spoofing.

If the IP spoofing Detection function returns true then controller instruct the data device to drop the packet from that host. Otherwise controller instructs to data device to forward the incoming packet to its destination if it is in the ipToMac dictionary otherwise flood the packet.

5   Implementation

Experiments are conducted for IP-spoofing attack to check the feasibility of proposed solution. The detail of this experiment is shown below.

5.1   Test Experiment

The test experiment is conducted on a ubuntu virtual machine with 1 core and 1 GB of RAM. The host machine is running on window 7 and has Intel Core i3 processor and 4 GB of RAM. For this experiment, Mininet tool is used that is installed on a ubuntu virtual machine. Mininet is an emulator that is used for creating and experiments. It creates a network with hosts, switches and controller according to a given topology. That controller can also be a remote controller. The hosts created in Mininet have Linux operating system. For graphics mode a Redhat machine is used and a secure connection is set up between two virtual machines. Here, to generate fake packet for attack the network tool Ostinato is used. In this tool any packet can be created and more than one packet can be sent to the target host (Fig. 3).

5.2   Experiment Setup

The topology used to check the proposed solution is with 20 hosts connected with switch. Here controller is used remotely that is a default POX controller with proposed solution. This topology created in Mininet emulator .15 hosts act as normal user and 5 hosts act as attacker. These 5 systems are compromised by an attacker. All the emulated links are with 10 Mbps bandwidth, 5 ms delay and 0% loss. All the hosts have CPU performance.

5.3   Test Scenario

To test the proposed solution an IP spoofing attack scenario is created. In this a network topology is created with 20 hosts. A controller is running remotely with proposed mitigation solution. Initially send icmp packet to the hosts in the network. Linux command

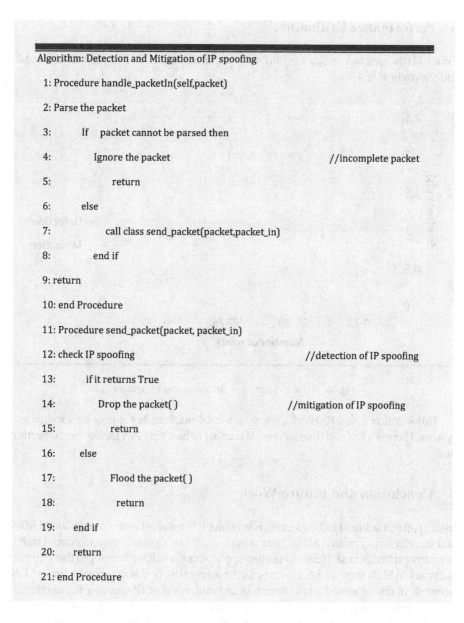

Algorithm: Detection and Mitigation of IP spoofing

1: Procedure handle_packetIn(self,packet)

2: Parse the packet

3:      If   packet cannot be parsed then

4:           Ignore the packet   //incomplete packet

5:               return

6:      else

7:           call class send_packet(packet,packet_in)

8:      end if

9: return

10: end Procedure

11: Procedure send_packet(packet, packet_in)

12: check IP spoofing   //detection of IP spoofing

13:      if it returns True

14:           Drop the packet( )   //mitigation of IP spoofing

15:               return

16:      else

17:           Flood the packet( )

18:               return

19:   end if

20:   return

21: end Procedure

Fig. 3. Algorithm for detection and mitigation of spoofing in SDN in LAN

"pingall" can be used. After that IP spoofing is performed by using a network tool Ostinato. Here fake IP packets are generated and send to the different hosts by the compromised host. Then this application detects the IP spoofing attack and in future, drops the
packet from that host.

6   Performance Evaluation

Time of Detection and mitigation of IP spoofing in SDN in LAN is measured in seconds and shown in Fig. 4

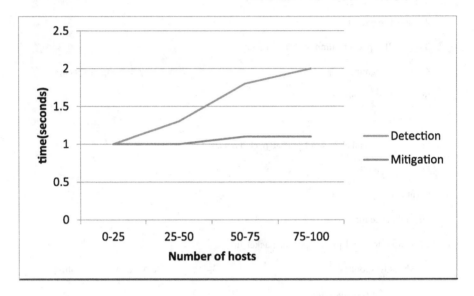

Fig. 4.   Time evaluation in detection and mitigation

This solution gives 100% efficiency on virtual machine but it may be vary on real system. There is a little bit time increase if number of hosts in LAN increases at detection time.

7   Conclusion and Future Work

In this paper, a solution to IP spoofing is presented. We started with a discussion of SDN and security issues related SDN. Then we discuss Firewall that is used to control traffic from any particular host. Here, some other applications also have developed that mitigate the attack in SDN network i.e. Dos attacks. Spoofing attacks is also serious issue in SDN network. In this proposed work, detection and mitigation of IP spoofing is presented.

This solution is presented for only one controller. So, it can be further developed for more than one controller. It can detect and mitigate IP spoofing in only LAN network. So, it also can be expend for WAN network.

Acknowledgement.   The author would like to thank Mr. Vipin Gupta for valuable help. This work is supported by U-net solution, Moga.

References

1. Mendon, M., Nunes, B., Nguyen, X.N., Obraczka, K.: A survey of software-defined networking: past, present and future of programming networks, p. 18. IEEE (2015)
2. Nick, F., Rexford, J., Zegura, E.: The road to SDN: an intellectual history of programmable network, p. 13 (2014)
3. Li, X., Yuan, D., Hu, H., Ran, J., Li, S.: DDs detection in SDN switches using support vector machine classifier. In: JIMET (2015)
4. Vizvary, M.: Mitigation of DDoS Attacks in Software Defined Networks (2015)
5. Shu, Z., Wan, J., Li, D., Vasilakos, A.V.: Imran, M: Security in software-defined networking: threats and countermeasures. Mob. Netw. Appl. **21**(5), 764–776 (2016)
6. Hande, Y.H, Kadhu, T, Hublikar, S, Kulkarni, S, Kasirsagar, A: DDos attacks in SDN networks. Int. Eng. Res. J. 2015
7. Open networking foundation SDN Architecture version 1.0., POX (2013). http://www.noxrepo.org/pox/about-pox/
8. Open networking foundation. www.mininet.org

Performance Modeling and ANFIS Computing for Finite Buffer Retrial Queue Under F-Policy

Madhu Jain and Sudeep Singh Sanga[(✉)]

Department of Mathematics, IIT Roorkee, Roorkee 247 667, India
drmadhujain.iitr@gmail.com,
sudeepiitroorkee@gmail.com

Abstract. This investigation is concerned with the performance prediction and admission control F-policy for the machine repair problem with retrial. To develop a Markov model, the steady state Chapman-Kolmogorov equations are constructed. The system state probabilities are obtained by using recursive method. Various performance measures are established explicitly in terms of steady state probabilities. To examine the effects of system parameters, the numerical simulation is performed by choosing a suitable illustration. The cost function is also framed to evaluate the optimal service rate and corresponding optimal cost. ANFIS soft computing technique is used to compare the numerical results obtained analytically and also by implementing ANFIS.

Keywords: F-policy · Retrial · Machine repair · Recursive technique · Cost optimization · ANFIS

1 Introduction

In real world, the formation of queues can be seen everywhere such as at railway reservation counters, in front of post office, doctor's clinic and many other places. The formation of queues and delay in service are important features for both customers as well as system organizers. To examine more practical retrial queues with finite population, the queueing scenario of machine repair shop can be seen a failed machine may try to get repair again and again after waiting sometimes in retrial orbit for the repairman. To be more specific, one can notice the computer repair shop where failed computers join the system for the repair job; in case when a repairman is busy, the failed computers may wait in the pool of blocked computers called retrial orbit and from there those failed computers seek for the repairing job and try again later. Whenever, the repairman becomes free, it takes the next failed computer for the repair, however, other failed computers may also try to get repair from the orbit. A vast survey on retrial queues can be seen in the study of Artalejo and Falin [1], Artalejo [2]. A finite capacity M/M/1 queue with retrial attempts for the retrial and normal queues was studied by Sherman and Kharoufeh [3]. They obtained the steady state joint distributions of the retrial size and the normal queue size when the server is in a busy or idle state.

Now-a-days our life completely depends on machines (e.g. computer, mobile, vehicles, etc.). But it is not always feasible to design perfect machines; it may failed at

K. Deep et al. (eds.), *Proceedings of Sixth International Conference on Soft Computing for Problem Solving*, Advances in Intelligent Systems and Computing 547, DOI 10.1007/978-981-10-3325-4_25

any time due to several reasons and this results in loss of revenue and goodwill of the concerned organization/system. To overcome these problems, significant works on the machine repair problem have been done by many researchers [4, 5]. Wang and Siwazlian [6], Wang [7], Jain [8] and many others have worked on machine repair problems and developed Markov models for the performance analysis of machining system by using queueing and reliability theory. Yang and Chiang [9] proposed an algorithm for the machining system using a threshold recovery policy. They formulated a cost function and optimized it by using particle swarm optimization (PSO) algorithm.

Control F-policy for the machining systems are broadly applicable in many real world industrial scenarios including the computer systems, communication networks, manufacturing units, inventory systems, etc. The other day-to-day realistic applications of F-policy for the machining system can be found in many machine repair shops namely automobile repair shop. Gupta [10] introduced the concept of F-policy to study Markovian single server model. According to F-policy, as the system reaches to its full capacity, no further units are permitted to enter in the system until the number of units again drops to a prefixed threshold value 'F'. In recent years, some research works have appeared on control F-policy for Markovian queueing models [11, 12]. Recently, Jain and Bhagat [13] considered transient analysis of finite retrial queue with delayed repair under F-policy. To solve the differential equations governing the model, they have used Runge-Kutta 4th order method and obtained the transient results for the performance matrices. Goswami [14] investigated the interrelationship between the randomized F-policy and randomized N-policy for the discrete-time queues with start-up time. The relationships between the discrete-time Geo/Geo/1/K queues with (p, F) and (q, N)-policies are established by a series of propositions. The benefit of this interrelationship is that the solution of one queue can be deduced from the other queue. They proposed genetic algorithm to determine the optimal values for the system parameters.

For developing the performance model of complex systems, the emerging hybrid soft computing technique ANFIS which is combination of neural networks and fuzzy logic, plays an important role. This hybrid soft computing technique ANFIS is also well suited in distinct areas of industries operating in machining environment. (cf. Lin and Liu [15]). Jain and Upadyaya [16] proposed a Markov model for the machining system by incorporating some realistic features such as N-policy, spares and multiple vacations. They had compared the results determined by matrix recursive method with the ANFIS results. The results obtained by Kumar and Jain [17] using SOR method have good match with the hybrid soft computing technique ANFIS results for the multi-component degraded machining system.

In the present paper, M/M/1/K model for the machining system operating under control F-policy by considering the constant retrial times is investigated. The noble and innovative features of present study are the provision of F-policy, state dependent rate and finite retrial orbit in the model formulation. To the best of authors' knowledge, so far there is no work available in the literature which is related to the machine repair problem with retrial attempts and control policy simultaneously. The rest of the article is constituted as follows. In Sect. 2, model description and notations to develop the model are presented. Section 3 provides the mathematical formulation and analysis of the model. Various system indices are established in the Sect. 4. The sensitivity analysis of the system performance with respect to different parameters is given in the

Sect. 5. Section 6 is developed to present ANFIS approach for obtaining the performance indices. Finally, Sect. 7 concludes the summary of the present study.

2 Model Description

Consider an M/M/1/K model with retrial attempts for the maintainability of multi components machining system operating under F-policy. The server renders the repair of one failed machine at a time by following the first come first served (FCFS) discipline.

The mathematical formulation of the Markov model is based on certain assumptions which are described as follows. The multi components machining system consists of K identical machines. The machine may fail independent of the other machines. The failure of the machines occurs in Poisson fashion with rate λ. The state dependent arrival rate λ_n is defined by $\lambda_n = (K - n)\lambda$. If the incoming failed machine finds the repairman free, it gets repaired following the exponential distribution with mean $1/\mu$. If the incoming failed machine finds the server (*i.e.* repairman) busy then it is forced to join the retrial orbit having finite capacity F. From the orbit it re-attempts with constant retrial rate γ again and again to seek the server free for the repair. When the system capacity becomes full, then a setup job is required to stop the failed machines from joining the system; the time required for the setup is assumed to be exponentially distributed with rate ε. Once the system capacity is full, the failed machines are not permitted to queue up in the system until the number of failed machines in the system is further cease to a prefixed threshold level 'F' of workload of failed machines and is said to follow F-policy for the admission of jobs. The service time during the period of F-policy is assumed to follow the exponential the distribution with mean $1/\mu'$. Further, it is considered that all the durations for which machines remain in different states are mutually independent.

For the mathematical formulation of the model, the following notations are used.

Let $N(\tau)$ and $Y(\tau)$ be the number of failed machines and states of the server respectively, at time τ. The server state $Y(\tau)$ is defined as follows:

$$Y(\tau) = \begin{cases} 0, \text{ the repairman is busy and the failed machines are forced to join the orbit;} \\ 1, \text{ the repairman is busy and the failed machine are permitted to enter in the system;} \\ 2, \text{ the repairman is busy and the failed machine are not permitted to enter in the system.} \end{cases}$$

To develop Markov model, define the system states probabilities at time τ for node (j, n), by $P_{j,n}(\tau) = \Pr ob\{Y(\tau) = j, N(\tau) = n\}$. It is noted that $\{Y(\tau), N(\tau) : \tau \geq 0\}$ is a bi-variate Markov process which is discrete in state space and continuous in time. The model shall analyze at steady state, *i.e.*, when $\tau \to \infty$ by denoting $P_{j,n} = \lim_{\tau \to 0} P_{j,n}(\tau)$.

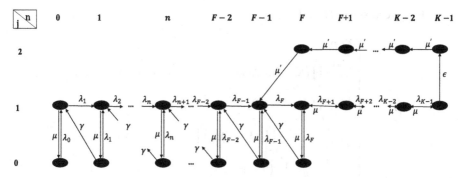

Fig. 1. Transition state diagram

3 Governing Equations

In view of transition rates for the system states (j, n) as depicted in Fig. 1, we formulate the steady state Chapman-Kolmogorov equations for the system states for three levels $Y(\tau) = j (j = 0, 1, 2)$ as follows:

(i) **For system state $j = 0$, $0 \leq n \leq F$.**

$$\lambda_0 P_{0,0} = \mu P_{1,0} \tag{1}$$

$$(\lambda_n + \gamma) P_{0,n} = \mu P_{1,n}; \ 1 \leq n \leq F. \tag{2}$$

(ii) **For system state $j = 1$, $0 \leq n \leq K - 1$.**

$$(\lambda_1 + \mu) P_{1,0} = \lambda_0 P_{0,0} + \gamma P_{0,1} \tag{3}$$

$$(\lambda_{n+1} + \mu) P_{1,n} = \lambda_n P_{0,n} + \gamma P_{0,n+1} + \lambda_n P_{1,n-1}; \ 1 \leq n \leq F - 2 \tag{4}$$

$$(\lambda_F + \mu) P_{1,F-1} = \lambda_{F-1} P_{0,F-1} + \gamma P_{0,F} + \lambda_{F-1} P_{1,F-2} + \mu' P_{2,F} \tag{5}$$

$$(\lambda_{F+1} + \mu) P_{1,F} = \lambda_F P_{0,F} + \lambda_F P_{1,F-1} + \mu P_{1,F+1} \tag{6}$$

$$(\lambda_{n+1} + \mu) P_{1,n} = \lambda_n P_{1,n-1} + \mu P_{1,n+1}; \ F + 1 \leq n \leq K - 2 \tag{7}$$

$$(\varepsilon + \mu) P_{1,K-1} = \lambda_{K-1} P_{1,K-2} \tag{8}$$

(iii) **For system state $j = 2$, $F \leq n \leq K - 1$.**

$$\mu' P_{2,n} = \mu' P_{2,n+1}; F \leq n \leq K - 2 \tag{9}$$

$$\mu' P_{2,K-1} = \varepsilon P_{1,K-1} \tag{10}$$

Solving recursively (1)–(10), we obtain

$$
P_{1,n} = \begin{cases} \frac{K\lambda}{\mu}P_{0,0}; & n = 0 \\ \frac{K\lambda^{n+1}}{\mu^{n+1}} \prod_{i=1}^{n}(K-i)\delta_i P_{0,0}; & 1 \leq n \leq F-2 \\ \frac{K\lambda^{n+1}}{\mu^{n+1}} \prod_{i=1}^{n}(K-i)\delta_i P_{0,0}; & n = F-1 \\ \frac{K\lambda^{n+1}}{\mu^{n+1}} \frac{\mu^{n+1-F}g(n)}{R} \prod_{i=1}^{n}(K-i)\delta_i \prod_{i=F}^{n}\frac{1}{\delta_i}P_{0,0}; & F \leq n \leq K-2 \\ \frac{K\lambda^{n+1}}{\mu^{n+1}} \frac{\mu^{n+1-F}}{R} \prod_{i=1}^{n}(n+1-i)\delta_i \prod_{i=F}^{n}\frac{1}{\delta_i}P_{0,0}; & n = K-1 \end{cases}
\tag{11}
$$

$$
P_{0,n} = \begin{cases} \frac{\lambda_1\lambda_0}{\mu\gamma}P_{0,0}; & n = 1 \\ \frac{K\lambda^n}{\mu^n} \prod_{i=1}^{n-1}(K-i)\delta_i \frac{\lambda_n}{\gamma}P_{0,0}; & 2 \leq n \leq F-2 \\ \frac{\mu}{\gamma+\lambda_{n+1}} \frac{K\lambda^{n+1}}{\mu^{n+1}} \prod_{i=1}^{n}(K-i)\delta_i P_{0,0}; & n = F-1 \\ \frac{\mu\lambda_n g(n)}{(\gamma+\lambda_n)R} \frac{K\lambda^n}{\mu^n} \prod_{i=1}^{n-1}(K-i)\delta_i P_{0,0}; & n = F \end{cases}
\tag{12}
$$

$$
P_{2,n} = \frac{\varepsilon K \lambda^K}{\mu'\mu^K} \frac{\mu^{K-F}}{R} \prod_{i=1}^{K-1}(K-i)\delta_i \prod_{i=F}^{K-1}\frac{1}{\delta_i}P_{0,0}; \quad F \leq n \leq K-1
\tag{13}
$$

where

$$
g(F) = \left(\mu^{K-F-2}(\mu+\varepsilon) + \varepsilon \sum_{j=1}^{K-F-2} \mu^{K-F-2-j} \prod_{i=1}^{j} \lambda_{K-i} \right),
$$

$$
R = \varepsilon \prod_{i=F+1}^{K-1}(\lambda_i) + \mu \frac{1}{\delta_F} g(F) \text{ and } \delta_i = \frac{\gamma+\lambda_i}{\gamma}
$$

Here $P_{0,0}$ is determined using the normalizing condition

$$
P_{0,0} + \sum_{n=1}^{F}P_{0,n} + \sum_{n=0}^{K-1}P_{1,n} + \sum_{n=F}^{K-1}P_{2,n} = 1
\tag{14}
$$

4 Performance Measures

To explore the performance of the system, various system indices namely average number of failed machines in the system $E[N_S]$, average number of failed machines in the queue $E[N_q]$ and in the retrial orbit $E[N_R]$ are obtained. Other indices such as throughput (TP), expected delay time (EDT), machine availability $(M.A.)$ and operative efficiency $(O.E.)$ are also obtained. To analyze the status of the server, long run probabilities P_I and P_{SB} have also been established. Various indices in terms of steady state probabilities determined in previous section, are as follows:

$$E[N_S] = \sum_{n=0}^{F} nP_{0,n} + \sum_{n=0}^{K-1} (n+1)P_{1,n} + \sum_{n=F}^{K-1} (n+1)P_{2,n} \qquad (15a)$$

$$E[N_q] = \sum_{n=0}^{K-1} nP_{1,n} + \sum_{n=F}^{K-1} nP_{2,n} \text{ and } E[N_R] = \sum_{n=0}^{F} nP_{0,n} \qquad (15b\text{--}c)$$

$$TP = \mu \sum_{n=0}^{K-1} P_{1,n} + \mu' \sum_{n=F}^{K-1} P_{2,n} \text{ and } EDT = \frac{E[N_s]}{TP} \qquad (15d\text{--}e)$$

$$M.A. = 1 - \frac{E[N_s]}{K} \text{ and } O.E. = 1 - \sum_{n=0}^{F} P_{0,n} \qquad (15f\text{--}g)$$

$$P_I = \sum_{n=0}^{F} P_{0,n} \text{ and } P_{SB} = \sum_{n=0}^{K-1} P_{1,n} + \sum_{n=F}^{K-1} P_{2,n} \qquad (15h\text{--}i)$$

Cost Function

For any machining system, it is a crucial issue to evaluate the optimal service rate which minimizes the total cost. For constructing the cost function, the cost elements associated with different activities are used. Our main aim is to evaluate the optimal value of μ that minimizes the expected total cost per unit time. We formulate cost function $TC(\mu)$ which gives the total cost per unit time, as follows:

$$TC(\mu) = C_I P_I + C_B P_{SB} + C_H E[N_q] + \mu C_F + \mu C_A + C_O E[N_R] \qquad (16)$$

Here per unit cost elements related to duration for which server being idle (busy) is $C_I(C_B)$, holding cost (cost incurred in retrial orbit) is $C_H(C_O)$, cost for providing service to the customer when the arrivals are not allowed (allowed) is $C_F(C_A)$.

5 Numerical Simulations

In this section, the numerical simulation results to analyze the influence of different parameters on various system indices is presented. The numerical results are obtained by coding the computer program in *MATLAB* software. For the illustration purpose, the default parameters are fixed as $K = 7$, $F = 4$, $\lambda = 3$, $\mu' = \mu = 8$, $\gamma = 1$, $\varepsilon = 1$.

To evaluate the optimal service rate (μ^*) and corresponding optimal cost $TC(\mu^*)$, three costs sets have taken as given in the Table 1.

Table 1. Cost sets with different cost elements

Cost set	C_I	C_B	C_H	C_F	C_A	C_o
I	10	10	120	10	5	90
II	5	5	125	10	5	90
III	10	10	122	14	5	90

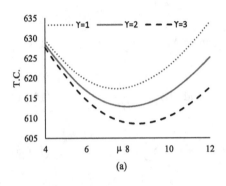

(a)

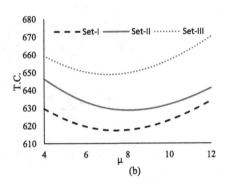

(b)

Fig. 2. (a–b) shows the variation in cost T.C. for varying different parameters (retrial rate, cost elements)

The optimal service rate 'μ^*' is evaluated by heuristic approach by plotting the curves for cost vs. 'μ' in Fig. 2(a–b). Table 2 displays the optimal service rate 'μ^*' and corresponding optimal cost $TC(\mu^*)$ for different cost elements C_I, C_B, C_H and C_F.

Based on numerical results, the trends of system performance indices is interpreted as follows:

(i) **Effect of arrival rate** (λ)

From Table 3, it is observed that as λ increases, the number of failed machines in the queue, operative efficiency, probability of server being busy and total cost per unit time of the system increase but the machines availability and probability of server being idle decrease. From Fig. 3(a), it is clear that as λ increases, the system size increases but changes are much remarkable for lower values of λ.

To demonstrate the sensitivity of performance indices with respect to different parameters we compute numerical results and display in Fig. 3(a–b) and Tables 3, 4 and 5.

(ii) **Effect of service rate** (μ)

In Table 4, we observe that as μ increases, the number of failed machines in the queue, operative efficiency, probability of server being busy and cost per unit time of the system decrease. On the other hand, machine availability and probability of server being idle increase with μ. It can be easily concluded that the performance in terms of better service in an economic way can be entranced by optimal choice of service rate μ.

Table 2. Effect of (λ, γ) on optimal $(\mu^*, TC(\mu^*))$

(λ, γ)	$(\mu^*, TC(\mu^*))$		
	Cost set-I	Cost set-II	Cost set-III
(1,1)	(7.342, 476.8671)	(7.762, 478.8706)	(6.039, 506.5275)
(2,1)	(7.244, 568.0753)	(7.735, 575.8664)	(6.113, 600.2447)
(3,1)	(7.477, 617.2652)	(8.094, 628.5784)	(5.851, 650.9819)
(3,2)	(8.013, 612.7879)	(8.643, 623.5170)	(6.301, 648.2109)
(3,3)	(8.510, 608.5357)	(9.154, 618.7555)	(6.703, 645.5289)

Table 3. Various performance measures for varying values of λ

λ	$E[N_q]$	$E[N_R]$	P_I	P_{SB}	M.A.	O.E.	T.C.
1	1.3203	2.0988	0.6242	0.3758	0.4578	0.3781	477.31
2	2.4352	1.6284	0.4282	0.5718	0.3396	0.5718	568.78
3	3.2110	1.1357	0.2920	0.7080	0.2821	0.7080	617.53
4	3.7497	0.7985	0.2038	0.7962	0.2434	0.7962	651.83
5	4.1295	0.5752	0.1465	0.8535	0.2154	0.8535	677.31

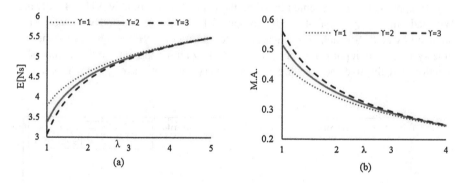

Fig. 3. (a–b) represents the variation in $E[N_s]$ and M.A. for varying λ

Table 4. Various performance measures for varying values of μ

μ	$E[N_q]$	$E[N_R]$	P_I	P_{SB}	M.A.	O.E.	T.C.
4	4.3530	0.4128	0.1064	0.8936	0.2112	0.8936	629.51
6	3.7408	0.7848	0.2015	0.7985	0.2480	0.7985	619.52
8	3.2110	1.1357	0.2920	0.7080	0.2821	0.7080	617.53
10	2.7813	1.4356	0.3703	0.6297	0.3099	0.6297	622.96
12	2.4370	1.6834	0.4360	0.5639	0.3321	0.5640	633.95

Table 5. Various performance measures for varying values of γ

γ	$E[N_q]$	$E[N_R]$	P_I	P_{SB}	M.A.	O.E.	T.C.
1	3.2110	1.1357	0.2920	0.7080	0.2821	0.7080	617.53
1.5	3.2172	1.0999	0.2854	0.7146	0.2853	0.7146	615.05
2	3.2237	1.0660	0.2789	0.7211	0.2883	0.7211	612.79
2.5	3.2304	1.0339	0.2726	0.7274	0.2910	0.7274	610.70
3	3.2371	1.0036	0.2665	0.7334	0.2935	0.7335	608.77

(iii) **Effect of retrial rate (γ)**

It is clear from Table 5 that as γ increases, the number of failed machine in the queue, operative efficiency, machine availability and probability of server being busy increase while probability of server being idle and total cost per unit time decrease. This trend of system indices with respect to retrial rate is quite obvious as retrial rate adds up more units in the queue orbit for the repair job.

6   Adaptive Neuro-Fuzzy Interface System (ANFIS)

Neuro-fuzzy technique can be easily applied for the numerical experiment to generate the numerical values of the system indices of the machining system with retrial. Here, ANFIS approach is implemented by using the neuro-fuzzy tool in MATLAB software. We made the comparison of results computed by recursive method with the results obtained by ANFIS which authenticates the feasibility of ANFIS controller for the real time systems. The input parameter γ is fuzzified by taking the trapezoidal membership function and the five membership values as very low, low, average, high, and very

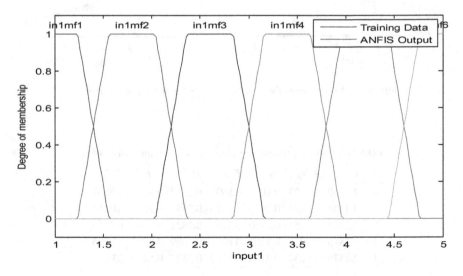

Fig. 4. Membership function for M.A., λ

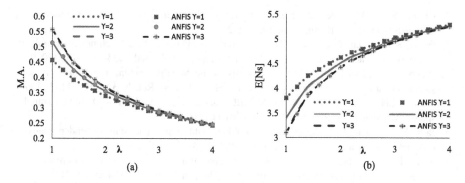

Fig. 5. (a–b) represents the variation in M.A. and $E[N_s]$ for varying λ

high. The membership function for γ is shown in Fig. 4. The ANFIS results for the machine availability and average number of failed machines in the system have been plotted by using tick marks in Fig. 5(a–b) along with the results obtained by using recursive method. The dotted and smooth lines depict the results obtained by recursive method. From these figures, it is clear that there are quite close results by both analytical and an ANFIS approaches. Based on numerical experiment, and trends of the results displayed in Fig. 5(a–b), it can be concluded that the results obtained by ANFIS are at par with the analytical results.

7 Conclusion

In this article, M/M/1/K model is investigated for the machining problem with retrial attempts by incorporating several realistic features such as F-policy, state dependent arrival rate and admission control policy. A recursive method to derive the steady state probabilities of the system in explicit form has been successfully implemented. Numerical illustrations are taken into consideration to analyze how different parameters affect the system performance measures. Our study may provide a valuable insight to the system designers and decision makers for the up gradation of existing systems. The numerical results compared with the results obtained by using ANFIS, demonstrate the feasibility of the scope and usefulness of neuro-fuzzy approach for the performance prediction of complex embedded systems.

References

1. Artalejo, J.R., Falin, G.I.: Standard and retrial queueing systems: A comparative analysis. Rev. Mat. Complutense **15**, 101–129 (2002)
2. Artalejo, J.R.: Accessible bibliography on retrial queues. Math. Comput. Model. **51**, 1071–1081 (2010)
3. Sherman, N.P., Kharoufeh, J.P.: An M/M/1 retrial queue with unreliable server. Oper. Res. Lett. **69**, 697–705 (2006)

4. Gupta, S.M.: Machine interference problem with warm spares, server vacations and exhaustive service. Perform. Eval. **29**, 195–211 (1997)
5. Wang, K.H., Chen, W.L., Yang, D.Y.: Optimal management of the machine repair problem with working vacation: newton's method. J. Comput. Appl. Math. **233**, 449–458 (2009)
6. Wang, K.H., Siwazlian, B.D.: Cost analysis of the M/M/R machine repair problem with spares operating under variable service rates. Microelectron. Reliab. **32**, 1171–1183 (1992)
7. Wang, K.H.: Cost analysis of the M/M/R machine-repair problem with mixed standby spares. Microelectron. Reliab. **33**, 1293–1301 (1993)
8. Jain, M.: An (m, M) machine repair problem with spares and state dependent rates: a diffusion process approach. Microelectron. Reliab. **37**, 929–933 (1997)
9. Yang, D.Y., Chiang, Y.C.: An evolutionary algorithm for optimizing the machine repair problem under a threshold recovery policy. J. Chin. Inst. Eng. **37**, 224–231 (2014)
10. Gupta, S.M.: Interrelationship between controlling arrival and service in queueing systems. Comput. Oper. Res. **22**, 1005–1014 (1995)
11. Yang, Y.D., Wang, K.H., Wu, C.H.: Optimization and sensitivity analysis of controlling arrivals in the queueing system with single working vacation. J. Comput. Appl. Math. **234**, 545–556 (2010)
12. Wang, K.H., Yang, D.Y.: Controlling arrivals for a queueing system with an unreliable server Newton-Quasi method. Appl. Math. Comput. **213**, 92–101 (2009)
13. Jain, M., Bhagat, A.: Transient analysis of finite F-policy retrial queues with delay repair and threshold recovery. Natl. Acad. Sci. Lett. **38**, 257–261 (2015)
14. Goswami, V.: Relationship between randomized F-policy and randomized N-policy in discrete-time queues. OPSEARCH **53**, 131–150 (2016)
15. Lin, Z.C., Liu, C.Y.: Application of an adaptive neuro-fuzzy inference system for the optimal analysis of chemical–mechanical polishing process parameters. Int. J. Adv. Manufact. Technol. **18**, 20–28 (2003)
16. Jain, M., Upadhyaya, S.: Threshold N-Policy for degraded machining system with multiple type spares and multiple vacations. Qual. Technol. Quant. Manag. **6**, 185–203 (2009)
17. Kumar, K., Jain, M.: Threshold N-policy for (M, m) degraded machining system with K-heterogeneous server, standby switching failure and multiple vacation. Int. J. Math. Oper. Res. **5**, 423–445 (2013)

Landslide Early Warning System Development Using Statistical Analysis of Sensors' Data at Tangni Landslide, Uttarakhand, India

Pratik Chaturvedi[1](✉), Shikha Srivastava[1], and Preet Bandhan Kaur[2]

[1] Defence Terrain Research Laboratory, DRDO, Metcalfe House, Delhi, India
`prateek@dtrl.drdo.com, shikha200591@gmail.com`
[2] Thapar University, Patiala, Punjab, India
`preetbandhan2@gmail.com`

Abstract. Rainfall induced landslides account for over 200 deaths and loss of over Rs.550 crores annually in Himalaya. Literature suggests sensors based site specific Early Warning System (EWS) to be feasible and economic to curtail losses due to landslides for high risk areas. Area selected for current study is Tangni landslide located in Chamoli district of Uttarakhand state, India due to high anticipated risk to the local community residing nearby. For realization of EWS, a near real time instrumentation setup was installed on the slope. The setup measures pore water pressure, sub-surface deformations, and surface displacements along with rainfall. Regression analysis models are developed using antecedent rainfall and deformation data which are further used to find out thresholds for sensors based on z-scores. In future using the results from the sensors installed in the field and laboratory characterizations, numerical analyses will be applied to develop a process based model.

Keywords: Landslide · Antecedent rainfall · Slope deformation · Prediction · Early Warning System (EWS)

1 Introduction

Mountains constitute different variety of rock structures and soils. Excessive weathering causes the slope material to disintegrate and decompose resulting in the fall of the debris down the slope under the effect of gravitational force. This fall of material from the mountain is termed as Landslide. The major reason that triggers debris flow (phenomena of Landslide) is heavy and prolonged rainfall. The other side factors causing landslides are deep excavations on slope for construction of roads, buildings, canal and mining without appropriate sewage of rain and ground water and disposal of debris [1]. Landslide occurrences are pretty much common in the Indian Subcontinent. The Indian Himalayas being the hilly terrain comprises of tectonically unstable geological formations subjected to landslides which lead to huge destruction of mankind and other utilities. In Himalayas, landslides are triggered primarily in the monsoon season. The annual average rainfall of Uttarakhand state where the study area lies is 1,580.9 mm. The mean

© Springer Nature Singapore Pte Ltd. 2017
K. Deep et al. (eds.), *Proceedings of Sixth International Conference on Soft Computing for Problem Solving*, Advances in Intelligent Systems and Computing 547,
DOI 10.1007/978-981-10-3325-4_26

rainfall during monsoon period (June to September) contributes 77.7% of annual rainfall and hence plays a significant role in triggering landslides [2].

There are two well-known approaches for the data analysis of parameters responsible for causing landslides: Statistical analysis and Process based analysis. Although in the present scenario, at the regional level lot of research is going on for calculating statistical threshold for probabilistic assessment of occurrence of landslide which is no doubt cost effective and accepted worldwide but to understand the kinematics and actual physical phenomenon of particular landslide lot of site specific data in terms of instrumentation setup and survey is required. According to the earlier research, physically based or process based approach can effectively be applied only over small sites [3]. In this study, the process based approach is applied and statistical analysis of the sensors' data is carried out for a specific site of Uttar-akhand. The landslide site named Tangni (Fig. 1) is located on NH-58 which connects Badrinath Dham, an ancient Hindu worship center and nearby hilly areas. Any kind of slope triggering can create the blockade for the traffic along this important hill route. Therefore, it is required to understand the physical processes involved in triggering of Tangni land-slide. Since rainfall is the most contributing factor for the landslide, an attempt has been made to set up a warning system on the basis of interaction between rainfall and landslide. Early warning systems (EWSs) may be helpful in reduction of the damages caused due to natural hazards like landslides. Landslide monitoring and warning can be achieved by monitoring rainfall intensity and changes in soil-properties [4]. This paper describes the relationship of site specific parameters triggering landslides such as regression of antece-dent rainfall with IPI sensor movement, predicting the future displacement of sensors based on the linear regression equation and then finding Z scores to set thresholds on the sensors values to generate alerts.

Fig. 1. Tangni landslide on NH-58

2   Background

Earlier research works have shown that in general, continuous rainfall in a landslide prone region is the main triggering factor of slope failures [5]. Antecedent or cumulative threshold of rainfall helps in predicting the landslides [6]. Studying the relationship in rainfall intensity and debris flow, researchers have found how an extreme rainfall event results in landslides near Darjeeling in the eastern Indian Himalayas [7]. It is also observed that steep slopes have landslips when rainfall intensity in the past 24 h reaches 130–150 mm or the antecedent rainfall for past 3 days is greater than 180 mm [8]. Numerical analyses of rainfall-induced slope failures have been done to study the controlling parameters [9–11] and the effects of cumulative rainfall on rainfall-induced slope failures [9, 12]. Deep seated landslides are the consequences of prolonged and less intense rainfall events [13]. As Tangni is a deep seated landslide, so this phenomenon may hold true for this landslide also. In these types of slope i.e. deep-seated landslides, antecedent rain plays conclusive role in predicting landslide. So, first hypothesis is that rainfall of past few days will play a significant role in triggering of Tangni landslide. Second hypothesis of this work is to develop a regression model between the antecedent rainfall and sensor displacement. The minimum monthly cumulative rainfall belongs to the month of December (1.26 mm) and the overall rainfall in the month of July 2013 (464.48 mm). This rainfall data is obtained from the NASA's Tropical Rainfall Measuring Mission (TRMM) data for 2012–2014 [19]. Approximately 20 landslide events have happened in the last 10 years near Tangni and for our research, we have worked on rainfall and movement data of 2012–2014.

3   Study Area: Tangni Landslide

Tangni landslide (Fig. 1) is located about 59 km from Joshimath at an altitude of about 1450 m near Tangni village with specific geographic location as Latitude 30° 27' 54.3" N and Longitude 79° 27' 26.3" E. The road direction and slide face direction is N60°. The uphill and downhill slopes are 650 m and 800 m respectively. The vegetation is totally absent on slide face, but margins are marked by mixed pine and oak forest. Some of the trees are observed to be tilting towards left side of the slide face which is funnel shaped and situated along a small stream which leads to considerable erosion of the slide surface. The width of road affected by slide is about 25 m. The slide is almost 1.2 km long and 0.6 km wide. The toe of landslide located below the road is the most active zone. It is a translation type of landslide. The annual rainfall of the area is around 1580 mm. The rainfalls show a peak in rainy season: the months of July and August display the maximum of monthly average rainfall of 426 mm each month [2].

4   Geology of the Study Area

The study area is having Precambrian lithology of Garhwal region. The study area lies in NW Lesser Himalaya. Main Central Thrust (MCT) and other faults are in vicinity of the study area. Thus, the area is tectonically vigorous. River Alaknanda and its tributaries are drained in that area. The rocks of the study area comprise of phyllites/slates and dolomites. The slope is continuous and is inclined at about 35° above road level and 42° below road level in dip direction of N10°. The right side of the slide area is consisting of jointed and fractured rocks. The left side of the study area consists of loose soil and debris material [14].

5   Methodology

As shown in Fig. 2, the research work can be described in the following steps:

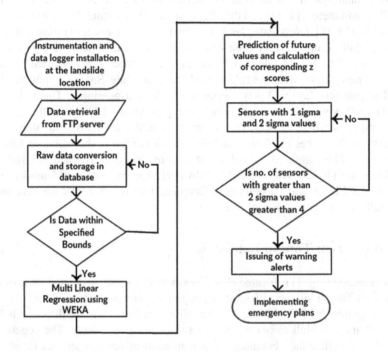

Fig. 2. Framework of landslide deformation model

5.1   Instrumentation Setup and Data Retrieval

To realize EWS, it is first necessary to install instruments which can give us the real time changes in slope. Monitoring system for landslide that has been used in current study includes field sensors, data acquisition systems, data communication and software for remotely collecting sensor data, processing and generating alerts. The field sensors used at the Tangni Landslide site are in-place inclinometers (IPI), vibrating wire piezometers, tipping bucket rain gauge and extensometers. IPIs consist of string of MEMS sensors installed in casing within a borehole. Multiple IPI's are linked together by Gauge tubes. Each IPI is further connected to the data logger which is power source of the sensors. Data logger records the data and communicates it to the server via GSM mode of communication. IPI sensor records the movement of slope in terms of voltage. Extensometers are used to measure the displacement of this slow moving landslide. Vibrating wire piezometers are used to measure pore water pressure to determine slope stability and safe rates of excavation. Piezometers are designed to convert the water pressure into the units of frequency signal (Hz). Tipping bucket rain gauge measures the amount of rain that has fallen. Rain falling in the pivot funnel shaped device drips into the calibrated bucket balanced on pivot.

Nine Boreholes using DTH drilling method were dug on the body of selected landslide to install piezometers and In-place-inclinometer. Each set of IPI consists of five Bi-axial sensors. Initial water level observations were taken by vibrating wire piezometers. Movements and deformation were also computed by determining the values by using "In-place" inclinometers. On the hypothesis that antecedent rainfall is responsible for landslide triggering, daily precipitation data (2012–2014) and IPI sensors data is collected in server from this remote location.

5.2 Raw Data Conversion, Filtering and Storage

The remote monitored data collected at the base station needs further conditioning to get some meaningful results. IPI data received in voltage terms is converted to engineering units (mm/m) using calibration factors from unique sensor specific sheet based on serial numbers. Conversion equations for IPI sensors are given by:

$$Deformation\,(mm/m) = b0 + b1 * V + b2 * V * V + b3 * V * V * V \tag{1}$$

Where $b0, b1, b2, b3$ are unit less calibration factors [14].

The total displacement value is calculated using the gauge length of 3 m.

$$Displacement\,(mm) = Deformation\,(mm/m) * Gauge\,Length\,(m) \tag{2}$$

The data received is erroneous because of factors like insignificant ground movements, high voltage power supply disturbances and untimely power cuts etc. The data received is filtered by using various assumptions:

(1) Eliminating out of range values by sensors (IPI data range - −173.8 mm/m to +173.8 mm/m),
(2) Sudden peaks and valleys in hourly data for an hour or two and
(3) Values obtained in between two hours were deleted.

Using nearest neighborhood method, the values of the missing hours is predicted keeping in view the factors like rainfall in that area for that particular hour. To ease the handling and analysis of data, the hourly day wise data is converted to average day wise data using Macro programming in MS-Excel. Thus, final data set contains averaged day wise data for A and B axis (25 sensors each).

5.3 Multi-linear Regression Using WEKA Software Tool

Waikato Environment for Knowledge Analysis (WEKA) is a open-source tool for data mining applications. Using WEKA tool, many of the machine-learning algorithms can be applied to the dataset [17].

Multi-linear regression analysis using WEKA attempts to model a relationship between two or more antecedent rainfall days (explanatory variables) and the displacement of sensor (responsive variable) by fitting a linear relationship between the variables [18]. This method is useful in predicting the sensor displacement for future days depending on the cumulative previous day's rainfall. Formally, the model for multiple linear regression, given n observations, is

$$Yi = Ci + Ci - 1 * Xp1 + Ci - 2 * Xp2 + Ci - 3 * Xp3 + \cdots \ldots + C1 * Xpi - 1 \tag{3}$$

Where, Yi represents the sensor displacement,
Xpi represents the cumulative rainfalls of p_i previous days and
Ci represents the multi linear equation coefficients,
$i = 1, 2, 3, \ldots n$.

5.4 Z-score Analysis

Z-score is statistical analysis of score's relationship to the mean in group of scores. It refers to number of standard deviations, a data point is away from the mean. Z score can be positive or negative which implies that Z score = 0 (score is same as mean.), Z score = 1 (data point is one standard deviation away from the mean), Z score = 2 (data point is two standard deviations away from the mean) [15]. Basic formula for Z score calculation is:

$$Zi = (xi - \mu)/s \tag{4}$$

Where s - standard deviation,
μ - mean of the data,
x_i - actual data point (displacement in consecutive days)

From the z-score, it is reported in the earlier research works that mod $|z|$ value greater than 2 is used for outlier detection in a dataset [20, 21]. These outlier values can be further used to set thresholds on the sensors' values as whenever multiple sensors show high variation from their usual pattern; this may imply high probability of the landslide occurrence.

After calculation of z-scores for all the sensors' data from Eq. 4, the change in sensors' value is compared to it. Change in displacement is basically divided in three classes low, medium and high on the basis of comparison with z-score. To maintain the simplicity only three warning levels (Table 1) for sensor displacement have been created. (Low level, medium level and high level alert) [16].

Table 1. Landslide warning threshold obtained from z-score.

Warning level	Colour	Minimum no. of sensors with greater than 2 sigma values
Low	Yellow	1
Medium	Orange	2–3
High	Red	4 or more

6   Results and Discussion

6.1   Plots of Sensors' Data and Rainfall

From the plots of sensors' data of different boreholes as shown in Figs. 3, 4 and 5, it can be clearly seen that the maximum displacement in the sensors happened during monsoon season (July, 2013–Aug, 2013 & July, 2014–Aug, 2014). This gives a very significant result that severity of landslide is very high during monsoon and precaution needs to be taken for those months. The analogy behind the oscillations in some parts can be related to human tooth as when the tooth is about to fall, it has maximum oscillations from its place before falling. Therefore, landslides can be deduced from the above plots (majorly in monsoon). This may be due to fact that slope became wet due to continuous rain in the month of July and thus became highly prone to landslide [2]. One interesting finding was that the part of landslide below the road i.e. the toe section was moving faster in comparison of the rest of the body which is consistent with literature regarding other multiple retrogressive landslides [14]. Figures 3, 4 and 5 below show the borehole wise plots (depth of IPI's in borehole vs. dates.)

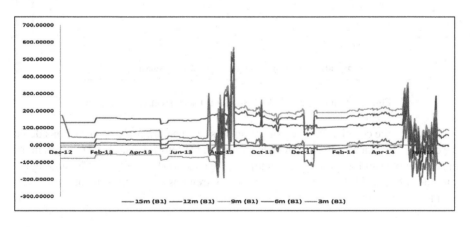

Fig. 3. Profile of Borehole-1(*A-axis*) IPI sensors at various depths

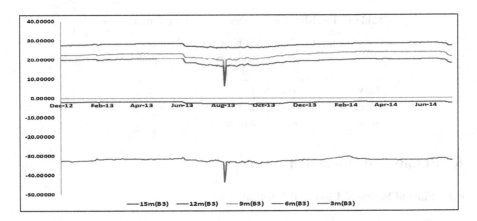

Fig. 4. Profile of Borehole-3(*A-axis*) IPI sensors at various depths

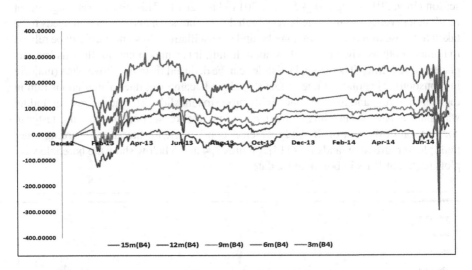

Fig. 5. Profile of Borehole-4(*B-axis*) IPI sensors at various depths

To verify the correlation between the sensors' data and rainfall, we have considered the data set (displacement and antecedent rainfall) for sensor located at 3 m depth in the third borehole (middle part of landslide) and it gives good correlation. Plot among the variables of data set (Fig. 6) shows satisfactory trend as sensor movement varies with the increase in rainfall in almost similar pattern.

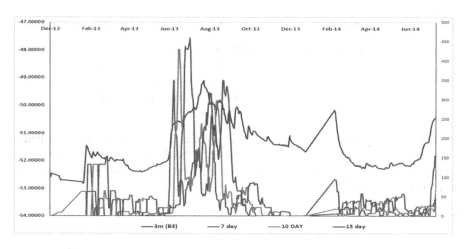

Fig. 6. Displacement and Antecedent rainfall (*7 day, 10 day, and 15 day*) plot of sensor at 3 m depth, borehole no. 3

6.2 Regression Model Development

Figure 7 represents the output screen of WEKA where linear regression classifier is applied on data set mentioned above. The above mentioned classifier is applied on the dataset by dividing the data to 60% in training and 40% in testing data as shown below:

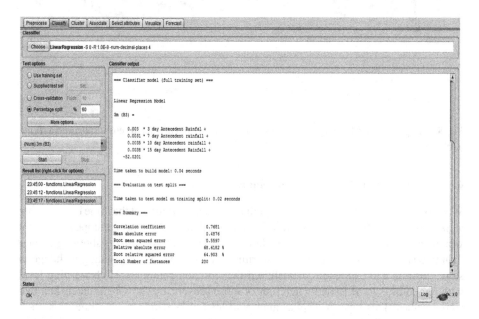

Fig. 7. Regression model of sensor at 3 m depth, third borehole using antecedent rainfall.

Developed model through the classifier gives following linear regression equation.

$$
\begin{aligned}
\textit{Displacement of the sensor at 3 m (B3)} = \ & (0.003) * 3\,\textit{day Antecedent Rainfall} \\
& + (0.0031) * 7\,\textit{day Antecedent rainfall} \\
& + (0.0035) * 10\,\textit{day Antecedent rainfall} \quad (5)\\
& + (0.0038) * 15\,\textit{day Antecedent Rainfall} \\
& + (-52.0201)
\end{aligned}
$$

Time taken by the classifier to build the model is 0.2 s and correlation coefficient value obtained is 0.7681 with mean absolute error 0.4876. From the linear regression equation, sensor displacement of future can be predicted. Below graph shows the predicted values (Fig. 8).

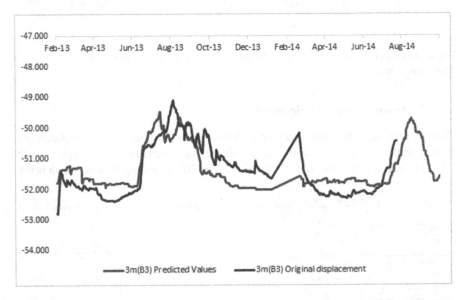

Fig. 8. Graph obtained for original sensor movement (*red*) and predicted sensor movement (*blue*) (Color figure online)

6.3 Z-score Based Thresholds Development

When only one sensor is showing greater than two sigma values, at that time low warning signal yellow in color is to be issued which implies that there has been a slope movement which may or may not lead into a landslide event. Two or three sensors showing greater than two sigma value imply medium warning (orange colored) suggesting a change in slope position and can lead to landslide if the situation continues. High warning level with red color is to be issued when four or more sensors are showing greater than two sigma values. This type of displacement will lead to landslide in any time in coming future (refer to Table 1). According to data of 25 sensors, if for a particular date 4 or more sensors are having $|z| >= 2$, this implies that there is high probability of landslide. Figure 9 shows the table with Z score values of sensors and last column represents the number of sensors which have greater than 2 sigma values. The landslide warnings have been shown as yellow color for low level warning, orange color for moderate warning level and red color for high level alert. The cells marked green are the |z| values greater than or equal to 2 sigma.

Date	15m (B1)	12m (B1)	9m (B1)	6m (B1)	3m (B1)	15m (B2)	9m (B2)	15m (B3)	12m (B3)	9m (B3)	6m (B3)	3m (B3)	3m (B4)	15m (B5)	12m (B5)	6m (B5)	No.of sensors with greater than 2 sigma values
04-08-2013	0.01	0.13	0.12	0.18	1.56	0.02	0.24	2.45	0.99	0.11	0.78	0.36	0.01	0.05	0.03	0.21	1
05-08-2013	0.01	0.04	1.17	0.53	3.21	0.48	0.40	6.14	5.14	0.11	0.79	1.86	1.03	0.03	0.05	1.32	3
06-08-2013	0.00	0.06	1.30	4.80	1.35	0.34	0.51	1.52	0.15	0.51	1.38	0.24	0.50	0.02	0.02	0.39	1
07-08-2013	0.00	0.02	1.44	1.13	5.66	0.23	0.05	0.22	0.94	0.16	2.44	0.34	0.48	0.08	0.05	0.02	2
08-08-2013	0.90	0.77	18.23	13.58	10.16	0.22	3.66	1.29	0.08	0.24	0.96	0.39	3.20	0.05	0.05	0.81	6
09-08-2013	0.15	1.24	6.05	11.01	3.47	0.00	1.17	3.73	1.94	0.05	1.00	0.62	2.89	0.92	0.02	0.22	5
10-08-2013	0.30	3.43	0.37	7.60	2.79	0.20	0.18	0.47	0.50	0.07	0.62	0.38	0.13	0.43	0.04	0.09	3
11-08-2013	0.86	0.89	6.98	0.06	2.85	0.64	1.12	0.89	0.34	0.06	0.37	0.54	0.89	0.50	0.06	1.37	2
12-08-2013	0.62	3.41	2.25	6.44	0.90	0.66	0.65	0.71	1.30	0.05	0.46	2.21	1.01	0.18	0.03	1.97	4
13-08-2013	3.66	1.12	2.02	2.29	0.89	0.05	0.36	1.30	0.98	0.07	0.59	0.56	0.30	0.56	0.03	2.44	5
14-08-2013	3.90	2.77	2.31	1.91	1.63	0.28	0.40	0.58	0.32	0.71	0.34	0.56	0.22	0.09	0.04	0.24	3
15-08-2013	0.68	0.95	1.21	0.44	0.03	0.01	0.53	0.53	0.67	18.70	0.66	0.20	0.59	0.66	0.04	0.40	1
16-08-2013	0.59	8.56	3.17	3.46	0.82	0.56	0.65	1.32	0.98	3.35	0.10	0.18	1.18	0.18	0.05	1.18	4
17-08-2013	3.87	5.59	0.54	4.63	0.05	0.91	0.44	0.58	0.58	8.82	0.74	0.33	0.23	0.38	0.04	0.23	4
18-08-2013	0.98	0.79	0.90	0.01	0.10	0.75	0.03	0.38	0.29	7.47	1.43	0.60	0.07	0.14	0.06	2.09	7
19-08-2013	0.42	0.33	0.29	1.62	0.04	0.19	0.23	0.42	1.04	0.10	0.35	0.26	0.17	0.08	0.03	2.77	1
20-08-2013	0.30	0.01	0.08	0.27	0.01	0.31	0.22	0.00	0.14	0.08	1.73	0.06	0.11	0.04	0.05	0.77	0
21-08-2013	0.15	0.01	0.60	0.11	0.32	0.43	0.10	0.54	0.80	0.01	2.92	1.52	0.42	0.37	0.04	0.65	1
22-08-2013	0.01	0.01	0.40	0.26	0.20	0.18	0.50	1.48	0.53	0.15	4.70	2.31	0.73	0.13	0.04	0.16	2
23-08-2013	0.57	0.64	1.16	0.03	0.12	0.73	0.57	0.31	1.44	0.59	4.32	2.07	0.35	0.98	0.22	0.09	2
24-08-2013	0.63	4.50	1.45	0.13	0.14	0.49	0.02	0.13	1.52	0.31	2.75	1.83	1.24	7.82	1.38	0.23	3
25-08-2013	0.14	2.21	0.44	0.40	0.13	2.37	0.37	2.04	3.30	0.24	1.71	0.74	1.39	0.01	4.38	0.48	5

Fig. 9. Table with Z-score values and number of sensors showing greater than 2 sigma values (*last column*). (Color figure online)

7 Conclusion

In order to minimize the risks imposed by the Tangni landslide, an Early Warning System has been implemented. As landslides are mostly triggered in monsoons, case study involves two monsoon seasons (2012–2014). EWS currently being used at the site has 25 IPI sensors, 4 vibrating wire piezometers and rain gauges. The data logger collects hourly data and uploads it on the ftp. This instrumentation setup is capable of providing necessary information to monitor the dynamics of unstable slope. The sensor movement predicted using the regression model is found to closely match the sensor movement in real. Using this technique, a near real time EWS has been generated with different warning signals. Using the regression model, displacement rates are predicted which are further used to generate alerts. This research work can be carried forward by developing a model of unsaturated soil slope stability variations when it is subjected to rain water infiltration.

Acknowledgement. The authors would like to thank Director, DTRL, DRDO and a team of scientists involved in the development of the instrumentation system for development of early warning system for landslide.

References

1. Rai, P.K., Mohan, K., Kumra, V.K.: Landslide hazard and its mapping using remote sensing and GIS. J. Sci. Res. **58**, 1–13 (2014)
2. Kanungo, D.P., Sharma, S.: Rainfall thresholds for prediction of shallow landslides around Chamoli-Joshimath region, Garhwal Himalayas, India. Landslides **11**(4), 629–638 (2014)
3. Martelloni, G., Segoni, S., Fanti, R., Catani, F.: Rainfall thresholds for the forecasting of landslide occurrence at regional scale. Landslides **9**(4), 485–495 (2012)
4. Chae, B.-G., Kim, M.-I.: Suggestion of a method for landslide early warning using the change in the volumetric water content gradient due to rainfall infiltration. Environ. Earth Sci. **66**(7), 1973–1986 (2012)
5. Brand, E.W., Premchitt, J., Philipson, H.B.: Relationship between rainfall and landslides in Hong Kong. Paper presented at the 4th International Symposium on Landslides, Toronto, Canada (1984)
6. Baum, R.L., Godt, J.W.: Early warning of rainfall-induced shallow landslides and debris flows in the USA. Landslides **7**(3), 259–272 (2010)
7. Dahal, R.K., Hasegawa, S.: Representative rainfall thresholds for landslides in the Nepal Himalaya. Geomorphology **100**(3), 429–443 (2008)
8. Froehlich, W., Gil, E., Kasza, I., Starkel, L.: Thresholds in the transformation of slopes and river channels in the Darjeeling Himalaya, India. Mt. Res. Dev. **10**, 301–312 (1990)
9. Gasmo, J.M., Rahardjo, H., Leong, E.C.: Infiltration effects on stability of a residual soil slope. Comput. Geotech. **26**(2), 145–165 (2000)
10. Tsaparas, I., Rahardjo, H., Toll, D.G., Leong, E.C.: Controlling parameters for rainfall-induced landslides. Comput. Geotech. **29**(1), 1–27 (2002)
11. Rahardjo, H., Lee, T.T., Leong, E.C., Rezaur, R.B.: Response of a residual soil slope to rainfall. Can. Geotech. J. **42**(2), 340–351 (2005)
12. Rahardjo, H., Leong, E.C., Rezaur, R.B.: Effect of antecedent rainfall on pore water pressure distribution characteristics in residual soil slopes under tropical rainfall. Hydrol. Processes **22**(4), 506–523 (2008)
13. Bonnard, C., Noverraz, F.: Influence of climate change on large landslides: assessment of long-term movements and trends. In: International Conference on Landslides: Causes, Impacts and Countermeasures (No. LMS-CONF-2001-002), pp. 121–138 (2001)
14. Chaturvedi, P., Jaiswal, B., Sharma, S., Tyagi, N.: Instrumentation Based Dynamics Study of Tangni Landslide near Chamoli, Uttrakhand. IJRAT **2**, 127–132 (2014)
15. Weinberg, S.L., Abramowitz, S.K.: Data Analysis for the Behavioral Sciences Using SPSS. Cambridge University Press, Cambridge (2002)
16. Intrieri, E., Gigli, G., Mugnai, F., Fanti, R., Casagli, N.: Design and implementation of a landslide early warning system. Eng. Geol. **147**, 124–136 (2012)
17. Hall, M., Frank, E., Holmes, G., Pfahringer, B., Reutemann, P., Witten, I.H.: The WEKA data mining software: an update. SIGKDD Explor. **11**(1), 10–18 (2009)
18. Todorovski, L., Ljubič, P., Džeroski, S.: Inducing Polynomial Equations for Regression, pp. 441–452. Springer, Heidelberg (2004)
19. Huffman, G.J., Adler, R.F., Bolvin, D.T., Gu, G., Nelkin, E.J., Bowman, K.P., Hong, Y., Stocker, E.F., Wolff, D.B.: The TRMM multi-satellite precipitation analysis: quasi-global, multi-year, combined-sensor precipitation estimates at fine scale. J. Hydrometeor. **8**, 38–55 (2007)
20. Zhu, A.X., Wang, R., Qiao, J., Qin, C.Z., Chen, Y., Liu, J., Zhu, T.: An expert knowledge-based approach to landslide susceptibility mapping using GIS and fuzzy logic. Geomorphology **214**, 128–138 (2014)
21. Hodge, V.J., Austin, J.: A survey of outlier detection methodologies. Artif. Intell. Rev. **22**(2), 85–126 (2004)

Face as Bio-Metric Password
for Secure ATM Transactions

Arun Singh[✉], Jhilik Bhattacharya, and Shatrughan Modi

CSED, Thapar University, Patiala, India
arunsingh121@gmail.com, {jhilik,shatrughan.modi}@thapar.edu

Abstract. Security in banking transactions is major concern. In this paper we propose a method to use face as a biometric password along with PIN number in ATM machines. For face recognition in controlled environment Wavelet, LBP and PCA are used. This can be used to perform more secure ATM transactions. Even after authorizing the access to a user the algorithm continuously monitors the user. If a user chooses to leave the ATM machine without completing his/her ongoing transaction or moves his/her head away from the camera for 10 s then his/her session will be logged out automatically and he/she will have to restart a new transaction. Our technique can easily be combined with regular ATM machines and people do not even require any further knowledge to use ATM machines. This technique can also be combined with banking websites to provide more safe and secure online transactions for online users.

Keywords: Face recognition · Biometric password · Secure ATM machines · Authentication using live camera · Face as biometric password

1 Introduction

Automated teller machine (ATM) is a device that allows the bank customers to carry out banking transactions such as, deposits, withdrawal, fast cash, transfers, balance enquiries and mini statement etc. It has its roots embedded in customer accounts and records of a banking institution [12]. With the increase of ATM frauds, new authentication techniques are developed to overcome security problems of personal identification numbers (PIN). New authentication techniques are mostly created with the aim to be better than password or PIN [3,4]. Security is certainly the most important aspect when designing authentication systems. Using face as a biometric password is a novel technique to provide better security at ATM. The goal of face recognition is to recognize a person based on his facial appearance [9]. Face recognition is a passive biometric compared to other biometrics. It does not require a subject to be near or in contact with the sensor. It has numerous applications including security and surveillance, e-commerce, video databases. It has been an attractive research area in computer science.

K. Deep et al. (eds.), *Proceedings of Sixth International Conference on Soft Computing for Problem Solving*, Advances in Intelligent Systems and Computing 547, DOI 10.1007/978-981-10-3325-4_27

In the last two decades, a number of algorithms were proposed and have been developed. Fisherfaces [1], Bayesian Eigenfaces [7], 3D Morphable Model [2] and Elastic Bunch Graph Matching [6] to name just a few popular ones. These techniques have shown very good results on a number of controlled training sets. A commercial face recognition system called as FRVT [10] showed that face recognition has been deteriorated when there were differences in lighting and pose between images of enrollment and recognition. Basically face recognition has proven to be very challenging because someones face can appear differently in different images. Eyeglasses, weight change, facial hair, makeup and aging are intrinsic factors that create differences in appearance. Some extrinsic factors such as brightness, direction and camera view point can lead to significant image variability. We believe for accurate and unconstrained face recognition, the lighting conditions must not vary. In ATMs lighting conditions does not vary. It is a controlled environment and can easily be used for face recognition accurately. Using high resolution images in database and HD camera at ATM machine can provide uncompromised face recognition environment.

Our research mainly focuses on using face recognition as a biometric password along with ATM PIN to perform a successful and secure transaction. In our technique we use facial features of a person as his biometric identity to recognize him at ATM machine. We continuously capture his face till he completes his transaction. If he leaves ATM just after entering his PIN and does not complete the transaction. Then it reverts back all the changes made by him during the transaction. Otherwise the system allows him to carry on with the ongoing process. This technique can be used to provide safe and secure environment for deposit and withdrawal of money at ATMs and can also protect elderly people (in case somebody took their ATM card and also know their ATM PIN).

This paper is organized into five sections. Section 2 describes problem statement. In Sect. 3 we describe our proposed algorithm which includes face detection, false face removal and face recognition. Section 4 depicts experimental results. In Sect. 5 we will conclude our work.

2   Problem Statement

There are so many instances in past, where security of ATMs have been breached. Most of the instances are with old age people or people who are new at ATM. Elderly people used to ask for help and robbers get a chance to snatch their ATM PIN. After helping robbers used to run away with their ATM cards and then they can easily withdraw the money before that person will approach bank authorities to block his/her ATM card. By adding face as a biometric password in the ATM machines, this will increase the authentication level for people and can easily solve this issue. This technique can also be applied in online banking websites to protect the online theft in case of compromised passwords.

3   Proposed Algorithm

Figure 1 depicts our algorithm to use face recognition in ATM machines:

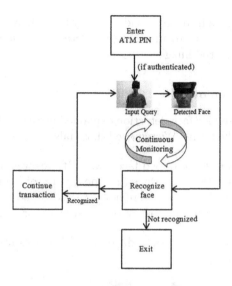

Fig. 1. Our algorithm

3.1   Face Detection

We applied face detection using Boosted Local Features [5] technique to detect the face position in an image. This technique used a learning algorithm to detect face position. A simplified version is discussed below:

There were total T possible features for a 303×303 pixel sub window. It was computationally expensive to calculate all the features hence a learning algorithm selected best features and used them for training the classifier. This algorithm constructed a strong classifier with linear combination of weighted weak classifiers t.

$$h(x) = sign(\Sigma_{t=1}^{T} \alpha_t h_t(x)) \tag{1}$$

Weak classifier was a threshold function. It was based on feature f_t

$$h_t(x) = \begin{cases} -s_t & \text{if} f_t < \phi_t \\ s_t & \text{otherwise} \end{cases}$$

The ϕ_t, s_t and coefficients α_t were determined in the training.

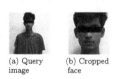

(a) Query   (b) Cropped
image   face

Fig. 2. Query image and cropped face

Figure 2(a) depicts a frame from video (query image). We applied the Boosted Local Features [5] technique to automatically detect face and we cropped the face of size 303×303 from the image.

3.2 False Face Removal

The algorithm we used for face detection explained in Sect. 3.1 also detected false faces sometimes in the frame. We removed those false faces by the following steps:

1. We detected eyes in the cropped face image. If eyes were also detected in the cropped face image than it reduced the possibility of being it as a false face.
2. In most of the false faces, eyes were not detected (in 95% cases). We removed those cropped faces from the list of detected faces for that frame.

Using the above steps we have reduced the false faces up to 98% (Fig. 3).

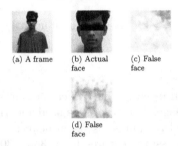

(a) A frame   (b) Actual face   (c) False face

(d) False face

Fig. 3. Actual face and false face

3.3 Face Recognition

First we captured frame from video (query image) and found the face in query image by using approach defined in Sect. 3.1. Then we applied Wavelet, LBP and PCA [11] on captured face image to recognize the person.

1. In face recognition, we resized query image into 303×303 dimension and performed single level two-dimensional wavelet decomposition.
2. We used the approximation and divided it into 31×31 blocks. After converting the approximation into 31×31 blocks, a 5×5 matrix was formed.
3. We calculated Local Binary Pattern [8] and Histogram of 5×5 matrix LBP For i, j: 1 to 5;

$$f(x) = \begin{cases} 1 & x_{ij} \geq p \\ 0 & \text{otherwise} \end{cases}$$

Here $f(x)$ is threshold function, p is center pixel and x_{ij} is neighboring pixel

4. After calculating LBP, we applied PCA and our algorithm for person identification which are discussed further in this section:
We had the training set of images $I_1, I_2, I_3, ..., I_m$

$$AverageFaceI_A = \frac{1}{m}\Sigma_{n=1}^{m} i_n \qquad (2)$$

Each face differed from average face by

$$\Theta_i = I_m - I_A \qquad (3)$$

Covariance matrix

$$C = \frac{1}{m}\Sigma_{n=1}^{m}\Theta_n\Theta_n^T \qquad = AA^T \qquad (4)$$

where $A = [\Theta_1, \Theta_2, \Theta_3, ..., \Theta_m]$

Now we found β, the principal component coefficient of covariance matrix C.

After finding β, we found ∇ and Ψ by using following formula:

$$\nabla = \beta \times \Theta^T \qquad (5)$$

$$\Psi = \nabla^T \Theta^T \qquad (6)$$

∇ and Ψ were the parameters calculated after applying wavelet, LBP and PCA on the database images.

5. The following algorithm was used for face recognition:
   (a) The test face T differed from average face by $\sigma = T - I_A$
   (b) $\epsilon = \nabla^T \times \sigma^T$
   (c) For $i = 1, 2, 3, ..., m$
   $$\Omega = \Psi(:, I)$$
   $$\Gamma(I) = norm(\Omega - \epsilon)$$
   The minimum value of $\Gamma(I)$ was an accurate match of query face.

4   Experimental Results

In this section we present the screenshots of our prototype tool in a scenario discussed below and after that the face recognition results are described.

A user named Arun entering his ATM PIN as shown in Fig. 4(a). The yellow color box represents that his face has been detected. After he enters his PIN the system will authenticate his ATM PIN from database and the face recognition algorithm recognizes his face. If his PIN and face matches with our records he gets directed to next screen shown in Fig. 4(b) where he can choose any one of different options viz; Fund Transfer, Deposit Cash, Mini Statement, Balance Enquiry, Withdraw Money etc. Suppose he selected Withdraw Money option and gets directed to Withdraw Money screen. He entered Rs. 5000/- to be withdrawn from his account but before pressing YES due to some reason he moved away.

276 A. Singh et al.

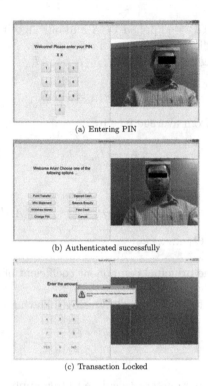

(a) Entering PIN

(b) Authenticated successfully

(c) Transaction Locked

Fig. 4. Different scenarios at ATM

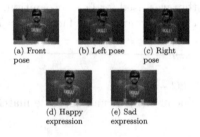

(a) Front pose   (b) Left pose   (c) Right pose

(d) Happy expression   (e) Sad expression

Fig. 5. Database images with different poses and expressions

The face recognition algorithm detected this and blocked the system and gave a warning message Arun! You are not in front of the camera. You will be logged out after 8 s as shown in Fig. 4(c).

Our gallery images were taken in a controlled imaging environment with a digital camera (Sony DSC-W320). Our database images have at least 4800 × 4900 resolutions. We took total 530 images of 106 individuals i.e. 5 images per person as shown in Fig. 5.

Table 1. Face recognition results

Gallery Images 530	Query Image		
	Recognized	UnRecognized	Error
	98.1%	1.9%	1.17%

Table 1 depicts the face recognition results of query image with our gallery images. There was 98.1% exact recognition of query faces with gallery images. In some cases query face was mismatched (i.e. matched with someone else face) in the gallery i.e. error rate (1.17%) while 1.9% was the unrecognition rate.

5 Conclusion

In this paper, we have proposed a new face recognition based authentication technique for ATMs. This technique can be used along with ATM PIN to provide another layer of security for ATM users. Although, face can be masked but it is a difficult endeavor for thieves to capture somebodys face image with ATM card. Face masking issue can also be avoided by tracking eyelid blink. This technique is developed to improve the existing ATM authentication technique by using face as a biometric. We have tested our technique on the database with 530 images of 106 individuals. The size of the database totally depends on the resolution of gallery images. The experimental results shows that our technique has reported 98.1% face recognition accuracy in a controlled environment. We have tested our system on DELL precision t5610 workstation and the time taken by our system to recognize the query image was 947 ms in average case and 670 ms in best case.

Conflicts of Interest. The authors declare that they have no conflict of interest with anyone.

References

1. Belhumeur, P.N., Hespanha, J.P., Kriegman, D.: Eigenfaces vs. fisherfaces: recognition using class specific linear projection. IEEE Trans. Pattern Anal. Mach. Intell. **19**(7), 711–720 (1997)
2. Blanz, V., Vetter, T.: Face recognition based on fitting a 3D morphable model. IEEE Trans. Pattern Anal. Mach. Intell. **25**(9), 1063–1074 (2003)
3. Coventry, L., De Angeli, A., Johnson, G.: Usability and biometric verification at the ATM interface. In: Proceedings of the SIGCHI Conference on Human Factors in Computing Systems, pp. 153–160. ACM, New York (2003)
4. Hayashi, E., Dhamija, R., Christin, N., Perrig, A.: Use your illusion: secure authentication usable anywhere. In: Proceedings of the 4th Symposium on Usable Privacy and Security, pp. 35–45. ACM, New York (2008)
5. Jones, M.J., Viola, P.: Face recognition using boosted local features. In: IEEE International Conference on Computer Vision (2003)

278   A. Singh et al.

6. Kepenekci, B.: Face recognition using gabor wavelet transform. PhD thesis, Middle East Technical University (2001)
7. Moghaddam, B., Wahid, W., Pentland, A.: Beyond eigenfaces: probabilistic matching for face recognition. In: Proceedings of Third IEEE International Conference on Automatic Face and Gesture Recognition, pp. 30–35. IEEE (1998)
8. Ojala, T., Pietikäinen, M., Harwood, D.: A comparative study of texture measures with classification based on featured distributions. Pattern Recogn. 29(1), 51–59 (1996)
9. Pandya, J.M., Rathod, D., Jadav, J.J.: A survey of face recognition approach. Int. J. Eng. Res. Appl. (IJERA) 3(1), 632–635 (2013)
10. Phillips, P.J., Grother, P., Micheals, R.J. et al.: FRVT 2002: Evaluation Report [r/ol] (2003)
11. Turk, M.A., Pentland, A.P.: Face recognition using eigenfaces. In: Proceedings of IEEE Computer Society Conference on Computer Vision and Pattern Recognition, CVPR 1991, pp. 586–591, June 1991
12. Wan, W.W.N., Luk, C.-L., Chow, C.W.C.: Customers' adoption of banking channels in Hong Kong. Int. J. Bank Mark. 23(3), 255–272 (2005)

SVM with Feature Selection and Extraction Techniques for Defect-Prone Software Module Prediction

Raj Kumar$^{(\boxtimes)}$ and Krishna Pratap Singh

Department of Information Technology, Indian Institute of Information Technology,
Allahabad, UP, India
kashwan92@gmail.com, kpsingh@iiita.ac.in

Abstract. In this paper, support vector machines with combinations of different feature selection and extraction techniques are used for the prediction of defective software module. It is tested on five NASA datasets. Correlation-based feature selection technique (CFS), principal component analysis (PCA) and kernel principal component analysis (KPCA or kernel PCA) techniques are used for feature selection and feature extraction. It has been shown that the CFS + SVM gives better prediction results and accuracy compare to PCA + SVM and KPCA + SVM.

Keywords: SVM · Software defect prediction · Feature selection

1 Introduction

Form various studies in software we can identify that most of the defects are generally found in some software modules. Due to these defective software modules, software failures can occur. Software failure affects the software company's reputation and also increase cost of the software. Therefore it is important for software development companies to reduce the cost as much as possible. With the help of a prediction model we predict if software is defective or not based on its properties and features and then we can notify the developers about their software being defective which will enable them to make defect free and good quality software and managing their effort and resources used in it. Early identification of defective modules reduced the software maintenance cost. As the software size increases, prediction of the defective software modules will perform a very important role to help to software developers to speed up the software and meet the time to market and customer requirements [1, 2]. Selection of features (attributes) which are more informative is an important part for better prediction and better accuracy. Successful feature selection is very helpful and has several advantages in many situations such as where number of features in datasets is very high or features are directly or indirectly correlated (may dependent to each other). Feature selection reduces dimensions which are employed to reduce the computation cost. Reduction of unwanted features and noises are performed for better classification (either module is defective or not). Correlation-based feature selection technique (CFS) [3] is used for feature selection and principal component analysis (PCA) [4] and kernel principal component analysis (KPCA or kernel PCA) [4, 7] are used for feature extraction.

© Springer Nature Singapore Pte Ltd. 2017
K. Deep et al. (eds.), *Proceedings of Sixth International Conference on Soft Computing for Problem Solving*, Advances in Intelligent Systems and Computing 547,
DOI 10.1007/978-981-10-3325-4_28

Subsequently, defectiveness of a module is predicted based on the selected features(attributes) using support vector machines (SVM). SVM is widely used for data classification for various application medical diagnosis, pattern recognition, Chinese character classification, text classification etc. Therefore, SVM may be very effective in prediction of software quality.

The purpose of this paper is to identifying defective software modules (methods in Object oriented language and functions in procedural language) with help of feature selection and extraction techniques (CFS, PCA and KPCA) and SVM.

The rest of the paper is organized: Sect. 2 discusses about the related work, Sect. 3 give basic introduction of SVM, Sect. 4 describes the feature selection and feature extraction techniques, Sect. 5 discusses the conducted empirical study, Sect. 6 shows the results of experiment and Sect. 7 concludes the paper and provides directions for future work.

2   Related Work

Elish and Elish [1] investigate the performance of SVM for the prediction of the defective software module. Moeyersomsa, et al. [2] used Association Rule Mining (ARM) to figure out the relationship between the software defect and software measures. NASA datasets were used for this purpose. Information Gain (IG) was calculated to find out the importance of the attributes. Selvaraj and Thangaraj [6] investigate the performance ofSVM with kernel functions (RBF, Polynomial) for the defect prediction and compare the performance of SVM with Naive Bayes and decision stump. Only KC1 dataset from PROMISE repository was used for this experiment. Chakraborty and Maulik [8] investigate the transdctive SVM (TSVM) using feature selection to identify the different type of cancers from microarray datasets. Qimeng et al. [10] proposed as transfer learning based neural network for software defect prediction. Yang et al. [9] proposed a learning to rank approach for software defect prediction by optimizing the performance of ranking.

3   Introduction of Support Vector Machine

SVM is well known supervised machine learning algorithm for classification [1]. It was initially designed for a linearly separable data and later on based on kernel concept designed for nonlinearly separable data. In literature, it is classified as linear and nonlinear SVM.

3.1   Linear SVM

In case of linear SVM [1], there can be two cases either linearly separable data or non-linearly separable data.

Case 1:- The separable case: In case of binary classification model, we have to model the function $f: R \rightarrow \{\pm 1\}$ using training data. Let us denotes that if $x \in C1$ then $y = 1$

and if x ∈ C2 then $y = -1; (x_i, y_i) \in R \, X\{\pm 1\}$. If it is possible for p-dimensional space then there will be a pair(w, b) such that:

$$w^T x_i + b \geq +1 \text{ For all } x_i \in C1$$
$$w^T x_i + b \leq -1 \text{ For all } x_i \in C2 \tag{1}$$

For all i = 1, 2, 3....n; w: normal to the hyperplane; b: bias term.
Both inequality constraints (1), can be written as:

$$y_i(w^T x_i + b) \geq 1 \text{ for all } x_i \in C1 U C2 \tag{2}$$

To optimize this we need to separate the data with maximal marginal hyperplane (MMH). Therefore, we have to minimize $||w||^2 = w^T w$.

$$\text{Minimize}_{(W,b)} \left\{ \frac{||w||^2}{2} \right\}, \text{ subject to } y_i(w^T x_i + b) \geq 1, \text{ i } = 1, 2, 3, \ldots, n \tag{3}$$

Where (x_i, y_i) is the training instances in training data and n is the number of such training instances. On applying Lagrangian duality techniques, we have:

$$MaximizeF(\lambda_i) = \sum_{i=1}^{n} \lambda_i - \frac{1}{2}||w||^2$$
$$= \sum_{i=1}^{n} \lambda_i - \frac{1}{2} \sum_{i=1}^{n} \sum_{j=1}^{n} \lambda_i \lambda_j y_i y_j x_i^T x_j \tag{4}$$

Subject to$\sum \lambda_i y_i = 0$ and$\lambda_i \geq 0$ for i = 1,2,...,n
Where λ_i- Lagrange multiplier.
Now for optimal hyperplane decision function is as follows:

$$f(x) = sign \sum_{j=1}^{n} \lambda_i y_i (x_i^T x) + b \tag{5}$$

Case 2:- The non-separable case: In case, unable to separate the data perfectly with linear SVM, a modification was proposed as to split the training data with a minimal error. In that case, some changes in minimization problem with a positive slack variable $\xi_i \geq 0$.

$$\text{Minimize}_{(w,b,\xi)} \left\{ \frac{1}{2}||w||^2 + C \sum_{i=1}^{n} \xi_i \right\}, \text{ subject to } y_i(w^T x_i + b) \geq 1 - \xi_i \tag{6}$$

where i = 1,2,...,n; C is a regularization parameter.

3.2 Nonlinear SVM

When the data is not separated by linear function such that the data is inseparable linearly we have to use mapping function $\emptyset$ which is nonlinear such that $x \rightarrow \emptyset(x)$, where $\emptyset: R^n \rightarrow R^m$ is the feature map which map the data into higher dimensional space [1]. Using kernel function we can transform the data from lower dimensional space to high dimensional feature space in which data may be linearly separable. So, the trick is to transform a low dimension non linearly separable data into high dimension, linearly separable data. Therefore, kernel functions play very important role for SVM. The kernel function $K\left(x_i, x_j\right)$ is evaluated as:

$$K\left(x_i, x_j\right) = \emptyset(x_i)^T \emptyset(x_j) \tag{7}$$

Therefore optimization problem of Eq. (5) becomes:

$$Maximize\left\{ \sum_{i=1}^{n} \lambda_i - \frac{1}{2} \sum_{i=1}^{n} \sum_{j=1}^{n} \lambda_i \lambda_j y_i y_j K(x_i, x_j) \right\}. \tag{8}$$

Subject to $\sum \lambda_i y_i = 0$ and $\lambda_i \geq 0$ for i = 1,2,...,n
Therefore finding the solution, the decision can be constructed as

$$f(x) = sign \sum_{j=1}^{n} \lambda_i y_i K(x_i^T x) + b \tag{9}$$

Following kernel functions are used in this study for prediction:

1. Linear: $K(x_p, x_q) = x_p^T x_q$
2. Gaussian: $K(x_p, x_q) = exp(-\gamma||x_p - x_q||^2)$
3. Sigmoid: $K(x_p, x_q) = tanh(\gamma(x_p^T x_q)-r)$

4   Feature Selection and Feature Extraction

Feature selection [12] is performed to identify the smaller number of attributes to be considered in final subset or which not to be consider in further study. For M number of features there will be 2^M possible subset. To find out the best subset for further study (or to use in classification), evaluation of all subset is good for small M, but as M is large it becomes very time consuming to evaluate all possible subsets. In such situations there are various feature selection techniques are applied. In this paper for feature selection correlation-based feature selection technique (CFS) is used. In correlation-based feature selection, features [3] [12] which are less correlated to each-other and high correlated to label-class are selected as subset of original set of features for further processing. CFS, first calculate the correlation matrix which give information about correlation between features.

Range of correlation value is from -1 to $+1$. On comparing with threshold value feature subset is selected from the original set of features.

Principal component analysis [4] is one of the efficient feature extraction techniques. The principle component can be understood as new axes of dataset that maximize variance along eigenvectors of covariance matrix. Kernel PCA is an extension of PCA [4, 7] for non-linear data. It uses kernel tricks for mapping the data into higher dimensions. Rbf KPCA is used for feature extraction. Therefore non-linear instances are mapped in higher dimensions using function$\emptyset$. So the sample element x can be written as $\emptyset(x)$ for higher dimension mapping.

For KPCA consider the following two steps:

1. Calculate the kernel matrix.

$$K(x_i, x_j) = exp(-\gamma||x_i - x_j||^2)$$

2. Eigendecomposition of kernel matrix.

$$\tilde{K} = K - 21_{\frac{1}{n}} K + 1_{\frac{1}{n}} K 1_{\frac{1}{n}} \tag{10}$$

where $1_{1/n}$ is a matrix with all elements $1/n$.

5 Empirical Evaluation

To evaluate the performance of SVM with different feature selection and feature extraction techniques various datasets are used. Detail about the data set is given in this Section.

5.1 Datasets

Datasets which we used are the mission critical NASA software projects [1, 2, 6] are given in Table 1. The datasets are publicly available from the repository of the NASA IV & V Facility Metrics Data Program [5]. Five datasets are used in our study. Each dataset has 21 independent variables at module level and one dependent Boolean variable (Label class): Defective, which gives information about module that particular module, is defective or not.

Table 1. Characteristics of datasets

Datasets	Language	No. of modules	%ND modules	Description
CM1	C	498	90.16	A spacecraft instrument of NASA
PC1	C	1109	93.05	Flight software for earth orbiting satellite
KC1	C ++	2109	84.54	Storage management for ground data
KC2	C ++	522	79.50	Processing of science data
KC3	Java	458	90.61	Collection, processing and delivery of satellite metadata

5.2   Independent and Dependent Variables

All 21 independent variables are static software metrics [1, 2, 5]. These metrics include the McCabe metrics, Halstead metrics (basic and derived), Line Count, and Branch Count as shown in Table 2. There may be some independent variables highly correlated, therefore, feature selection is done using correlation based feature selection to find out the best predicators. Table 3 contains the features selected using CFS for different datasets. Dependent variable is a Boolean variable: *Defective*, which give information about the module that it is defective module or non-defective module. In this study we predict that a module is defective or not. We don't consider the type of defect or number of defect in this study.

Table 2. Module level metrics (Independent variables)

Type	Metrics	Description
McCabe	V(g)	Cyclomatic complexity
	IV (g)	Design complexity
	EV (g)	Essential complexity
	LOC	Total lines of code
Derived Halstead	N	Total number of operators and operands
	V	Halstead volume
	L	Program length
	I	Intelligent count
	B	Effort estimation
	D	Difficulty
	T	Time to write program
	E	Effort to write program
Line count	LOComment	Number of line of comments
	LOCode	Number of line of statement
	LOCodeandComment	Number of line of code and comments
	LOBlank	Number of line of blank
Basic Halstead	UniqOp	Number of unique operator
	UniqOpnd	Number of unique operands
	TotalOp	Number of total operator
	TotalOpnd	Number of total operator
Branch	BranchCount	Total number of branch count

Table 3. Best subset of feature of each dataset using CFS

Dataset	Best features from original feature set using CFS
CM1	LOC, I, EV(g), LOCodeandComment, LOBlank,LOComment
PC1	L, I, EV(g), E, T, LOCodeandComment, LOBlank,LOComment
KC1	V(g), LOC, LOCodeandComment, L, UniqOpnd
KC2	L,LOCodeandComment,LOComment, UniqOp
KC3	LOCodeandComment,LOComment, D, EV(g)

5.3 Prediction Performance Measures

To measure the correct classification rate we used the confusion matrix [1, 5]. Accuracy is measured using Eq. (11) (Table 4).

$$\text{Accuracy} = \frac{TP + TN}{TP + TN + FP + FN} \tag{11}$$

Table 4. Confusion matrix

		Predicted	
		Not defective	Defective
Actual	Not defective	TN	FP
	Defective	FN	TP

TN tells that module is not defective and classifier predicts no defect whereas **TP** tells that module is defective and classifier predict defect. Similarly **FP** tells that module is not defective and classifier predict defect whereas **FN** tells that module is defective and classifier predict no defect.

5.4 Experimental Setup

For SVM we used the following parameters sets: Kernel: {'rbf', 'sigmoid', 'linear'}, C: {1,2,3} and gamma: {0.3,0.2,0.1,0.01,0.001, 0.0001}. We used 5-fold cross validation except KPCA, where dataset is randomly divides data into 5 bins, For 5 times, and 4 bins are selected for training of the model and remaining one is used for testing purpose. Each time a new bin is selected for testing purpose. In case of kernel PCA we use 5-fold, 10- fold and 20-fold cross validation.

To check the type of kernel, value of C and gamma for which the support vector machine gives best results we have to run our program for different kernels, different values of C and gamma from the sets parameters.

6   Results and Discussion

The performances of SVM with combinations with different feature selection and extraction techniques for different datasets are shown in Tables 5, 6, 7, 8 and 9. In case of PCA, NX: it shows that the N eigenvectors with the highest eigenvalues are taken in consideration.

Table 5. Different Hyperparameters for PC1-dataset for best accuracy

Dataset: PC1		Kernel	C	Gamma	Accuracy
Gen.		Rbf	1	0.1	90.16
Correlation		rbf	1	0.001	93.41
Linear PCA	2x	rbf	1	0.3	93.15
	5x	sigmoid	1	0.3	91.42
	7x	rbf	1	0.2	93.15
	21x	sigmoid	1	0.3	91.42
Kernel PCA	5-Fold	rbf	1	0.3	93.23
	10-Fold	rbf	2	0.3	93.51
	20-Fold	rbf	2	0.3	93.79

Table 6. Different Hyperparameters for KC1-dataset for best accuracy

Dataset: KC1		Kernel	C	Gamma	Accuracy
Gen.		rbf	1	0.001	84.21
Correlation		linear	2	0.3	84.68
Linear PCA	2x	linear	1	0.3	84.40
	5x	rbf	1	0.3	84.07
	7x	linear	1	0.2	84.12
	21x	rbf	1	0.001	84.21
Kernel PCA	5-Fold	rbf	1	0.2	84.54
	10-Fold	rbf	1	0.2	84.54
	20-Fold	rbf	1	0.2	84.54

Table 7. Different hyperparameters for CM1-dataset for best accuracy

DATASET: CM1		Kernel	C	Gamma	Accuracy
Gen.		Sigmoid	3	0.1	90.36
Correlation		Rbf	1	0.2	90.56
Linear PCA	2x	Rbf	1	0.3	90.16
	5x	Rbf	1	0.3	89.14
	7x	Rbf	1	0.3	90.16
	21x	Sigmoid	1	0.3	90.36
Kernel PCA	5-Fold	Rbf	1	0.3	90.17
	10-Fold	Rbf	1	0.3	90.17
	20-Fold	Rbf	1	0.3	90.224

6.1 Results for PC1

In the Table 5 different values of kernel, C and gamma are taken with different models (CFS + SVM, PCA(2x, 5x, 7x, 21x) + SVM, KPCA + SVM) for best accuracy of PC1 dataset. It is observed that KPCA(20-Fold) + SVM gives the best result for the PC1 dataset with kernel: rbf, C: 2 and gamma: 0.3 with accuracy 93.79%. KPCA(10-fold) + SVM and CFS + SVM also give comparable results.

6.2   Results for KC1

Different hyperparameter's values of SVM for best accuracy for the KC1 datasets are shown in Table 6

It is observed that CFS + SVM gives the best accuracy for the KC1 dataset with kernel: linear, C: 2 and gamma: 0.3 with accuracy 84.68%. After that KPCA + SVM give the second highest accuracy 84.54% with hyperparameter kernel: rbf, C:1 and gamma: 0.2.

6.3   Results for CM1

In the Table 7 different values of kernel, C and gamma are taken with different models (CFS + SVM, PCA(2X, 5X, 7X, 21X) + SVM, KPCA + SVM) for best accuracy for CM1 dataset. From the Table 7, it is observed that the best accuracy of CM1 dataset is given by CFS + SVM for hyperparameter kernel: rbf, C:1 and gamma: 0.2 with accuracy 90.56%. PCA (21X) + SVM also give the good result.

6.4   Results for KC3

Different values of hyperparameters of SVM for best accuracy for KC3 dataset are shown in Table 8. It is observed that the best accuracy for KC3 dataset is given by CFS + SVM model with accuracy 90.83% for hyperparameters – linear kernel, C = 1 and gamma = 0.3.

Table 8.   Different hyperparameters for KC3-dataset for best accuracy

Dataset: KC3		Kernel	C	Gamma	Accuracy
Gen.		rbf	1	0.3	90.61
Correlation		linear	1	0.3	90.83
Linear PCA	2x	sigmoid	1	0.0001	90.61
	5x	sigmoid	1	0.3	90.61
	7x	sigmoid	1	0.3	90.61
	21x	Rbf	1	0.3	90.61
Kernel PCA	5-Fold	Rbf	1	0.3	90.61
	10-Fold	rbf	1	0.3	90.63
	20-Fold	rbf	1	0.3	90.64

6.5   Results for KC2

Different values of hyperparameters of SVM for best accuracy for KC3 dataset are given in Table 9. It is shown that the best accuracy for KC2 dataset is given by PCA (2x) + SVM with hypermeters: kernel: 'linear'; C: 2; gamma: 0.3 with accuracy 83.88%.

Table 9. Different hyperparameters for KC2-dataset for best accuracy

Dataset: KC2		Kernel	C	Gamma	Accuracy
Gen.		sigmoid	1	0.3	79.50
Correlation		sigmoid	1	0.01	83.51
Linear PCA	2x	linear	2	0.3	83.88
	5x	sigmoid	1	0.0001	83.51
	7x	sigmoid	3	0.0001	83.32
	21x	sigmoid	1	0.3	79.50
Kernel PCA	5-Fold	rbf	1	0.2	79.50
	10-Fold	rbf	1	0.2	79.51
	20-Fold	rbf	1	0.2	79.53

Proposed combinations of methods have been simulated on various datasets and performances of methods are summarized in the Table 10. It is observed that CFS + SVM give best results compare to all others. KPCA + SVM have better results compare to PCA + SVM

Table 10. Accuracy for different datasets based on different feature selection model

Datasets	Accuracy (mean)								
	Gen.	Correlation	PCA				Kernel PCA		
			2x	5x	7x	21x	5-Fold	10-Fold	20-Fold
KC1	84.21	84.68	84.40	84.07	84.12	84.21	84.54	84.54	84.54
PC1	90.16	93.41	93.15	91.42	93.15	91.42	93.23	93.51	93.79
CM1	90.36	90.56	90.16	89.14	90.16	90.36	90.17	90.17	90.224
KC3	90.61	90.83	90.61	90.61	90.61	90.61	90.61	90.63	90.64
KC2	79.50	83.51	83.88	83.51	83.32	79.50	79.50	79.51	79.53

7   Conclusion

In this paper we evaluate the performance of SVM with feature selection (CFS) and feature extraction techniques (PCA and KPCA) using five NASA datasets. From the results we can conclude that SVM with correlation-based feature selection technique (CFS + SVM) give the best result among all models. Overall accuracy of CFS + SVM model varies between 83.51% to 93.41%. KPCA + SVM also give better results for some datasets or at least comparable with other methods. KPCA + SVM are required less time for computation compare to other methods. For PC1 dataset KPCA + SVM give the best result with accuracy 93.79% (20-FOLD). Overall accuracy of KPCA + SVM model varies between 79.50% to 93.79%. In case of PCA + SVM, PCA with top two eigenvectors gives the best results for all PCA feature extraction. Overall accuracy of PCA(2x) +SVM model varies between 83.88% to 93.15%.

References

1. Elish, K.O., Elish, M.O.: Predicting defect-prone software modules using support vector machines. J. Syst. Softw. **81**(5), 649–660 (2008)
2. Zafar, H., Rana, Z., Shamail, S., Awais, M.M.: Finding focused itemsets from software defect data. In: 2012 15th International Multi Topic Conference (INMIC), pp. 418–423. IEEE (2012)
3. Hall, M.A., Smith, L.A.: Feature subset selection: a correlation based filter approach. In: International Conference on Neural Information Processing and Intelligent Information Systems (1997)
4. Wang, Q.: Kernel principal component analysis and its applications in face recognition and active shape models. arXiv preprint arXiv:1207.3538 (2014)
5. http://www.nasa.gov/centers/ivv/home/index.html
6. Selvaraj, P.A., Thangaraj, P.: SVM for Software Defect Prediction. Int. J. Eng. Technol. Res. **1**(2), 68–76 (2013)
7. Schölkopf, B., Smola, A., Müller, K.R.: Nonlinear component analysis as a kernel eigenvalue problem. Neural Comput. **10**(5), 1299–1319 (1998)
8. Chakraborty, D., Maulik, U.: Identifying cancer biomarkers from microarray data using feature selection and semi supervised learning. IEEE J. Trans. Eng. Health Med. **2**, 1–11 (2014)
9. Yang, X., Tang, K., Yao, X.: A learning-to-rank approach to software defect prediction. IEEE Trans. Reliab. **64**(1), 234–246 (2015)
10. Cao, Q., Sun, Q., Cao, Q., Tan, H.: Software defect prediction via transfer learning based neural network. In: 2015 First International Conference on Reliability Systems Engineering (ICRSE), pp. 1–10. IEEE (2015)
11. Hall, M.A., Lloyd A.S.: Feature subset selection: a correlation based filter approach. In: International Conference on Neural Information Processing and Intelligent Information Systems, pp. 855–857 (1997)

Retrial Bulk Queue with State Dependent Arrival and Negative Customers

Charan Jeet Singh[1](✉), Madhu Jain[2], Sandeep Kaur[3],
and Rakesh Kumar Meena[2]

[1] Department of Mathematics, Guru Nanak Dev University,
Amritsar 143005, India
cjsmath@gmail.com
[2] Department of Mathematics, Indian Institute of Technology Roorkee,
Roorkee 247667, India
drmadhujain.iitr@gmail.com, rakeshmeena3424@gmail.com
[3] Department of Applied Sciences, Khalsa College of Engineering
& Technology, Amritsar 143002, India
sandeepsaini49@gmail.com

Abstract. In this investigation, the single server retrial queue is studied under the assumption that the arrival of the positive customers occur in bulk with state dependent rates. The server may breakdown during the essential and optional services due to arrival of negative customers. The combined supplementary variable and generating function approach is used to analyze the mathematical model and to find the queueing characteristics. The cost function has been constructed to determine the optimal number of parameters involved for providing the desired efficiency to the system. The numerical results for various performance indices are presented to examine the system behavior. The Adaptive Neuro Fuzzy Inference System (ANFIS) technique which is the combination of neural network and fuzzy logic has been used to design a fuzzy inference model for the retrial queueing system. The neuro fuzzy based numerical results are generated for the mean queue length.

Keywords: Unreliable server · Negative customers · Retrial queue · Optional service · Adaptive Neuro Fuzzy Inference System (ANFIS)

1 Introduction

In the congestion situations of digital communication systems and high tech-production systems, the queueing modeling of retrial phenomenon can be widely used for the performance evaluation of system characteristics. In the recent past, the modeling of retrial queues has formulated and analyzed mathematically by many researchers under different assumptions. The contributions of Artalejo and Corral [1], Wu and Lian [2], Choudhury and Ke [3], Yang *et al.* [4] and many others are worth-mentioning in the direction of extensive survey on the retrial queue. The situations with negative customers can be observed in the computer systems which may fail with the arrival of the virus. Due to negative customers, the service of positive customers may be affected and

© Springer Nature Singapore Pte Ltd. 2017
K. Deep et al. (eds.), *Proceedings of Sixth International Conference on Soft Computing for Problem Solving*, Advances in Intelligent Systems and Computing 547,
DOI 10.1007/978-981-10-3325-4_29

the flow of arrivals of the customers may be dependent on the server's current positions. The adverse impact of the growth of negative customers on the efficiency of queueing system can be experienced in many congestion situations. The performance analysis of queueing system with negative customers and optional service are investigated by many researchers. The detailed account on the service system with negative customers can also be found in the research works of Rajadurai and Chandrasekaran [5], Bagyam et al. [6], Gao and Wang [7], etc.

The neuro fuzzy technique is a powerful soft computing technique, which has successfully applications in various fields including manufacturing/ production systems, telecommunication systems, hardware and software systems, automobiles, and many more to study the performance characteristics. In the modern era, the soft computing has facilitated the task of computation of performance indices for complex systems in a big way. The neuro fuzzy technique came into existence due to combination of two commonly used soft computing approaches namely fuzzy logic and neural network. The combination of these approaches provides a more versatile technique including adaptive neuro-fuzzy inference system (ANFIS). For malware detection in an information security system, Altaher et al. [8] presented a neural fuzzy classifier system based on adaptive neuro-fuzzy inference system (ANFIS). The ANFIS technique is employed to predict the performance of flexible and degraded manufacturing systems by Lin and Liu [9], Jain et al. [10] and Jain and Upadhyaya [11], and others. Recently, Jain and Bhagat [12] used ANFIS technique to facilitate some approximate results and compared it with the analytical results obtained for the unreliable server $M^X/G/1$ retrial queueing model with vacation.

The present study deals with the retrial bulk arrival queuing system with negative customers. The arrivals of positive customers are state dependent and the customers can opt the optional service after getting the essential service. The queue size distribution and cost function are established by using the supplementary variable and generating function methods to explore the behavior of the system. The adaptive neuro fuzzy based soft computing technique is also used to find the numerical results for the mean queue length of the system.

2 Model Description

To develop the queueing model of service system with negative customers, consider the congestion situation governed by the non-Markovian model with general distributed functions associated with service $M_i(x)$, delayed to repair $D_i(x)$, repair $R_i(x)$ and retrial process $\Gamma(x)$. The requisite notations used in the model formulation are given in Table 1. The supplementary variables are associated with the general distributed elapsed times of different states of the server along with the assumptions that $M_i(x)$, $\Gamma(x)$, $D_i(x)$ and $R_i(x)$ are continuous at $x = 0$; with $\Gamma(0) = 0$, $\Gamma(\infty) = 1$, $M_i(0) = 0$, $M_i(\infty) = 1$, $D_i(0) = 0$, $D_i(\infty) = 1$, $R_i(0) = 0$ and $R_i(\infty) = 1$; $i = 1, 2$.

Table 1. Notations used for model formulation

$\lambda_1, \lambda_2, \lambda_3$	The arrival rates of positive customers who join the system in bulk of size X with probability mass function $P(X = j) = c_j$, $j \geq 1$, when server is in idle state, in busy state and delayed/repair state, respectively
$E(X), E(X^2)$	The first two moments of X
σ	Arrival rate of negative customer
r_0	The probability to get the optional service, $\bar{r}_0 = 1 - r_0$
p	The joining probability of customers from the retrial orbit (if any) after finishing the service or repair of the failed server, $\bar{p} = 1 - p$
q	The probability that the customer from retrial orbit with random variable ϖ and distribution function $\Gamma(x)$, give-up his attempt first, when new positive customer arrives, and return to his position of the retrial queue, $\bar{q} = 1 - q$
$N(t)$	The orbit size of the customers at time t
$M_1(M_2)$	The general distributed random variable for essential (optional) service with distribution function $M_1(x)(M_2(x))$ and first two moments $\beta_1^{(j)}(\beta_2^{(j)})$, $j = 1, 2$
$D_i(R_i)$	The random variable of delayed time (repair time) of the server while failed during the service with distribution function $D_i(x) R_i(x)$ and respective first two moments $\gamma_i^{(j)}(g_i^{(j)})$, $j = 1, 2$; $i = 1, 2$
$A_n(x, t), P_n^i(x, t)$	The transient state probability corresponding to n retrial customers and n customers with probability function K_x in the system respectively, with elapsed service time x at time t; $i = 1, 2$
$D_n^i(x, t), R_n^i(x, t)$	The probability of n customers in the system with elapsed delayed to repair and repair time respectively, at time t
$\Gamma^0(t), M_i^0(t)$	The respective retrial time and service time at time t
$D_i^0(t), R_i^0(t)$	At time t, the delayed time for the repair and repair time of failed server during any stage of the service, respectively
$\tilde{F}(s)$	The Laplace-stieltjes transforms of the functions of random variable $F(x)$ associated with different server's states
$E(I), E(H), E(C)$	The expected length of idle period, completion period, cycle

3   Steady State Equations

The following governing equations are constructed by using the probability reasoning:

$$\lambda_1 P_0^0 = \left[\int_0^\infty \mu_2(x) P_0^2(x) dx + \bar{r}_0 \int_0^\infty \mu_1(x) P_0^1(x) dx \right] + \int_0^\infty R_0^1(x) \pi_1(x) dx + \int_0^\infty R_0^2(x) \pi_2(x) dx \tag{1}$$

$$\frac{d}{dx} A_n(x) + [\lambda_1 + K(x)] A_n(x) = 0; \quad n \geq 1, x > 0 \tag{2}$$

$$\frac{d}{dx}P_n^i(x) + [\lambda_2 + \sigma + \mu_i(x)]P_n^i(x) = \lambda_2 \sum_{j=1}^{n} c_j(1 - \delta_{n,0})P_{n-j}^i(x); \ i = 1, 2, \ n \geq 0, \ x > 0$$

(3)

$$\frac{d}{dx}D_n^{(i)}(x) + [\lambda_3 + \eta_i(x)]D_n^i(x) = \lambda_3 \sum_{j=1}^{n} c_j(1 - \delta_{n,0})D_{n-j}^i(x); \ i = 1, 2, \ n \geq 0, \ x > 0$$

(4)

$$\frac{d}{dx}R_n^i(x) + [\lambda_3 + \pi_i(x)]R_n^i(x) = \lambda_3 \sum_{j=1}^{n} c_j(1 - \delta_{n,0})R_{n-j}^i(x); \ i = 1, 2, \ n \geq 0, \ x > 0$$

(5)

The normalizing condition is

$$P_0^0 + \sum_{n=1}^{\infty} \int_0^{\infty} A_n(x)dx + \sum_{i=1}^{2} \sum_{n=0}^{\infty} \left[\int_0^{\infty} P_n^i(x)dx + \int_0^{\infty} D_n^i(x)dx + \int_0^{\infty} R_n^i(x)dx \right] = 1 \quad (6)$$

$$P_0^0 = \rho_1(\rho_2 + \rho_3)^{-1} \tag{7}$$

4 Model Solution

The present non-Markovian model is analyzed by using supplementary variable technique with probability generating function approach.

(i) The stability condition of the system

$$s_1 < 1 - \bar{p}q(1 - \tilde{\Gamma}(\lambda_1))E(X). \tag{8}$$

(ii) The marginal PGF of the system states

$$A(z) = \rho_1 \bar{p}\lambda_1(z - X(z)\tau(z))\tilde{\Gamma}^*(\lambda_1)[U(z)(\rho_2 + \rho_3)]^{-1} \tag{9}$$

$$P^1(z) = \rho_1\lambda_1(1 - X(z))(1 - \bar{p}q(1 - \tilde{\Gamma}(\lambda_1)))\tilde{M}_1^*(\psi_1(z) + \sigma)[U(z)(\rho_2 + \rho_3)]^{-1} \tag{10}$$

$$P^2(z) = \rho_1 r_0 \lambda_1(1 - X(z))(1 - \bar{p}q(1 - \tilde{\Gamma}(\lambda_1)))\tilde{M}_1(\psi_1(z) + \sigma)\tilde{M}_2^*(\psi_1(z) + \sigma)[U(z)(\rho_2 + \rho_3)]^{-1} \tag{11}$$

$$D^1(z) = \rho_1 \sigma \lambda_1(1 - X(z))(1 - \bar{p}q(1 - \tilde{\Gamma}(\lambda_1)))\tilde{M}_1^*(\psi_1(z) + \sigma)\tilde{D}_1^*(\psi_2(z))[U(z)(\rho_2 + \rho_3)]^{-1} \tag{12}$$

$$D^2(z) = \rho_1 r_0 \sigma \lambda_1 (1 - X(z))(1 - \bar{p}q(1 - \tilde{\Gamma}(\lambda_1)))\tilde{M}_1(\psi_1(z) + \sigma)\tilde{M}_2^*(\psi_1(z) + \sigma))\tilde{D}_2^*(\psi_2(z))[U(z)(\rho_2 + \rho_3)]^{-1}$$
(13)

$$R^1(z) = \rho_1 \sigma \lambda_1 (1 - X(z))(1 - \bar{p}q(1 - \tilde{\Gamma}(\lambda_1)))\tilde{M}_1^*(\psi_1(z) + \sigma)\tilde{D}_1(\psi_2(z))\tilde{R}_1^*(\psi_2(z))[U(z)(\rho_2 + \rho_3)]^{-1}$$
(14)

$$R^2(z) = \rho_1 \sigma r_0 \lambda_1 (1 - X(z))(1 - \bar{p}q(1 - \tilde{\Gamma}(\lambda_1)))\tilde{M}_1(\psi_1(z) + \sigma)\tilde{M}_2^*(\psi_1(z) + \sigma)\tilde{D}_2(\psi_2(z))R_2^*(\psi_2(z))[U(z)(\rho_2 + \rho_3)]^{-1}$$
(15)

Also

$$\lambda_e = \lambda_1(P_0^{(0)} + A(1)) + \lambda_2(\sum_{i=1}^{2} P^i(1)) + \lambda_3(\sum_{i=1}^{2} D^i(1) + \sum_{i=1}^{2} R^i(1))$$
$$= \lambda_1 [(1 - s_1 - E(X))(p + \bar{p}\,\tilde{\Gamma}(\lambda_1)) + (1 - \bar{p}q(1 - \tilde{\Gamma}(\lambda_1)))(E(X) + s_1)](\rho_2 + \rho_3)^{-1}$$
(16)

$$\rho = \lambda_e E(X)[\tilde{M}_1^*(\sigma)(1 + \sigma(\gamma_1^{(1)} + g_1^{(1)})) + r_0 \tilde{M}_1(\sigma)\tilde{M}_2^*(\sigma)(1 + \sigma(\gamma_2^{(1)} + g_2^{(1)}))]$$
(17)

where

$$\overline{M_i(x)} = 1 - M_i(x), \overline{\tilde{M}_i(\sigma)} = 1 - \tilde{M}_i(\sigma) ; i = 1, 2.$$

$$\psi_1(z) = \lambda_2(1 - X(z)), \psi_2(z) = \lambda_3(1 - X(z)), \Delta = [1 - \bar{p}qE(X)(1 - \tilde{\Gamma}(\lambda_1)) - s_1]$$
$$\tau(z) = \tilde{M}_1(\psi_1(z) + \sigma)(\bar{r}_0 + r_0\tilde{M}_2(\psi_1(z) + \sigma)) + \sigma[(1 - \tilde{M}_1(\psi_1(z) + \sigma))\tilde{D}_1(\psi_2(z))\tilde{R}_1(\psi_2(z))$$
$$+ r_0\tilde{M}_1(\psi_1(z) + \sigma)(1 - \tilde{M}_2(\psi_1(z) + \sigma))\tilde{D}_2(\psi_2(z))\tilde{R}_2(\psi_2(z))][(\psi_1(z) + \sigma)]^{-1}$$

$$s_1 = [E(X)\tilde{M}_1^*(\sigma)][\lambda_2 + \sigma\lambda_3(\gamma_1^{(1)} + g_1^{(1)})]$$
$$+ [r_0E(X)\tilde{M}_1(\sigma)\tilde{M}_2^*(\sigma)][\lambda_2 + \sigma\lambda_3(\gamma_2^{(1)} + g_2^{(1)})]$$
$$\tilde{M}_i^*(\sigma) = \overline{\tilde{M}_i(\sigma)}\sigma^{-1}, \rho_1 = [1 - \bar{p}qE(X)(1 - \tilde{\Gamma}(\lambda_1)) - s_1], \rho_2 = (1 - s_1 - E(X))(p + \bar{p}\tilde{\Gamma}(\lambda_1))$$

$$\rho_3 = E(X)(1 - \bar{p}q(1 - \tilde{\Gamma}(\lambda_1)))\{1 + \lambda_1[\tilde{M}_1^*(\sigma)(1 + \sigma(1 + (\gamma_1^{(1)} + g_1^{(1)}))) + r_0\tilde{M}_1(\sigma)\tilde{M}_2^*(\sigma)\times$$
$$(1 + \sigma(1 + (\gamma_2^{(1)} + g_2^{(1)})))]\}$$
$$U(z) = \tau(z)(1 + \bar{p}q(1 - \tilde{\Gamma}(\lambda_1))(X(z) - 1)) - z$$

5 System Characteristics

The performance measures based on the queue characteristics are derived to explore the behavior of the system.

(i) Long run system state probabilities

$$P_I = [1 - s_1 - \bar{p}q(1 - \tilde{\Gamma}(\lambda_1))][\rho_2 + \rho_3]^{-1} \tag{18}$$

$$P_N = [\bar{p}\lambda_1 \tilde{\Gamma}^*(\lambda_1)(s_1 + E(X) - 1)][\rho_2 + \rho_3]^{-1} \tag{19}$$

$$P_{M_1} = [\lambda_1(1 - \bar{p}q(1 - \tilde{\Gamma}(\lambda_1)))\tilde{M}_1^*(\sigma)E(X)][\rho_2 + \rho_3]^{-1} \tag{20}$$

$$P_{M_2} = \lambda_1 r_0(1 - \bar{p}q(1 - \tilde{\Gamma}(\lambda_1)))\tilde{M}_1(\sigma)\tilde{M}_2^*(\sigma)E(X)[\rho_2 + \rho_3]^{-1} \tag{21}$$

$$P_{D_1} = [\sigma\lambda_1(1 - \bar{p}q(1 - \tilde{\Gamma}(\lambda_1)))\tilde{M}_1^*(\sigma)\gamma_1^1 E(X)][\rho_2 + \rho_3]^{-1} \tag{22}$$

$$P_{D_2} = [\sigma r_0\lambda_1(1 - \bar{p}q(1 - \tilde{\Gamma}(\lambda_1)))\tilde{M}_1(\sigma)\tilde{M}_2^*(\sigma)\gamma_2^1 E(X)][\rho_2 + \rho_3]^{-1} \tag{23}$$

$$P_{R_1} = [\sigma\lambda_1(1 - \bar{p}q(1 - \tilde{\Gamma}(\lambda_1)))\tilde{M}_1^*(\sigma)g_1^1 E(X)[\rho_2 + \rho_3]^{-1} \tag{24}$$

$$P_{R_2} = [\sigma r_0\lambda_1(1 - \bar{p}q(1 - \tilde{\Gamma}(\lambda_1)))\tilde{M}_1(\sigma)\tilde{M}_2^*(\sigma)g_2^1 E(X)][\rho_2 + \rho_3]^{-1} \tag{25}$$

(ii) Mean orbit size (L_0)

$$L_0 = \rho_1(\lambda_3(\rho_2 + \rho_3))^{-1}[N_1''(1)N_2'(1) - N_1'(1)N_2''(1)][2(N_2'(1))^2]^{-1} \tag{26}$$

Where

$$N_1'(1) = -[\lambda_3 E(X)(1 - \bar{p}q(1 - \tilde{\Gamma}(\lambda_1))) + \lambda_3(1 - s_1 - E(X))(p + \bar{p}\tilde{\Gamma}(\lambda_1))$$
$$+ \lambda_1\lambda_3(1 - \bar{p}q(1 - \tilde{\Gamma}(\lambda_1)))E(X)\{\tilde{M}_1^*(\sigma)(1 + \sigma(\gamma_1^{(1)} + g_1^{(1)})) + r_0\tilde{M}_1(\sigma)\tilde{M}_2^*(\sigma)(1 + \sigma(\gamma_2^{(1)} + g_2^{(1)}))\}]$$

$$N_1''(1) = -\lambda_3[2s_1 E(X) + E(X^{(2)})](1 - \bar{p}q(1 - \tilde{\Gamma}(\lambda_1))) + \lambda_3[E(X^{(2)}) + 2s_1 E(X) + \tau''(1)](p + \bar{p}\tilde{\Gamma}(\lambda_1))$$
$$+ \lambda_1(1 - \bar{p}q(1 - \tilde{\Gamma}(\lambda_1)))\{-2\lambda_2\lambda_3(E(X))^2(\tilde{M}_1'(\sigma) + \tilde{M}_1^*(\sigma))\delta^{-1}(1 + \sigma(\gamma_1^{(1)} + g_1^{(1)}))$$
$$- \tilde{M}_1^*(\sigma)[\lambda_3 E(X^{(2)})(1 + \sigma(\gamma_1^{(1)} + g_1^{(1)})) + \sigma(\lambda_3 E(X))^2(\gamma_1^{(2)} + 2\gamma_1^{(1)}g_1^{(1)} + g_1^{(2)})]$$
$$+ 2r_0\lambda_2\lambda_3(E(X))^2(-(\tilde{M}_2'(\sigma) + \tilde{M}_2^*(\sigma))\sigma^{-1}\tilde{M}_1(\sigma) + \tilde{M}_2^*(\sigma)\tilde{M}_1'(\sigma))(1 + \sigma(\gamma_2^{(1)} + g_2^{(1)}))$$
$$- r_0\tilde{M}_2^*(\sigma)\tilde{M}_1(\sigma)b_1b_2[\lambda_3 E(X^{(2)})(1 + \sigma(\gamma_2^{(1)} + g_2^{(1)})) + \sigma(\lambda_3 E(X))^2(\gamma_2^{(2)} + 2\gamma_2^{(1)}g_2^{(1)} + g_2^{(2)})]\}$$

$$N_2'(1) = -[1 - s_1 - \bar{p}qE(X)(1 - \tilde{\Gamma}(\lambda_1))], \quad N_2''(1) = \tau''(1) + 2s_1\bar{p}qE(X)(1 - \tilde{\Gamma}(\lambda_1)) + \bar{p}qE(X^{(2)})(1 - \tilde{\Gamma}(\lambda_1))$$

$$\tau''(1) = (2\sigma^{-2}(\lambda_2 E(X))^2 + \sigma^{-1}\lambda_2 E(X^{(2)}))[1 - \tilde{M}_1(\sigma)(\bar{r}_0 + r_0\tilde{M}_2(\sigma))]$$
$$+ 2\lambda_2\sigma^{-1}(E(X))^2[\lambda_2\tilde{M}_1'(\sigma) + \lambda_3\overline{\tilde{M}_1(\sigma)}(\gamma_1^{(1)} + g_1^{(1)}) - r_0\lambda_2\tilde{M}_1'(\sigma)\overline{\tilde{M}_2(\sigma)} + r_0\lambda_2\tilde{M}_1(\sigma)\tilde{M}_2'(\sigma)$$
$$+ r_0\lambda_3\tilde{M}_1(\sigma)\overline{\tilde{M}_2(\sigma)}(\gamma_2^{(1)} + g_2^{(1)})] + 2\lambda_2\lambda_3\tilde{M}_1'(\sigma)(E(X))^2(\gamma_1^{(1)} + g_1^{(1)})$$
$$+ \overline{\tilde{M}_1(\sigma)}[(\gamma_1^{(2)} + g_1^{(2)} + 2\gamma_1^{(1)}g_1^{(1)})(\lambda_3 E(X))^2 + (\gamma_1^{(1)} + g_1^{(1)})\lambda_3 E(X^{(2)})]$$
$$- 2r_0\lambda_2\lambda_3(E(X))^2(\gamma_2^{(1)} + g_2^{(1)})[\tilde{M}_1'(\sigma)\overline{\tilde{M}_2(\sigma)} - \tilde{M}_1(\sigma)\tilde{M}_2'(\sigma)]$$
$$+ r_0\tilde{M}_1(\sigma)\overline{\tilde{M}_2(\sigma)}[(\gamma_2^{(2)} + g_2^{(2)} + 2\gamma_2^{(1)}g_2^{(1)})(\lambda_3 E(X))^2 + (\gamma_2^{(1)} + g_2^{(1)})\lambda_3 E(X^{(2)})]$$

(iii) **Mean system size (L_s)**

$$L_s = L_0 + \sum_{i=1}^{2} P^i(1) \tag{27}$$

6 Cost Analysis

The cost function per unit time associated with different server's activities is described to find the optimum cost of the system. The cost elements per unit time used to frame the corresponding function are holding cost (C_H) per day, the cost (C_O) per day of server being in idle state, start up cost (C_S) per day, (C_{B_i}) cost per customer when the server is busy with essential/optional service, delayed repair cost (C_{D_i}) per day when the server is broken down during essential/optional service, repair cost (C_{R_i}) per day incurred on the server failed during essential/optional service. The optimal cost depends on the efficient outputs of the system and it can be determined by using the following cost function.

$$TC = C_H L_s + C_O(P_0^0 + P_N) + [E[C]]^{-1} [C_S + \sum_{i=1}^{2} \{C_{B_i} E[S_i] + C_{D_i} E[D_i] + C_{R_i} E[R_i]\}] \tag{28}$$

where

$$E[C] = E(I) + E(H) = (\rho_2 + \rho_3)[\lambda_e E(X)(1 - s_1 - \bar{p}q(1 - \tilde{\Gamma}(\lambda)))]^{-1} \tag{29}$$

7 Numerical Illustration

The present section describes the numerical simulation for the studied model. For computation purpose, it is assumed that the flow of the arrivals in groups is geometrical distributed with first two moments $E(X) = d(1-d)^{-1}$, $E(X^2) = d(1+d)(1-d)^{-2}$, respectively. The essential/optional service time is considered according to l−Erlangian distribution with first and second moments as $\beta_i^{(1)} = \mu_i^{-1}$ and $\beta_i^{(2)} = (l+1)(l\mu_i^2)^{-1}$; $i = 1, 2$. It is assumed that the delayed time follows the exponential distribution with parameters γ_i and first two moments are taken as $\gamma_i^{(1)} = (\gamma_i)^{-1}$, $\gamma_i^{(2)} = 2(\gamma_i^2)^{-1}$ by setting the parameter $\gamma_i^{(1)} = g_i^{(1)}/2$. Further, we assume that the repair time follows the gamma distribution $G(2, g_i)$ with parameter g_i with first two moments as $g_i^{(1)} = 2(g_i)^{-1}$, $g_i^{(2)} = 6(g_i^2)^{-1}$ by taking the parameter $g_i^{(1)} = \beta_i^{(1)}/3$. The distribution of retrial time is considered as exponential distributed with parameter ϖ, such that $\tilde{\Gamma}(\lambda) = \varpi(\varpi + \lambda)^{-1}$. The following default parameters are taken for the computation purpose.

$E(X) = 2, \mu = 4, \mu_1 = \mu, \mu_2 = 3\mu, r_0 = 0.6, p = 0.25, q = 0.45, \theta = 0.05, \varpi = 4, l = 2, \sigma = .05, \lambda_1 = 1.4, \lambda_2 = 1.2,$
$\lambda_3 = 1.0, C_H = \$ 10/day, C_S = \$ 500/day, C_{B_i} = \$ 50/customer, C_{D_i} = \$ 20/day, C_{R_i} = \$ 30/day, C_O = \$ 10/day.$

The numerical results are displayed in Tables 2, 3, 4, 5 and 6 and graphs 1–6 to explore the trends of performance indices by varying different parameters. The various performance indices of the system based on queueing and reliability characteristics are affected by various system parameters as demonstrated in tables and graphs. To improve the system performance, the queueing indices by varying the various system descriptors are explored. Tables 2, 3 and 4 exhibit the effects of arrival rates of positive customers (λ), negative customers (δ) and service phases (l) respectively, for other fixed parameters on the system states (Fig. 1).

The graphical representations of these numerical results obtained by using neuro-fuzzy results in adaptive neuro-fuzzy inference system (ANFIS) approach are done. The soft computing method, based on ANFIS provides the reasonable good approximation as clear from the graphs. In Fig. 2, the parameter σ is considered as linguistic variable for the fuzzy system and the membership function for the same is

Table 2. Effect of arrival rates of negative customers on various performance indices

σ	P(Idle state)	P_{M_1}	P_{M_2}	P_{D_1}	P_{D_2}	P_{R_1}	P_{R_2}	L_s
0.1	0.3701	0.5202	0.1028	0.0022	0.0001	0.0043	0.0003	21.36
0.5	0.3913	0.4853	0.0913	0.0101	0.0006	0.0202	0.0013	16.44
0.9	0.4101	0.4543	0.0814	0.017	0.001	0.0341	0.002	13.53
1.7	0.4421	0.4021	0.0656	0.0285	0.0015	0.057	0.0031	10.25
2.9	0.4797	0.3417	0.0488	0.0413	0.002	0.0826	0.0039	7.81

Table 3. Effects of service phases and probability of optional service on mean queue size

μ	$r_0 = 0.3$			$r_0 = 0.5$			$r_0 = 0.7$		
	$l = 1$	$l = 2$	$l = 5$	$l = 1$	$l = 2$	$l = 5$	$l = 1$	$l = 2$	$l = 5$
4.0	14.21	13.83	13.6	19.01	18.58	18.32	27.89	27.48	27.21
4.1	12.81	12.47	12.26	16.6	16.21	15.97	23.03	22.61	22.34
4.2	11.68	11.37	11.18	14.77	14.41	14.19	19.66	19.27	19.02
4.3	10.75	10.46	10.29	13.32	13	12.8	17.2	16.83	16.6
4.4	9.97	9.71	9.54	12.16	11.86	11.67	15.31	14.98	14.77

Table 4. Effect of probabilities of retrial customers on performance indices

p	P(Idle state)	P_{M_1}	P_{M_2}	P_{D_1}	P_{D_2}	P_{R_1}	P_{R_2}	L_s
0.1	0.3881	0.5076	0.1009	0.0011	0.0001	0.0021	0.0001	33.39
0.3	0.3602	0.5308	0.1055	0.0011	0.0001	0.0022	0.0001	19.94
0.5	0.3312	0.5548	0.1103	0.0012	0.0001	0.0023	0.0002	13.91
0.7	0.3011	0.5798	0.1152	0.0012	0.0001	0.0024	0.0002	10.45
0.9	0.2698	0.6058	0.1204	0.0013	0.0001	0.0025	0.0002	8.18

Table 5. Effects of arrivals of positive customers on total cost (*TC*)

λ	μ = 4.0			μ = 4.1			μ = 4.2		
	l = 1	l = 2	l = 5	l = 1	l = 2	l = 5	l = 1	l = 2	l = 5
1.55	416.47	413.01	410.9	425.65	422.36	420.37	435.17	432.04	430.14
1.60	414.67	410.92	408.64	423.15	419.6	417.44	432.19	428.81	426.76
1.65	**414.19**	**410.13**	**407.65**	421.58	417.73	415.39	429.84	426.19	423.98
1.70	415.72	411.31	408.61	**421.46**	**417.29**	**414.75**	**428.53**	424.58	422.18
1.75	420.32	415.52	412.58	423.53	419.01	416.24	428.81	**424.53**	**421.93**

Table 6. Effects of arrivals of negative customers on total cost (*TC*)

λ	μ = 4.0			μ = 4.1			μ = 4.2		
	ϖ = 4.0	ϖ = 4.1	ϖ = 4.2	ϖ = 4.0	ϖ = 4.1	ϖ = 4.2	ϖ = 4.0	ϖ = 4.1	ϖ = 4.2
1.6	410.92	410.81	410.75	419.6	419.67	419.78	428.81	429.04	429.29
1.65	**410.13**	**409.7**	**409.34**	417.73	417.56	417.43	426.19	426.22	426.28
1.7	411.31	410.44	409.67	**417.29**	**416.76**	**416.31**	424.58	424.33	424.13
1.75	415.52	413.99	412.63	419.01	417.98	417.08	**424.53**	**423.89**	**423.34**
1.8	424.38	421.87	419.62	424	422.26	420.71	426.84	425.65	424.61

chosen as the Gaussian function. The linguistic values of input parameter for these figs are (i) very low, (ii) low, (iii) average, (iv) high and (v) very high. It is clear from the Fig. 2 that ANFIS results shown by tick marks are quite close to those as obtained analytically and depicted by curves. Figure 3 depicts L_S vs. *p* for different service time distributions. It is noted that L_S decreases for lower value of *p* but as *p* approaches to 1, the decreasing trend becomes negligible. The total cost depends on the cost incurred on different activities of the system. To determine the optimum cost and examine the effects of different parameters on the cost function, the different cost elements are taken as default values as C_H = \$10/day, C_S = \$500/day, C_{B_i} = \$50/customer, C_{D_i} = \$20/day, C_{R_i} = \$30/day, C_O = \$10/day. The convexity nature of curve of total cost

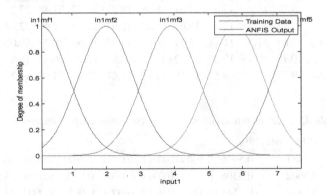

Fig. 1. Membership function

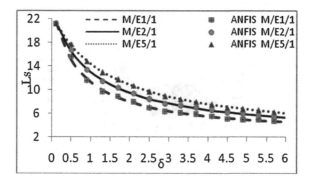

Fig. 2. ANFIS L_s vs. σ

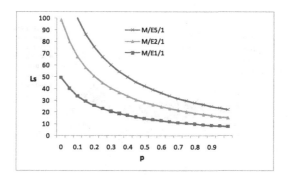

Fig. 3. L_s vs. p

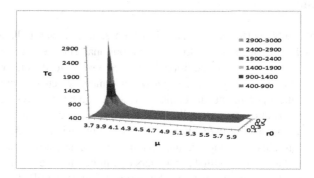

Fig. 4. Effects of r_0 and μ on TC

(TC) shown in Figs. 4, 5 and 6 yields the optimal values of the TC for the parameters $(\mu*, r_0*) = (4.5, 0.8), (\sigma*, \lambda*) = (0.05, 1.65), (\lambda*, \varpi*) = (1.75, 5.0)$,   respectively, along with corresponding optimal total costs \$444.30, \$410.13 and \$405.80.

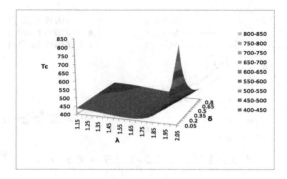

Fig. 5. Effects of λ and σ on *TC*

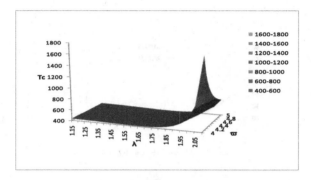

Fig. 6. Effects of λ and ϖ on *TC*

8 Conclusion

In the present study, performance analysis of unreliable server bulk arrival retrial queueing model with negative customers and optional services is considered. By incorporating the some realistic additional features viz. state dependent arrivals of positive customers and arrival of the negative customers which interrupt the service environment of the system, the developed queueing model portrays more versatile scenarios and can be effectively used due to improve the system performance in many congestion situations. The cost function provides useful insights to the system engineers and decision-makers to upgrade the system at reasonable cost. The cost analysis can be further used to obtain optimal parameters by minimizing the cost associated with various activities of the service system. The applicability of ANFIS for complex real time service system with unreliable server and many other noble features has been validated by taking suitable example. To realize the application of present model, we cite the situation experienced in computer networks due to virus, which affects the efficiency of the service system. The investigation studied can be further extended by incorporating *N*-policy, vacation schedule, etc.

References

1. Artalejo, J.R., Gomez-Corral, A.: Retrial Queueing Systems. Springer, Heidelberg (2008)
2. Wu, J., Lian, Z.: A single-server retrial G-queue with priority and unreliable server under Bernoulli vacation schedule. Comput. Ind. Eng. **64**, 84–93 (2013)
3. Choudhury, G., Ke, J.C.: An unreliable retrial queue with delaying repair and general retrial times under Bernoulli vacation schedule. Appl. Math. Comput. **230**, 436–450 (2014)
4. Yang, D.Y., Chang, F.M., Ke, J.C.: On an unreliable retrial queue with general repeated attempts and J optional vacations. App. Math. Mod. **40**, 3275–3288 (2016)
5. Rajadurai, P., Chandrasekaran, V.M.: Analysis of an $M^{[x]}/G/1$ unreliable retrial G-queue with orbital search and feedback under Bernoulli vacation schedule. OPSEARCH **53**(1), 197–223 (2016)
6. Bagyam, J.E.A., Chandrika, K.U., Rani, K.P.: Bulk arrival two phase retrial queueing system with impatient customers, orbital search, active breakdowns. Int. J. Comput. Appl. **73**(11), 13–17 (2013)
7. Gao, S., Wang, J.: Performance and reliability analysis of an $M/G/1$-G retrial queue with orbital search and non-persistent customers. Eur. J. Oper. Res. **236**(2), 561–572 (2014)
8. Altaher, A., Almomani, A., Ramadass, S.: Application of adaptive neuro-fuzzy inference system for information security. J. Comput. Sci. **8**(6), 983–986 (2012)
9. Lin, Z.C., Liu, C.Y.: Application of an adaptive neuro-fuzzy inference system for the optimal analysis of chemical–mechanical polishing process parameters. Int. J. Adv. Manuf. Technol. **18**, 20–28 (2001)
10. Jain, M., Maheshwari, S., Baghel, K.P.S.: Queueing network modeling of flexible manufacturing system using mean value analysis. Appl. Math. Model. **32**, 700–711 (2008)
11. Jain, M., Upadhyaya, S.: Threshold N-policy for degraded machining system with multiple type spares and multiple vacations. Qual. Technol. Quant. Mange. **6**, 185–203 (2009)
12. Jain, M., Bhagat, A.: $M^X/G/1$ retrial vacation queue for multi-optional services, phase repair and reneging. Qual. Technol. Quant. Mange. **13**, 263–288 (2016). doi:10.1080/16843703.2016.1189025

Wearable Haptic Based Pattern Feedback Sleeve System

Anuradha Ranasinghe[1(✉)], Kaspar Althoefer[2], Prokar Dasgupta[3], Atulya Nagar[1], and Thrishantha Nanayakkara[4]

[1] Liverpool Hope University, Liverpool, UK
{dissana,nagara}@hope.ac.uk
[2] Queen Mary, University of London, London, UK
k.althoefer@kcl.ac.uk
[3] MRC Centre for Transplantation, DTIMB & NIHR BRC, London, UK
prokar.dasgupta@kcl.ac.uk
[4] Kings College London, London, UK
thrish.antha@kcl.ac.uk

Abstract. This paper presents how humans trained in primitive haptic based patterns using a wearable sleeve, can recognize their scaling and shifting. The wearable sleeve consisted of 7 vibro-actuators to stimulate subjects arm to convey the primitive haptic based patterns. The focus of this study to understand (1) whether the human somatosensory system uses primitive patterns that can be modeled using Gaussian like functions to represent haptic perceptions, (2) whether these primitive representations are localized (cannot be shifted along the skin) and magnitude specific (cannot be scaled). These insights will help to develop more efficient haptic feedback systems using a small number of templates to be learnt to encode complex haptic messages.

Keywords: Haptics · Human-robot interactions · Vibro-actuator arrays · Guidelines and algorithm

1 Introduction

Haptics would be the best way to convey messages in critical tasks to provide spatial information [1]. Some of the studies demonstrated that haptic perceptions can be used to assist humans in navigation in unfamiliar environments [2,3]. Therefore, it is important to understand how humans perceive haptic feedback patterns using primitive haptic based templates. The results would give us an insight as to how to convey messages to humans when they work in noisy or uncertain environments like factory or search and rescue.

There have been some studies demonstrating that haptic feedback can be used to navigate humans [3–5]. Vibro-actuators have been used for different purposes to convey messages to humans. For example, the study in [5] presented an active belt which is a wearable tactile display that can transmit multiple

© Springer Nature Singapore Pte Ltd. 2017
K. Deep et al. (eds.), *Proceedings of Sixth International Conference on Soft Computing for Problem Solving*, Advances in Intelligent Systems and Computing 547,
DOI 10.1007/978-981-10-3325-4_30

directional information in combination with a GPS directional sensor and 7 vibro-actuators. Furthermore, in some studies vibro-tactile displays have used to improve the quality of life in different ways such as reading devices for those with visual impairments [6] to provide feedback of body tilt [7], balance control and postural stability [8], and navigation aid in unfamiliar environment [9]. Until now there have been many studies that used vibro-actuator belts for different purposes. However, this paper attempts to understand how to use vibro-tactile actuator arrays to understand representation of distributed haptic feedback which can be used to convey messages to humans.

Amplitude has been widely used to stimulate the human skin in most of the previous studies [10–13]. However, we argue that frequency would be better for persistent perception. The monotonic nature in amplitude could effect humans' responses.

It has been shown that humans learn movements through flexible combination of primitives that can be modeled using Gaussian like functions [14]. Therefore, in this paper we try to understand whether human somatosensory system also uses primitive patterns that can be modeled using Gaussian like functions to represent haptic perception. These insights will help to develop more efficient haptic feedback systems using a small number of templates to be learnt to encode complex haptic messages.

Basic two scientific questions are tested in this paper. (1) whether the human somatosensory system uses primitive patterns that can be modeled using Gaussian like functions to represent haptic perceptions, (2) whether these primitive representations are localized (cannot be shifted along the skin) and magnitude specific (cannot be scaled).

Therefore, to test those questions, the experiments were carried out to study humans' ability (1) to generalize (scaling/shifting) the trained primitive vibro-actuator array templates (the Gaussian template (T), shifted right (TR), shifted left (TL), half Gaussian (THA), and shrink (THS)), and (2) to recognize trained these templates and their inverse even they played randomly. The results of this paper would give an idea as to how humans construct the cutaneous feedback in different messages/scenarios. The results would help us to understand what the sensitive geometrical shapes are when we need to code haptic messages to humans in noisy crowded areas such as factory, search and rescue via cutaneous feedback. The possible applications could be human-robot interactions in uncertain environments, noisy situations like factories to convey messages to the humans.

2   Materials and Methods

The results of three experiments would answer the following scientific questions: Experiment 1: How humans generalize a Gaussian template in scaling and shifting, and Experiment 2: How humans can recognize trained templates even they are presented in a random order given by a set of discrete vibro-actuators on the arm.

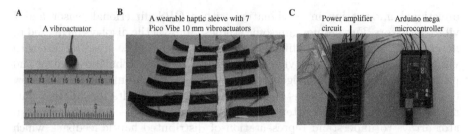

Fig. 1. Hardware design for wearable haptic sleeve: (A) Pico Vibe 10 mm vibro-actuator, (B) A wearable vibro-tactile actuator arrays with 7 Pico Vibe 10 mm vibro-actuator motors, and (C) Arduino Mega motherboard and power amplifier circuit to generate different intensity patterns.

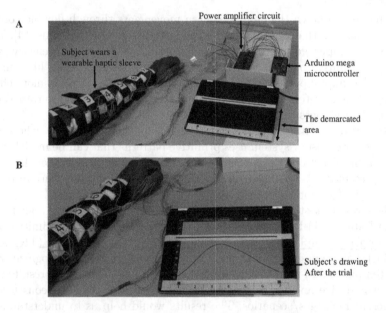

Fig. 2. An experimental trial: (A) The subjects wear a wearable haptic sleeve with 7 Pico Vibe 10 mm vibration motor. The drawing area is demarcated and used hardware is shown, and (B) The subject drew the intensity felt from the vibro-tactile actuator arrays was during the trial on ipad. Draw free app (Apple Inc.) software is used as a drawing tool.

Pico Vibe 10 mm vibration motor - 3 mm type (Precision Micro-drives) in Fig. 1A was used to make wearable haptic based pattern feedback system as shown in Fig. 1B. There are 7 Pico Vibe 10 mm vibro-actuators arranged in equal distance (7 cm) in the array as shown in Fig. 1B. The 7 Pico Vibe 10 mm vibro-actuators are attached to the seven belts which can be adjusted by strapping securely to arm of the different subjects. The different intensities for the vibrators are generated by Arduino Mega motherboard and the amplitude was modulated by a simple power amplifier circuit as shown in Fig. 1C.

2.1 Haptic Primitive Templates Generation

The standard Gaussian function was used to generate templates. The templates were generated by $y = gaussmf(x, [sig, c])$ by MATLAB 2014b, where $sig = std$, and c is the center of the distribution. The sig for pattern T, TR, TL, and THA are 1 and 0.5 for THS respectively. Moreover the amplitude of the THA was maintained at half of the rest of the templates.

2.2 Experimental Procedure

Subjects wore the haptic based pattern feedback sleeve as shown in Fig. 2A. Subjects were asked to keep the arm outstretched during the experiments. The smooth curves of Fig. 3 were selected as templates to generate different stimulation patterns. During a trial, all vibro-actuators vibrate simultaneously. Single

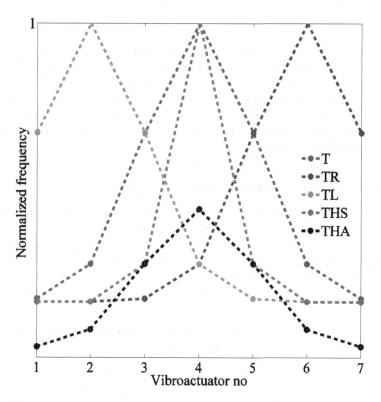

Fig. 3. The templates for experiment 1 and experiment 2: T- Gaussian Template. The standard deviation 1 for T, TR, TL, and THA. The standard deviation 0.5 was used for THS.

Table 1. The intensities (Hz) for different templates in experiment 1 and experiment 2

	Vib. 1	Vib. 2	Vib. 3	Vib. 4	Vib. 5	Vib. 6	Vib. 7
T	103.33	140.60	281.96	400.00	281.96	140.60	103.33
TR	100.00	100.10	103.33	140.60	281.96	400.00	281.96
TL	281.96	400.00	281.96	140.60	103.33	100.10	100.00
THS	100.00	100.10	140.60	400.00	140.60	100.10	100.00
THA	51.66	70.30	140.98	200	140.98	70.30	51.66

Table 2. The order for different templates in experiment 1 and experiment 2

	Exp 1: order	Exp 2: order
T	1-5, 11-14, 20-23, 29-32, 38-41	1-5, 21, 29, 32, 34, 38, 41
TR	6, 10, 18, 27, 33, 45	6-10, 23, 27, 31, 33, 39, 42
TL	7, 17, 24, 35, 37, 44	16-20, 22, 26, 28, 35, 37, 44
THS	9, 16, 26, 28, 34, 42	THS was dropped in Ex2
THA	8, 15, 19, 25, 36, 43	11-15, 24, 25, 30, 36, 40, 43

trial ran for an average of 80 ms. During the first five trials, the template (T) in Fig. 3 was played. Before playing each template, subjects were shown the printed template. Subjects were asked to draw a smooth curve representing what they perceive on an ipad sketching app (Draw free app (Apple Inc.)) after each trial as shown in Fig. 2B. A drawing area on the ipad was clearly demarcated to match the size of the printed template as shown in Fig. 2B. Just after the drawing, the next stimulation was given.

2.3 Data Processing and Statistical Analysis

The same available pencil in Draw free app (Apple Inc.) was used for drawing throughout the experiments. Get Data Graph Digitizer version 2.6 was used to digitize the data (16 digits) on drawn lines. To obtain the regression coefficients, the respective template was generated by MATLAB 2014a with the exact length of the drawing curve for each trial. The regression coefficients were calculated between humans sketch data (raw data) and the respective templates.

2.4 Experiment 1: To Understand How Humans Generalize a Gaussian Pattern in Scaling and Shifting

Eight healthy naive subjects (6 - male, 2 - female) age between 24 to 39 participated in the experiment 1. The experiment 1 was conducted to test how humans generalize a primitive template pattern (T) with respect to scaling and shifting. The shifting was done by left shift (TL) and right shift (TR), not up or down and scaling was done by shrinking (THS), and half in magnitude (THA) as shown in Fig. 3.

In experiment 1, subjects were asked to wear the vibro-actuator belt. Subjects were trained and shown only printed template T. They were informed that the intensities of the vibro-actuators in the belt are directly proportional to the height of the template T during the stimulation. Subjects were told that they are supposed to draw a smooth curve with heights directly proportional to the intensities of the vibro-actuators after the stimulation. During the experiment, subjects were trained for only template T. Therefore, at the beginning of the experiment, template T was played 5 times. Then subjects were informed that trained and untrained templates would be played pseudo randomly during the experiment. Therefore, after first five training trials, TL, TR, THS, and THA patterns were played randomly. However, there are training blocks in between other templates to train template T as shown in Table 2. During the training blocks, subjects were informed that the printed template was shown prior to the trial. For all training sessions (when T played) the visual cue was provided. At the end of each trial subjects were asked to draw what they felt during the trial. Pattern T repeated four times like in a block as shown in Table 2. Likewise four blocks of templates were played during the experiments after first five trials. The rest of four intensity patterns (TR, TL, THA, and THS) played six times each during the experiment randomly as shown in Table 2. Therefore, subjects participated in 45 trials during the experiments. For more clarity, the frequencies and the trial number and respective played patterns are shown in Tables 1 and 2 respectively.

2.5 Experiment 2: To Understand How Humans Can Recognize Trained Templates When They Are Presented in a Random Order

The second experiment was conducted to test how subjects recognize all trained haptic feedback patterns when they are presented in a random order. Six healthy naive subjects (4 - male, 2 - female) aged 24 to 28 participated in the experiment 2. All instructions in experiment 1 were given to the subjects. However, not only the template T, but also the templates TR, TL, and THA were shown to the subjects. Since the subjects were not able to distinguish the pattern THS from other patterns in experiment 1, the pattern THS was dropped and only patterns TL, TR, THA, and T were considered for the experiment 2. During the experiment, first 20 trials were designed to train the subjects to learn the patterns T, TR, TL, and THA. Each training pattern was played 5 times. During those 20 training trials, subjects were shown the pattern to be played. Finally the subjects were asked to draw a smooth curve representing the vibro-actuator intensity pattern they felt on an ipad screen. During the testing session, the four training templates were played in pseudo random order to achieve counter balancing. Each template was played 6 times making the total number of trials experienced by each subject to be 44.

3   Results

3.1   Experiment 1

The sketched raw data for the pattern T, TL, TR, THS, and THA in experiment 1 are shown in Fig. 4. The template patterns are shown by black dashed line. The sketched data in Fig. 4 were regressed by respective templates shown by black dashed line for each template. The average regression coefficients are shown in Fig. 5. The results in Fig. 5A show high correlationship between the data and template only when the template T was played. The regression coefficients of the first five trials in Fig. 5A confirms that this training block sets up a baseline.

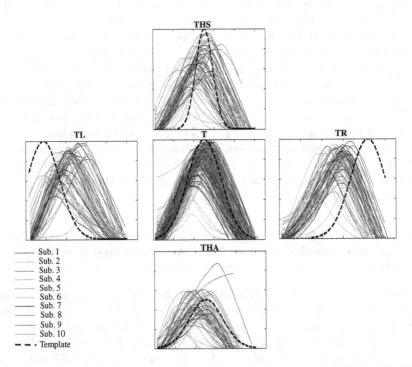

Fig. 4. The raw data for experiment 1: The sketched data for pattern T, TL, TR, THS, and THA for all trials. The all templates are shown by black dashed line.

Moving to the regression coefficients in templates TL, TR, THA, and THS regression coefficient values are relatively low when the sketched data are regressed with respective templates as shown in Fig. 5B. Moreover, it is noticed that some regression values less than 0 for templates TR and TL in Fig. 5B. The deviation can be noticed in humans' sketched data in Fig. 4 for templates TR and TL. However, the low and negative regression coefficients and higher variability values in Fig. 5C suggest that subjects were not able to shift the pattern (TL and TR) they are trained in. This might come from the fact that the memory

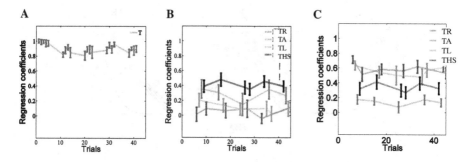

Fig. 5. Experiment 1: Average regression coefficients (In this experiment subjects were trained only for template T): (A) Regression coefficients for training template T over the trials, (B) Testing session regression coefficients for templates TL, TR, THS, and THA when the data are regressed with its respective templates, and (C) Regression coefficients for templates TL, TR, THS, and THA when the data are regressed with template T.

of the pattern T interferes with subjects' perception as shown in raw data in Fig. 4. For more clarity, the data were regressed with template T as shown in Fig. 5C (Note that in Fig. 5B the regression was done against the actual pattern that was played).

3.2   Experiment 2

The sketched raw data for the pattern T, TL, TR, and THA in experiment 2 are shown in Fig. 6. The templates are shown by black dashed line. The sketched data in Fig. 6 were regressed against respective templates. The average regression coefficients are shown in Fig. 7 when subjects were trained and tested in a random order of patterns T, THA, TL, and TR in experiment 2. In general, in Fig. 7 all regression coefficients have improved with respect to Fig. 5B for all templates. The average regression coefficients of training session are higher for T and THA with respect to TL and TR as shown in Fig. 7. It implies that subjects have a better ability to recognize scaled template than shifted ones as noticed in experiment 1. Moreover, it can be seen in sketched data in Fig. 6 too. Interestingly, subjects can recognize these four templates when they are played in random order, provided they were trained earlier. Therefore, those results show that subjects can recognize trained primitive patterns when vibro-actuator array generates different stimulations.

4   Discussion

This paper presents experimental evidence of the capabilities and limitations of the human somatosensory system to distinguish and recognize a class of primitive haptic feedback pattern presented after prior training. Three consecutive

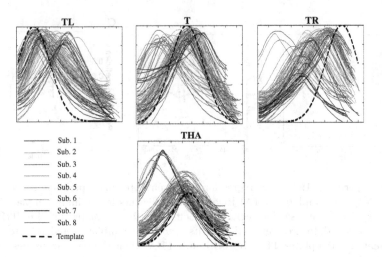

Fig. 6. The raw data for experiment 2(a): The sketched data for pattern T, TL, TR, and THA for all trials in experiment 2. The templates are shown by black dashed line. The template THS in experiment 1 was dropped in experiment 2.

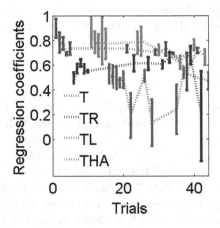

Fig. 7. (A) Experiment 2: Average regression coefficients when data regressed with respective template in Fig. 3. Standard deviation of the regression coefficients are shown by error bars.

experiments that provide insights into how humans recognize trained cutaneous feedback patterns as well as their shifts and scales. The results in this paper provide new insights into an important area of tactile input to inform shape recognition. The subjects' drawings of stimulus waveforms were captured and regression coefficients were used to understand humans' ability to recognize the given stimulations by different templates a via vibro-actuator array.

It would be interesting to test whether subjects could decode the messages from the wearable haptic sleeve when the user is mobile and active. It will inform to develop a vocabulary to be used in a haptic language for special information and to design training protocols.

A deeper understanding of humans' ability to generalize and recognize trained templates are important to convey some messages via vibro-actuator stimuli when humans have to work in noisy environments. Moreover, those trained templates could be considered as primitives of a haptic based language. Those primitives and subjects' responses would give an idea as to how they can be used to increase humans' perception when they are in noisy environments like factory, search and rescue.

Results of experiment 1 show that even though subjects were able to scale they find it difficult shift a trained template T. This results suggest that trained template T interfere for shifting. The raw data in experiment 1 shows higher deviation from the played template. Even for the template T in Fig. 5 has a higher deviation from the trained template T. This variability could come from physiological factors like muscle tension and psychological factors like attention.

However, results in experiment 2 show that average regression coefficients have improved with respect to experiment 1. These improvements show that subjects have an ability to recognize the same distributed haptic pattern (in this case T) even when it is shifted along the arm. Here TL is more precise and less strongly shifted to the center of the TR. It could come from that more sensitivity bias in shoulder than elbow (please note vibro-actuators labeled from 1 to 7 in Fig. 3 are from upper arm to lower arm.) as shown in raw data in Fig. 6. It would be useful to further investigate the distribution of sensitivity to vibro-tactile feedback along the human arm. Even though the same procedure was used for training for all experiments, the low regression coefficients in Fig. 5 and Fig. 7 suggest that even if recognition of the tactile patterns were good, performance would still be poor if drawing the visual representation was difficult.

The results explain as to how to use cutaneous feedback to the humans to convey some messages when humans working in noisy environments. It would help them to mentally construct the message by their training experiences as noticed in the results in experiments. The findings would give an insight about how special haptic memory is represented in the brain, how they could be linearly combined, and how humans can be trained for using multiple haptic patterns to decode complex messages. Those findings would be used to design guidelines/algorithm to convey messages to human in uncertain/noisy situations.

Acknowledgments. The authors would like to thank UK Engineering and Physical Sciences Research Council (EPSRC) grant no. EP/I028765/1 and grant no. EP/NO3211X/1, the Guy's and St Thomas' Charity grant on developing clinician-scientific interfaces in robotic assisted surgery: translating technical innovation into improved clinical care (grant no. R090705), Higher Education Innovation Fund (HEIF), and Vattikuti foundation.

References

1. Hale, K.S., Stanney, K.M.: Deriving haptic design guidelines from human physiological, psychophysical, and neurological foundations. IEEE Comput. Graph. Appl. **24**(2), 33–39 (2004)
2. Gilson, R.D., Redden, E.S., Elliott, L.R.: Remote tactile displays for future soldiers, technical report, DTIC Document (2007)
3. Jones, L.A., Lederman, S.J.: Human hand function. Oxford University Press (2006)
4. Gilson, R.D., Redden, E.S., Elliott, L.R.: Remote tactile displays for future soldiers, University of Central Florida, Orlando (2007)
5. Tsukada, K., Yasumura, M.: ActiveBelt: belt-type wearable tactile display for directional navigation. In: Davies, N., Mynatt, E.D., Siio, I. (eds.) UbiComp 2004. LNCS, vol. 3205, pp. 384–399. Springer, Heidelberg (2004). doi:10.1007/978-3-540-30119-6_23
6. Bliss, J.C., Katcher, M.H., Rogers, C.H., Shepard, R.P.: Optical-to-tactile image conversion for the blind. IEEE Trans. Man Mach. Syst. **11**(1), 58–65 (1970)
7. Wall III, C., Weinberg, M.S., Schmidt, P.B., Krebs, D.E.: Balance prosthesis based on micromechanical sensors using vibrotactile feedback of tilt. IEEE Trans. Biomed. Eng. **48**(10), 1153–1161 (2001)
8. Priplata, A.A., Niemi, J.B., Harry, J.D., Lipsitz, L.A., Collins, J.J.: Vibrating insoles and balance control in elderly people. Lancet **362**(9390), 1123–1124 (2003)
9. Rupert, A.H.: An instrumentation solution for reducing spatial disorientation mishaps. IEEE Eng. Med. Biol. Mag. **19**(2), 71–80 (2000)
10. Van Erp, J.B.: Guidelines for the use of vibro-tactile displays in human computer interaction. In: Proceedings of Eurohaptics, pp. 18–22. IEEE (2002)
11. Stepanenko, Y., Sankar, T.S.: Vibro-impact analysis of control systems with mechanical clearance and its application to robotic actuators. J. Dyn. Syst. Meas. Control **108**(1), 9–16 (1986)
12. Benali-Khoudja, M., Hafez, M., Alexandre, J.M., Khedda, A., Moreau, V.: VITAL: a new low-cost vibro-tactile display system. In: IEEE International Conference on Robotics and Automation, vol. 1, pp. 721–726 (2004)
13. Zaitsev, V., Sas, P.: Nonlinear response of a weakly damaged metal sample: a dissipative modulation mechanism of vibro-acoustic interaction. J. Vibr. Control **6**(6), 803–822 (2000)
14. Thoroughman, K.A., Shadmehr, R.: Learning of action through adaptive combination of motor primitives. Nature **407**(6805), 742–747 (2000)

Job Scheduling Algorithm in Cloud Environment Considering the Priority and Cost of Job

Mohit Kumar[✉], Kalka Dubey, and S.C. Sharma

Indian Institute of Technology Roorkee, Roorkee, India
mohit05cs33@gmail.com, kalka.dubey267@gmail.com,
subhash1960@rediffmail.com

Abstract. Distribution of work load (job) among the virtual machine is one of the challenging issues in cloud environment. It is very difficult to predict the execution time of job in cloud computing. So Cloud job scheduler should be dynamic in nature and distribute the job among the virtual machine in such a manner, no virtual machine should be in overloaded or ideal condition. We proposed an algorithm considering the priority of jobs and cost of resource. Job priority and cost of resources is major issue to establish a cloud environment for enterprises. For better quality of service and utilization of resources IBA algorithm is suited for this purpose. Result shows that IBA minimize the idle time of resources but IBA does not provide the guarantee for handling job priority and cost of the resource. So there is a need of job scheduling algorithm that considers job priority and recourse cost.

Keywords: Load balancing · Virtual machine · Job scheduling

1 Introduction

Today cloud computing is one of the most emerging evolutions in technology. The main aim of this technology is providing the type of services on the pay per use basis like software as a service, infrastructure as a service etc. There are some challenging problem exist in cloud computing like security problem in cloud computing (distribute denial of service attack, economics denial of service attack, data leakage problem etc.) [6] And job (task) scheduling problem (there are various quality of service parameter exist like deadline of job, cost of cloud resource, priority of job etc. we have to consider these parameter and find out the optimum results). We have proposed and implement a job scheduling algorithm considering priority of task for job in this paper. Cloud services successful attract the customers which are either the small firm or large enterprises towards its side. The basic concept of cloud is try to minimize the burden of user and provided the flexibility to take over the load of user to its cloud. For distribution of job in cloud computing technology job scheduling is one of the major factor. To fully utilize the cloud resources and distribute the job to the resources, job scheduling algorithm is used for this purpose.

© Springer Nature Singapore Pte Ltd. 2017
K. Deep et al. (eds.), *Proceedings of Sixth International Conference on Soft Computing for Problem Solving*, Advances in Intelligent Systems and Computing 547,
DOI 10.1007/978-981-10-3325-4_31

There are two types of scheduler work in cloud environment: first one is local scheduler that works with in virtual machine i.e. manage all the task of a single virtual machine and all the virtual machine contain a local job scheduler, second one is called global scheduler or meta scheduler. User submits all through the web interface in cloud environment after that all the jobs goes to meta scheduler because meta scheduler contain the information about all the virtual machine. Meta scheduler distributed these jobs to existing virtual machine [1]. After this the role of local scheduler comes in the existence processing element is provided by this scheduled for completion of jobs. A cloud contain more than one data center. A data center contains pool of scalable resources. Hardware, software or processing element all just types of various resources. One or more than one Virtual machine (VM) a data center contains. A virtual machine have more than one processing element for completion of any job these processing element executed the job which is submitted by the cloud user.

Architecture of the cloud metascheduler is shown in Fig. 1.

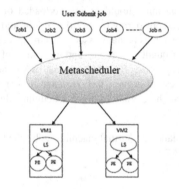

Fig. 1. Metascheduler architecture

This architecture has mainly three important parts.

- In First part user submitted their job for completion to the cloud user.
- In second part metascheduler perform its part the user job is mapped to various virtual machines according to the scheduling algorithm. This part will play one of the most important roles in job scheduling.
- In the last part local scheduler comes in the existences, allocation of processing element is done in this part.

2 Related Works

There are two type of job scheduling algorithm one is static algorithm and second is dynamic algorithm. Static algorithm are those algorithm in which variation of work load is very less and all the job information (required number of processor, required memory etc.) is known at static time like first come first serve, round robin, backfill algorithm using balanced spiral method etc. so these type of algorithm will not work well in cloud environment because you cannot predict the job in cloud environment at compile time. We need dynamic job scheduling algorithm for scheduling in cloud environment. Job scheduling problem is a NP complete problem in computer science that cannot be solved easily in polynomial time. So author's are using different technique to solve the job scheduling problem, some authors are using conventional approach to solve the problem of job scheduling while some are using heuristic and meta heuristic approach to solve the job scheduling problem in cloud environment. Therefore there are lots of algorithm has been proposed in past to solve the problem of job scheduling.

A. Mualem and D. Feitelson [2] proposed and algorithm for job scheduling considering the parameter average response time and execution time using backfilling approach. A. Suresh and P. Vijayakarthick [3] proposed an algorithm for job scheduling using the meta scheduler technique and consider the parameter processing time but this algorithm will not work well because this algorithm was a static algorithm. Improved Backfilling Algorithm is shown in Fig. 2. S. Yi et al. [4] proposed an algorithm for parallel job scheduling considering the resource utilization as a parameter. This algorithm work well for clusters and don not give the guarantee to work well in cloud environment. X. Xu and N. Ye proposed an algorithm to reduce the waiting time of job [5]. T. Ma and R. Buyya proposed an algorithm which is based on priority and non priority of job for gird computing [7]. P. Gupta and N. Rakesh proposed an algorithm for web server in cloud environment that reduce the waiting time and increase the response time of job request [8]. D. G. Feitelson et al. [9] proposed an algorithm for reducing the execution time of job.

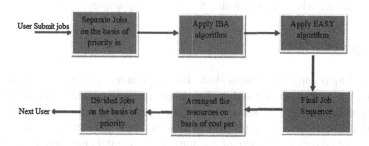

Fig. 2. Architecture of proposed method

Chen et al. [10] proposed an algorithm for reducing the make span time of task (job) and increase the average resource utilization ratio. Authors select that job first who have minimum task length and allocate to that virtual machine who can execute in minimum time. This algorithm reduce the make span time and increase the utilization of resource but this algorithm is a static algorithm and do not consider the deadline of task. There is some conventional dynamic algorithm proposed by the authors for job scheduling in cloud environment. J. Bhatia et al. [11] proposed a dynamic algorithm for task scheduling and load balancing in cloud environment that calculate the load on virtual machine (node) and find out the response time of each node so that cloud task scheduler take the decision next task is allocate to which virtual machine. K. Dubey et al. [12] proposed a dynamic algorithm that is based upon the priority of task using the combined approach of IBA and EASY algorithm. There are various dynamic job scheduling algorithm also exist in cloud environment but the entire algorithm have some pros and cons. Different algorithm choose the different parameter to find out the results like response time, average waiting time, make span time, throughput, resource utilization ratio etc. A single algorithm cannot fulfill the entire objective at a time.

There are some author's using the soft computing technique specially metaheuristic approach like honey bee behavior, particle swarm optimization, ant colony optimization etc. to solve the problem of job scheduling. D. Babu and P. Venkata [13] proposed an algorithm for load balancing in cloud environment in which honey is consider as resource and honey bee is consider as job. This algorithm optimizes the response time and execution time parameter. F. Ramezani and F. Khadeer hussain [14] proposed an algorithm which is based on particle swarm optimization technique. This algorithm reduces the execution time and transfer time parameter.

Algorithm IBA:

1. Sort the jobs according to the required number of PE's.
2. Select the largest job and placed it last position in the list.
3. Place the job which is second largest in the right side of the pool.
4. Then, check for the next job (do until all the jobs are finished)
   (a) If the sum of left pool < sum of right pool, then put the job in the left pool otherwise in the right pool.
5. Concatenate both the left pool and the right pool.
6. The sequence which we get in step 5 is the optimal sequence.

3   Proposed Model

Job scheduling is one of challenging issue in cloud computing. Various jobs scheduling algorithm had been proposed and implemented. The main requirements of job scheduling algorithm are to maximum utilization of resources and minimize the cost of resources per user considering the priority of jobs. Many job scheduling algorithm had been exist for take care of these entire requirement separately but there is no single job scheduling algorithm that fulfill the entire requirement. So we proposed and develop an algorithm that considers all these required parameter.

A. *Architecture of proposed model*

In our proposed approach, users submit their job with required processing element (PE's) for completion of job with priority of each job. Now the first step is to separate the job according to the priority of job in different queues. All the same priority job store in a one queue with first come first serve basis, after that all jobs are arranged according to priority in the queue.

After the first step various queue $q1, q2, q3 \ldots qn$ is created. These queues is form according to priority means $q1$ contain all the job which have the priority 1, $q2$ contains all the jobs which have the priority 2 and same is for all queue.

Jobs are grouped with priority now need to optimizing the recourses utilization apply IBA and EASY algorithm on the queue of jobs and after this final job sequence is formed. Now jobs are arranged according to allotment of processing element execution. Resources is classified according to the minimum cost to execute of a single job which is calculated as

$$C_j = (ET_j / PR) * C_r$$

where

C_j = Cost of a job
ET_j = Execution time of a job
PR = Processing power of resource
C_r = Cost of Resource

Minimum cost of a job resource is selected and assigned the jobs from Final job sequence list until the resource capacity is more than the length of jobs in the list, when the length of jobs is more than the resource capacity stop assigning the jobs on this resource and selected next resource in list and started assigning jobs to this resource until same condition is not occurred, continue assigning the jobs to resources until the job sequence list is empty.

B. *Description of the algorithm*

We take following assumptions in our proposed algorithm:-

- There is no relation between the jobs means the execution of one job is not dependent on the other job.
- User need to necessary define the priority and processing element requirement for completion of that job in advance.
- Execution time for job is taking a constant for result and analysis purpose.

The proposed algorithm process is as follows:

1. Jobs are submitted by the user along with priority and processing element requirement.
2. Jobs are separated on the basis of priority same priority job are placed in same queue so there are equivalent queue are maintained as number of priorities provided by users.
3. Once the priority queue is formed IBA algorithm is applied on these queues separately.
4. After applying IBA algorithm all queues are combined to form a intermediate sequence.
5. Now applying the EASY algorithm on this intermediate sequence for better utilization of resources.
6. Final List of job according to their priority and best utilization of resources submitted to the resources.

7. Arranged the resources on the basis of their cost of a job to execute. Resource which has minimum cost to execute of a job comes first and the resource which is having maximum cost comes last in order.
8. Assigned jobs to resources from the list, when the jobs length is more than the capacity of resource then assigned jobs to next resource in the list and this process continue until job length is reached to zero.

4   Simulation Tool and Experimental Results

The Cloudsim tool kit here we use as a simulator tool to test our proposed model with other job scheduling algorithm. We use cloudsim 2.1.1 version. Cloudsim tool kit is open source simulator where prebuilt class is present we need to inherited these classes for our experiment result. Cloudsim simulator contains the datacenter that represents the processing capacity of all virtual machine as shown in Fig. 3. A datacenter contain the many number of host an each host contain the number of virtual machine, it's totally depends upon the configuration of host and virtual machine. Broker work as a job scheduler in cloudsim. It selects the job and allocate to those virtual machine that can execute in minimum time and utilize the resource effectively. Cloudsim is basically used for job scheduling [15] so we implement the proposed algorithm on cloudsim tool.

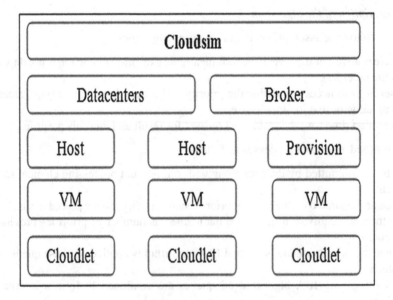

Fig. 3. Basic architecture of Cloudsim

Proposed model performance is tested on the following parameter

- Utilization of resources
- Time to process job
- Cost of the resources

We compare our proposed model with other job scheduling algorithm FCFC, EASY and IBA algorithm with execution time and Average resource usage (Fig. 4).

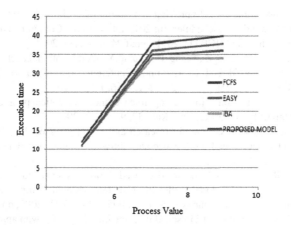

Fig. 4. Analysis of algorithm with respect to execution time

We take assumption that time to process all job is same.

After analysis the experimental result of cloudsim tool kit we found that our proposed model have better performance than FCFS and EASY algorithm but Approximately equal with IBA, IBA does not consider priority and cost of to process the job so overall our proposed model have shown better result.

5 Conclusion and Future Work

Cloud computing technology have various challenges such as task scheduling, utilization of resources, minimization of cost to perform user work, data leakage etc. Our proposed algorithm used the concept backfilling while considering user's priority among the job and cost consideration of resources. Simulation result showed that the proposed algorithm gives better result compare to other backfill algorithm like EASY and IBA and FCFS. Backfilling algorithms are successful to increases utilization of resources but do not consider priority and other issues.

We used cloudsim as our simulator to simulate our result and we find that in some cases high priority jobs continuously executed while the low job priority jobs need to wait for execution in some cases there is a possibility of starvation. We also consider in future that in our algorithm assume that all jobs have same execution time for analysis but in real scenario jobs have different execution time.

References

1. Peixoto, M., Santana, M., Estrella, J., Tavares, T., Kuehne, B., Santana, R.: A metascheduler architecture to provide QoS on the cloud computing. In: 2010 IEEE 17th International Conference on Telecommunications (ICT 2010), pp. 650–657, April 2010

2. Mu'alem, J.A., Feitelson, D.: Utilization, predictability, workload and user runtime estimates in scheduling the ibm sp2 with backfilling. IEEE Trans. Parallel Distrib. Syst. **12**(6), 529–543 (2001)
3. Suresh, A., Vijayakarthick, P.: Improving scheduling of backfill algorithm using balanced spiral method for cloud metascheduler. In: 2011 International Conference on Recent Trends in Information Technology (ICRTIT 2011), pp. 624–627, June 2011
4. Yi, S., Wang, Z., Ma, S., Che, Z., Liang, F., Huang,Y.: Combinational backfilling for parallel job scheduling. In: 2010 2nd International Conference on Education Technology and computer (IECTC 2010), vol. 2, pp. v2-112–v2-116, June 2010
5. Xu, X., Ye, N.: Minimization of job waiting time variance on Identical parallel machine. IEEE Trans. Syst. Man Cybern. Part C Appl. Rev. **37**, 917–927 (2007)
6. Yang, J., Chen, Z.: Cloud computing research and security issues. In: 2010 International Conference on Computational Intelligence and Software Engineering (CiSE 2010), pp. 1–3 (2010)
7. Ma, T., Buyya, R.: Critical-path and priority based algorithms for scheduling workflows with parameter sweep tasks on global grids. In: 17th International Symposium on Computer Architecture and High Performance Computing 2005, SBAC-PAD 2005, pp. 251–258 (2005)
8. Gupta, P., Rakesh, N.: Different job scheduling methodologies for web application and web server in a cloud computing environment. In: 2010 3rd International Conference on Emerging Trends in Engineering and Technology (ICETET 2010), pp. 569–572 (2010)
9. Feitelson, D.G., Rudolph, L., Schwiegelshohn, U.: Parallel job scheduling — a status report. In: Feitelson, D.G., Rudolph, L., Schwiegelshohn, U. (eds.) JSSPP 2004. LNCS, vol. 3277, pp. 1–16. Springer, Heidelberg (2005). doi:10.1007/11407522_1
10. Chen, H., et al.: User-priority guided min-min scheduling algorithm for load balancing in cloud computing. In: National Conference on Parallel Computing Technologies, Bangalore, pp. 1–8 (2013)
11. Bhatia, J., et al.: HTV dynamic load balancing algorithm for virtual machine instances in cloud. In: International Symposium on Cloud and Services Computing, Mangalore, pp. 15–20 (2012)
12. Dubey, K., et al.: A priority based job scheduling algorithm using IBA and EASY algorithm for cloud metaschedular. In: International Conference on Advances in Computer Engineering and Applications, Ghaziabad, India, pp. 66–70 (2015)
13. Babu, D., Venkata, P.: Honey bee behavior inspired load balancing of tasks in cloud computing environments. Appl. Soft Comput. **13**(5), 2292–2303 (2013)
14. Ramezani, F., Khadeer Hussain, F.: Task-based system load balancing in cloud computing using Particle Swarm Optimization. Int. J. Parallel Program. **42**(5), 739–754 (2013)
15. Buyya, R., Ranjan, R., Calheiros, R.: Modeling and simulation of scalable cloud computing environments and the cloudsim toolkit: challenges and opportunities. In: International Conference on High Performance Computing Simulation 2009, HPCS 2009, pp. 1–11, June 2009

Computational and Parametric Analysis of Parabolic Trough Collector with Different Heat Transfer Fluids

Rupinder Singh[✉], Yogender Pal Chandra, and Sandeep Kumar

Mechanical Engineering Department, Thapar University,
Patiala 147004, Punjab, India
rupinderratan4@gmail.com

Abstract. Solar energy is abundantly available on earth. The temperature of heat source needs to be high for the higher efficiency; be it energy production through thermodynamic cycle or heat extraction using heat transfer media, and solar energy concentration devices helps in achieving high temperature. Parabolic trough collector (PTC) has its own advantage with concentration ratio upto 215 times with reasonable cost and operational convenience, especially when low to medium range temperature heating is required. Present work is focused on the experimental and computational study on PTC with different heat transfer fluids towards identifying a suitable heat transfer fluid and flow parameters towards achieving higher heat collection and transfer efficiency. Most decisive thermo-physical entity such as heat transfer fluid (HTF) and its property *i.e.* flow rate is varied and its influence on the thermal efficiency, heat transfer and net effective temperature gain is analysed with the numerical model and results validated with experimental work. For numerical study, computational fluid dynamics (CFD) approach is taken using ANSYS–fluent software package. The experimentation results are in good agreement with the numerical model and suggest that with the flow rates of different HTF maintained within 3–5 LPM, the temperature gain can be achieved between 3–6 °C in a single pass with a maximum efficiency of 59.7%.

Nomenclature

CFD   Computational Fluid Dynamics
CSP   Concentrated Solar Power
HTF   Heat Transfer Fluid
PTC   Parabolic Trough Collector
TES   Thermal Energy Storage
Re   Reynolds Number
ρ Density
u   Velocity in x-direction
δ_{ij} Intermolecular Distance
T   Temperature
g_i Gravitational Acceleration
λ, C_P Constant
S_T Momentum Source

© Springer Nature Singapore Pte Ltd. 2017
K. Deep et al. (eds.), *Proceedings of Sixth International Conference on Soft Computing for Problem Solving*, Advances in Intelligent Systems and Computing 547,
DOI 10.1007/978-981-10-3325-4_32

1 Introduction

Concentrated solar power (CSP) has been a subject of great interest worldwide for several decades with Spain in its leading rein [1]. Concentrating solar energy employs a complicated scheme of mirrors or mirror finished curved metallic plates with a sun tracking arrangement, heat transfer fluids to transport the necessary thermal energy to the power block where it can be fed to power turbine to generate electricity using thermodynamic cycle, or can be used for thermal application.

Parabolic trough collector (PTC) is a mature solar thermal technology and is used widely for various industrial heating application and is very useful while working below 400 °C [2]. PTC comprises of a mirror finish curved metallic plate bent into parabolic shape so as to linearly focus all the incident radiation onto the absorber tube. In some large application, many of such sheets are put together to form a long arrays of troughs. The receiver/absorber system which is typically a metal pipe enclosed in an evacuated glass tube to reduce the convection losses, is mounted on the focus of the collector. To increase the thermal effectiveness and minimize the losses, anti-reflective coating is also applied on metal pipe.

Concentration ratio is defined as ratio of the collector's aperture area to the receiver area [3]. Characteristically, concentrating collectors can own different concentration ratio henceforth can operate at different temperatures. It should be noted that efficiency of the heat transfer or power production through thermodynamic cycle is a direct function of the operating temperature. However, practically, material of concentrating system, heat transfer fluid, storage system and the type of power cycle are sole decisive factors [4].

Primarily, the function of the collector is to focus the incident solar radiations onto the receiver/absorber system which is a heat exchanger device from where heat is carried away by heat transfer fluid (HTF). Heat transfer fluid is one of the critical constituent for the overall performance of the CSP. HTF can also be used as thermal energy storage (TES) device, storing heat in insulated TES tanks. Most suitable characteristics of the HTF includes thermal stability, high boiling point, high thermal conductivity, low melting point, high specific heat capacity for energy storage purpose and low viscosity [5]. Depending on the working temperature and end use, suitable HTF is used. HTF's could be assorted into six groups namely air or other gases; water/steam; Thermal oil; organics; molten-salts; liquid metals. Different researchers has used different types of materials for heat absorbing tube, such as copper, stainless steel, brass, etc. along with different type of absorbing fluid materials such as, Nano fluid, thermal oil, molten salt, water, air, etc. [6–8].

Air can be used for a wide range of temperatures and can be heated up to 700 °C at atmospheric pressure [5] and further can be used for steam generation while exchanging its heat with water in heat exchanger. Low specific heat, thermal conductivity and density of air result in low thermal density which makes it comparatively uncommon HTF choice in large size solar thermal plants. Water both as HTF and working fluid (called direct steam generation, DSG), complications are simplified many folds and the result is improved cycle efficiency and decreased power cost [9]. Most primary of thermal oils includes synthetic oils, silicone oils and mineral oils. Thermal stabilities of these oils break at around 400 °C and that's the reason they are used for low temperature applications [6]. On the other hand, organics like Biphenyl/Diphenyl oxide or TherminolVp-1^{TM} in common jargon is a big success in commercial CSP plants. TherminolVp-1, a universally accepted organic oil, is a eutectic agglomeration of two very stable compounds Biphenyl ($C_{12}H_{10}$) and Diphenyl Oxide ($C_{12}H_{10}O$) [5] and has very favorable working range for practically any PTC operations such as industrial process heat and power generation. Furthermore, a molten salt makes an excellent HTF's due to most favorable thermal stability at elevated temperatures (above 1000 °C). Molten salts are most pioneering technology used by state of the arts CSP plants based in France, United States and Spain. Most of its typical salts are based on nitrates or nitrites such as $NaNO_3$ (60 wt. %) & KNO_3 (40 wt. %). Liquid metals have not been used yet in CSP technology, though; they show very good thermo physical characteristics which makes them suitable for a HTF.

The present paper uses conventional water based and oil based HTF's such as water, endethylene glycol, ThermnolVp-1^{TM} and Mythol-Therm 500. Secondly, experimental work is followed with theoretical modeling of receiver tube for various mass flow rates of HTFs with the use of CFD based ANSYS- Fluent software. To further illustrate, mass flow rate of each fluid was varied in range of 0.043–0.091 kg/s.

2 Experimental Setup

The experimental setup is a single axis tracking, 1.22 m long and single trough test rig situated at Thapar University, Patiala, Punjab. The major components of test facility includes stainless steel collector with mirror finish curved metallic sheet having arch

Fig. 1. Experimental test rig (TU)

Table 1. System specification.

Total aperture area of solar field	8175 m^2
Length of PTC	121.9 cm
Focal length	60.7 cm
Aperture width	197 cm
Outer diameter of absorber	3.1 cm
Inner diameter of absorber	2.8 cm
Outer diameter of glass envelop	4.5 cm
Inner diameter of glass envelop	3.9 cm
Absorber tube material	Copper
Absorber glass material	Silicate glass
Rim angle	81°
Concentration ratio	50

length 1.83 m and focal length of 0.613 m, copper absorber tube with absorptance of 0.89, diameter of 0.03 m and length 1.2 m. Absorber tube is air filled annulus with glass envelope having transmittance of 0.95 and emittance of about 0.84. An east to west tracking system powered by stepper motors is also tuned up with the PTC to effectively track the sun movement for enhanced solar energy collection. A 28 L stainless steel storage tank is incorporated with the system with glass wool insulation. Figure 1 depicts the schematic of test rig while, Table 1 demonstrates its specification.

3 Numerical Method

For experimental validation, computational fluid dynamics (CFD) is used which entails the geometric modeling of receiver in ANSYS geometric modeler. After successful modeling, the model is imported for meshing and specific named selection is performed to differentiate each part of receiver such as glass envelop, air annulus, and inlet and outlet. Energy conservation equation is used in ANSYS-Fluent to solve the heat conduction and heat transfer problem. To simulate the turbulent flow condition, k-ε model is turned on. Radiation heat exchange on receiver is simulated by invoking solar S2S model. Furthermore, for comparison of thermal efficiency of receiver, 4 types of working fluids were used which are categorized under conventional and high grade industrial acquainted water based and oil based HTFs. Table 2 describes the material and their thermo-physical properties assigned in material section of Fluent.

3.1 Governing Equations

For fluid flow, Navier-Stokes equation is invoked which takes into account for continuity, energy and momentum equations. Furthermore, to define heat transfer in fluid, energy conservation equation which includes convection in fluid and conduction in absorber tube is invoked in the solution procedure. For radiation heat exchange,

Table 2. Thermo physical properties of material used

Material	K (W/m-k)	ρ (kg/m³)	C_p (J/kg-k)	μ (kg/m-s)
Air	0.0242	1.125	1006.43	0.00001789
Water	0.622	1000	4180	0.001003
Mythol-Therm500	0.102	868	2150	0.002430
Ethylene Glycol	0.254	1110	2430	0.001587
TherminolVp-1	0.117	1068	2270	0.004570
Copper	387.6	8978	381	–

radiation model is invoked. Employing turbulent flow RNG k-ε model for which turbulent intensity is calculated by the flowing relation [10].

$$k_{in} = 0.16(\text{Re}_{Dh})^{-\frac{1}{8}} \times 100\% \tag{1}$$

Following are the principle equations solved for current problem by numerical approach of CFD

Continuity equation:

$$\frac{\partial \rho}{\partial t} + \frac{\partial}{\partial x_i}(\rho u_j) = 0 \tag{2}$$

Momentum equation:

$$\frac{\partial}{\partial t}(\rho u_i) + \frac{\partial}{\partial x_i}(\rho u_i u_j) = \frac{\partial}{\partial x_j}\left[-p\delta_{ij} + \mu\left(\frac{\partial u_i}{\partial x_j} + \frac{\partial u_j}{\partial x_i}\right)\right] + \rho g_i \tag{3}$$

Energy equation:

$$\frac{\partial}{\partial t}(\rho c_p T) + \frac{\partial}{\partial x_i}(\rho u_i c_p T) - \frac{\partial}{\partial x_j}\left(\lambda \frac{\partial T}{\partial x_j}\right) = s_T \tag{4}$$

3.2 Geometry and Grid Generation

After properly designing the receiver tube in design modeler of ANSYS, it is exported for mesh generation where its domain is discretized into finitude of smaller sub-elements. Figure 2(a), (b) and (c) describe the modelled geometry, isometric and front view respectively of grid generated in the modelled receiver.

3.3 Boundary Conditions

Boundary conditions are needed to be identified which includes no slip condition for fluid flow over wall; Pressure outlet conditions accounting for fully grown viscous flow

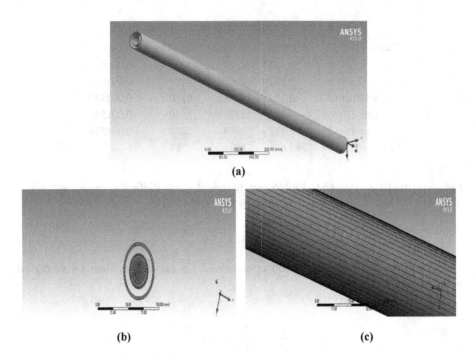

Fig. 2. (a) Physical model of receiver, grid generation in (b) isometric view of receiver, and (c) front view of receiver.

of dissimilar and viscid HTF used in experimentation; non zero heat flux condition at outer surface of absorber tube; constant heat flux around the absorber. Finally, radiation model is invoked and internal emissivity of receiver tube and glass material is set to 0.32 and 0.84 respectively. Appropriate cell zone conditions are given to emulate the receiver well closely to various documented cases in literature [11].

4   Results and Discussions

Experiment was performed for two velocities of HTF *i.e.* 0.066 m/s and 0.11 m/s (3 and 5 LPM respectively). As explained above, variant of fluids were used to compare the existing state of affairs to the new state of the art fluids so as to explain the increase in efficiency in setup. Firstly, conventional fluids *i.e.* water and Mythol–Therm 500 are experimented for the two described velocity and afterwards, high end fluids *viz.* Ethylene glycol and TherminolVp-1 is tested upon for the same parameters.

4.1   Water Based Heat Transfer Fluids

Figure 3 demonstrates the efficaciousness of the conventional receiver in terms of thermal efficiency and outlet temperature for different velocities of fluid flow. To illustrate, water leads to lowest thermal efficiency at flow rate of 0.11 m/s *i.e.* 46.7%, while it increases to 52.7% at 0.066 m/s. However, the increase in thermal efficiency in case of Ethylene Glycol is comparatively less *i.e.* 43.9% with fluid velocity of 0.066 m/s and 45.2% with fluid velocity of .011 m/s. Furthermore, maximum temperature rise was in a single pass was observed in Ethylene Glycol *i.e.* 4.64 K with the flow velocity 0.066 m/s, and 2.92 K with flow velocity 0.11 m/s. With water, maximum temperature rise was 2.89 K with flow velocity of 0.066 m/s and 1.39 K with 0.11 m/s.

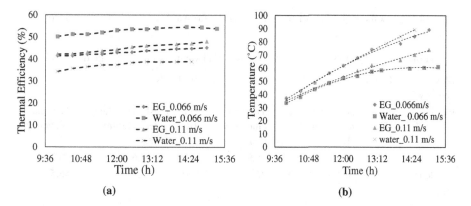

(a)   (b)

Fig. 3. Influence of mass flow rate of water based HTFs on (a) Thermal efficiency, and (b) Outlet temperature.

In addition, Fig. 4 presents the temperature profile of receivers on both water and Ethylene Glycol. This simulation was carried out for the same fluid velocity as in experimentation. Simulation results are compared with the experimental results and presented in Tables 3 and 5. Thermal efficiency with water having flow velocity 0.11 and 0.066 m/s is evaluated as 55.9% and 50.2% respectively. The mean error between experimental and modeling result is about 6%. While this error is only 2.5% for Ethylene Glycol (refer Tables 3 and 5).

4.2   Oil Based Heat Transfer Fluids

Same operational procedure was adopted for oil based fluids such as Mythol–Therm 500 and TherminolVp-1. Figure 5 presents the result of simulation. It can be inferred from the graphs that TherminolVp-1 has highest thermal efficiency i.e. 56.1% with fluid velocity of 0.066 m/s and 53.4% with fluid velocity of 0.11 m/s. While its 33% with fluid velocity of 0.066 m/s and 32.2% with fluid velocity of 0.11 m/s for Mythol–Therm 500. Furthermore, Mythol–Therm 500 doesn't show such an appreciable change in thermal efficiency with fluid flow velocity though the net temperature rise was far

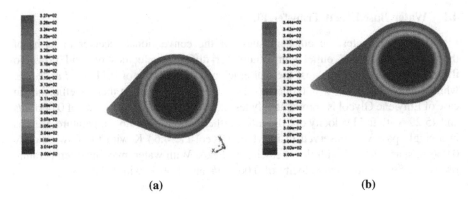

(a)   **(b)**

Fig. 4. Isothermal zones of receiver for HTFs (a) water, and (b) Ethylene Glycol.

Table 3. CFD analyses of both water based HTFs with two different fluid velocities

Flow velocity	0.066 m/s			0.11 m/s		
Fluids	ΔT (K)	Absorber tube temperature (K)	Thermal efficiency (%)	ΔT (K)	Absorber tube temperature (K)	Thermal efficiency (%)
Water	3.4	320.7	55.9	1.89	317.6	50.2
Ethylene Glycol	4.9	334.8	46.4	3.02	330.4	46.5

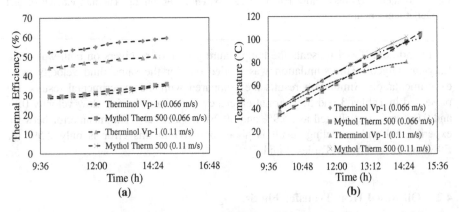

(a)   **(b)**

Fig. 5. Influence of mass flow rate of oil based HTFs on (a) thermal efficiency, and (b) outlet temperature.

better than both water, Ethylene Glycol and TherminolVp-1. This fact is further corroborated by Fig. 5(b). Figure 6 explains the isothermal zones of receiver for both HTFs, and its results are presented in Table 4.

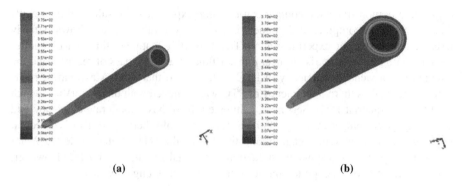

(a)   (b)

Fig. 6. Isothermal zones for oil based HTFs (a) TherminolVp-1, and (b) Mythol–Therm 500

Table 4. CFD analyses of both oil based HTFs with two different fluid velocities

Flow velocity	0.066 m/s			0.11 m/s		
Fluids	ΔT K)	Absorber tube temperature (K)	Thermal efficiency (%)	ΔT (K)	Absorber tube temperature (K)	Thermal efficiency (%)
Mythol Therm 500	5.8	362.2	34.2	3.4	356.1	32.9
Therminol Vp-1	6.6	367.8	59.7	4.1	376.4	56.6

Table 5. Comparative stance of modeling and experimentation results

Flow velocity (m/s)	Fluids	CFD Model		Experimental		% Error
		ΔT (K)	Thermal efficiency (%)	ΔT (K)	Thermal efficiency (%)	
0.066	Water	3.4	55.9	2.89	52.7	6.7
	Ethylene Glycol	4.9	46.4	4.64	43.9	2.5
	Therminol Vp-1	6.6	59.7	5.83	56.1	6
	Mythol Therm 500	5.8	34.2	5.61	33.0	3.5
0.11	Water	1.9	50.2	1.39	46.7	6.9
	Ethylene Glycol	3.0	46.5	2.92	45.2	2.7
	Therminol Vp-1	4.2	56.6	3.59	53.4	5.6
	Mythol Therm 500	3.4	32.9	3.32	32.2	2.1

Table 5 demonstrates the comparison between experimental results and the results obtained through computational model of the receiver tube. Result shows a fairly perfect agreement in the experimentation. The distinctive feature of this stance includes the maximum temperature rise in heat transfer fluid in a single pass of receiver and the maximum approachable efficiency of PTC pertaining to that fluid. Maximum temperature rise as well as thermal efficiency of PTC was achieved with therminolVp-1 which is oil based industrial HTF. Secondly, Ethylene Glycol has a moderate efficiency and reasonably good temperature rise. Furthermore, Mythol–Therm 500 also has second most significant temperature gain among the rest of the HTFs (refer Table 5). Water has been evaluated as second most favorable thermal efficiency of PTC. However, thermal efficiency of PTC pertaining to utilization of it is ungainly low.

5 Conclusion

Two groups of HTF were used for the current experimental setup namely conventional and non-conventional water based and oil based fluids respectively. Conventional category includes water as water based and Mythol–Therm 500 as oil based fluid. While, non-conventional fluids are more pushed towards leading industrial usage and includes, Ethylene Glycol as water based and TherminolVp-1 as oil based fluids. Experiments were performed for HTFs at different set of mass flow rate to calculate mean outlet temperature and thermal efficiency of PTC. In addition, computation fluid dynamics approach was adopted to numerically emulate the model and hence to convey numerical validity of the experimentation. Computational model entails proper and realistic geometric and boundary conditions and hence its approaches and checks for veracity of experimentation results. Results showed that TherminolVp-1 is a vanguard of oil based HTF for current setup and increases both thermal efficiency of PTC and outlet temperature of fluid. However, as far as water based fluid is concerned, water has favorable thermo-physical properties if it is used for low temperature applications due to its boiling constraints.

References

1. Antonelli, M., Baccioli, A., Francesconi, M., Desideri, U., Martorano, L.: Electrical production of a small size Concentrated Solar Power plant with compound parabolic collectors. Renewable Energy **83**, 1110–1118 (2015)
2. Zhang, H.L., Baeyens, J., Degreve, J., Caceres, G.: Concentrated solar power plants: review and design methodology. Renew. Sustain. Energy Rev. **22**, 466–481 (2013)
3. Duffie, J.A., Beckman, W.A.: Solar engineering of thermal process. Gear Team (2013)
4. Segal, A., Epstein, M.: Optimized working temperatures of a solar central receiver. Sol. Energy **75**, 503–510 (2003)
5. Tian, Y., Zhao, C.Y.: A review of solar collectors and thermal energy storage in solar thermal application. Appl. Energy **104**, 538–553 (2013)
6. Modi, A., Haglind, F.: Performance analysis of a Kalina cycle for a central receiver solar thermal power plant with direct steam generation. Appl. Therm. Eng. **65**, 201–208 (2014)

7. Beretta, D., Loveless, F.C., Nudenberg, W.: Use of synthetic hydrocarbon oils as heat transfer fluids. US Patent 4239638 (1980)
8. Goods, S., Bradshaw, R.: Corrosion of stainless steels and carbon steel by molten mixtures of commercial nitrate salts. J. Mater. Eng. Perform. **13**, 78–87 (2004)
9. Feldhoff, J.F., Benitez, D., Eck, M., Riffelmann, K.: Economic potential of solar thermal power plants with direct steam generation compared with HTF plants. Sol. Energy **132**, 41–100 (2010)
10. Wilcox, D.C.: Turbulence modelling for CFD. DCW Industries Inc. (1998)
11. Dudley, V., Kolb, G., Sloan, M.: Test results: SEGS LS2 solar collector. Report of Sandia National Laboratories, Sandia 94-1884 (1994)

Automatic Location of Blood Vessel Bifurcations in Digital Eye Fundus Images

Thanapong Chaichana[1(✉)], Zhonghua Sun[2(✉)], Mark Barrett-Baxendale[1], and Atulya Nagar[1]

[1] Biomechanical Electronics Research Group, Department of Mathematics and Computer Science, Liverpool Hope University, Hope Park, Liverpool, UK
{chaicht,barretm,nagara}@hope.ac.uk
[2] Department of Medical Radiation Sciences, School of Science, Curtin University, Perth, WA, Australia
z.sun@curtin.edu.au

Abstract. Retinal blood vessels are linked with hypertension and cardiovascular disease. It is generally known that vascular bifurcation is mainly involved in varying blood flow velocity as well as its pressure. This paper presents an efficient method for automatic location of blood vessel bifurcations in digital eye fundus images. The proposed algorithm comprised of three main steps: image enhancement, fuzzy clustering, and searching vascular bifurcation. The purposed algorithm revealed successful detection of bifurcations upon test images. Results showed improved diagnostic accuracy in identifying bifurcations with use of the proposed algorithm and encourage its use for further applications such as image registration, personal identification and pre-clinical scanning of retina diagnosis.

Keywords: Bifurcation · Retinal image · Fuzzy clustering · Imaging algorithm · Automatic location · Fundus image

1 Introduction

Bifurcation is a common connection in vascular network, basically a connected form of vascular tree structure in retinal image. Blood vessels within the eye supply the blood to retina. Fundus image is a photograph that captures the base of eyeball and principally provides basic information to characterise human eye condition. The main features in retinal image include blood vessels, optic nerve, macula and bifurcation [1–3]. In addition, many eye diseases do not show any signs and symptoms, and they may be painless with no any change in the vision and it can be noticed until the condition is detected at an advanced stage. Retinal image analysis is an important research context to the development of computer-assisted diagnosis and biomedical imaging analysis due to a large amount of future data storage technology and clinical informatics [4]. It requires pre-clinical scanning software prototype to review fundus images for retina diagnosis that can provide quick information and feedback on retinal main features. Those details can be used to assist ophthalmologist in term of diagnosis and treatments for reducing time,

© Springer Nature Singapore Pte Ltd. 2017
K. Deep et al. (eds.), *Proceedings of Sixth International Conference on Soft Computing for Problem Solving*, Advances in Intelligent Systems and Computing 547,
DOI 10.1007/978-981-10-3325-4_33

prevention, and making decision for further planning and or surgery. Retinal images representing healthy and eye disease are shown in Fig. 1.

Previous works on identifying vascular bifurcations in fundus images have focused on characterising the vascular junctions as well as vascular crossover-sections. Fatepuria et al. [7] proposed the windows matching techniques to search locations of vascular crossover and bifurcations. They reviewed the accuracy of their proposed algorithm against with an estimation of total numbers of crossovers and bifurcations. The results presented with the test image, have shown 82% of success rate in average.

Calvo et al. [8] presented many windows of crossover and bifurcation patterns to match with those forms in retinal images. They tested their developed algorithm with retinal images obtained from the varpa retinal images for authentication database [9]. Results of identifying crossovers and bifurcations were measured with many classic parameters sensitivity, to determine the specificity, and accuracy. They concluded that their algorithm has shown 93.6% of accuracy in the identified locations. They also addressed that this detection system of crossovers and bifurcations in fundus images is needed to improve on classification technique for more accurate detection of crossover locations.

Azzopardi and Petkov [10] introduced a combination of shifted filter responses (COSFIRE filters) to automatically detect vascular bifurcations in segmented retinal blood vessels from fundus images. They proposed an algorithm by creating trained windows which were used to filter selected bifurcation in different filtering levels and oriented directions, and finally to integrate all filtering results then defined as detected or undetected location of bifurcations. This algorithm was evaluated as 97.88% of an average precision. The technique is quite in line with "directional filtering in edge detection" proposed by Paplinski [11]. They concluded that COSFIRE filters to auto-matically search bifurcations can be used to detect any similar bifurcation patterns from trained prototype patterns of bifurcations and in fact the filtered output is a computed weighted geometric denoted of blurred and shifted Gabor filter results.

Waheed et al. [12] presented vascular and non-vascular features including blood vessels, bifurcated points, exudate points, luminance, contrast and structure from retinal images for person identification. The biometric system is stable and reliable to identify a person from his/her eye fundus image. Recognition algorithm was proposed in two stages differently, one was using features in retinal image and the other one was using photographic details of retinal image. They concluded that vascular based method is used to improve efficiency of retinal recognition and non-vascular based method is intended to reduce time complexity of recognition system. This biometric system can be improved by integrating retinal pathologies.

Welikala et al. [4] introduced automated system to assess image quality of large numbers of fundus images stored in United Kingdom Biobank. They proposed an algo-rithm to detect retinal images (big data) which are good to clarity conditions across 800 images that are saved in the database, and their assessment condition is paid on structural attributes of segmented vascular, bifurcation and branching trees. Their main aim was to use the imaging quality for future epidemiological studies. However, the segmented results of blood vessels were required to compare with ground truth data (manually hand label blood vessels) for obtaining the classic factors of true and false positive detected.

They concluded that the proposed software algorithm was effective in grading retinal images with less time consuming than the manual grading technique, and the algorithm computed completely in around 22 h, whilst manual processing completed in 567 h. Thus, this software can be used for grading quality of entire retinal images (136,000 images) in UK Biobank and other large retinal datasets.

This work proposes a new searching technique to locate and identify bifurcations in retinal images. The proposed algorithm consisted of three steps which include image enhancement, fuzzy k-median clustering, and searching vascular bifurcation. Figure 1 reveals the flowcode of the current technique.

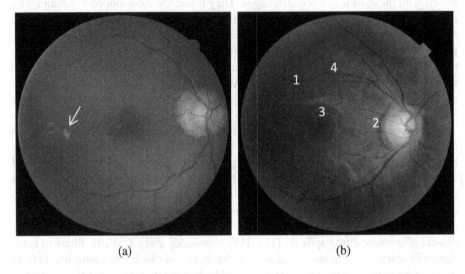

(a)   (b)

Fig. 1. Fundus images: (a) right eye disease [5], arrow points to an area of exudates and this can cause blur/dark spots in the vision; (b) Healthy Eye of a young male adult [6], and main features are: (1) blood vessels, (2) optic nerve, (3) macula and (4) bifurcation.

2   Retinal Imaging Data

Fundus images used in this study were obtained from three sources including the digital retinal images for vessel extraction (DRIVE) database [5], the structured analysis of the retina (STARE) database [13] and the fundus images of Rangsit University (FIRSU) database [6]. The DRIVE dataset contained both healthy and diseased eye images. These were 40 retinal images which were divided into two groups: training and test datasets, with each set including 20 images equally. The STARE dataset was recorded to have approximately 400 raw images including healthy and diseased conditions. In addition, FIRSU dataset was composed of 10 images of healthy volunteers who have participated in recording digital eye fundus at Robotics Laboratory, Biomedical Engineering, Rangsit University, Thailand, in March 2015. Hence, in this work the test images for automatic locating algorithm were randomly selected from these three sources (Fig. 2).

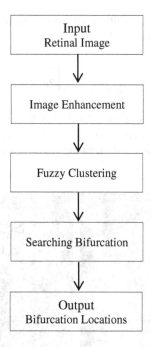

Fig. 2. Flowcode of the proposed algorithm.

3 Proposed Algorithm

3.1 Image Enhancement

A typical fundus image is recorded in Red, Green and Blue (RGB) colour space by using digital charge-coupled device (CCD) camera attached to the fundus photographic device. Each of RGB compositions is the 2-dimensional arrays of grayscale image; composited G plane of retinal image in Fig. 1(b) which can be expressed in matrix form as:

$$F = \begin{bmatrix} f_{0,0} & \cdots & f_{0,N-1} \\ \vdots & \ddots & \vdots \\ f_{M-1,0} & \cdots & f_{M-1,N-1} \end{bmatrix}$$

where F is a 2 dimensional image (G plane of retinal image), matrix F of size $M \times N$.

N is shape width (N columns). M is shape height (M rows). f is pixel value of an image.

Hence, Green channel was selected as grayscale image which is G plane representing mostly the luminosity values of the pixels in the median region as shown in our previous study [1]. Then, the blood vessels in retinal image were enhanced by using image convolution approach. A matched filter kernel was designed to convolve with grayscale image and the equation denoted as [1]:

$$h(x, y) = -e^{\left(\frac{-x^2}{2\sigma^2}\right)}$$

where $h(x, y)$ is a 2 dimensional matched filter kernel, matrix of size 16×15. x is column. y is row. σ is standard derivative which was setup value of 2.

An example of retinal image #40_training [5] was processed with image enhancement, and the output is shown in Fig. 3.

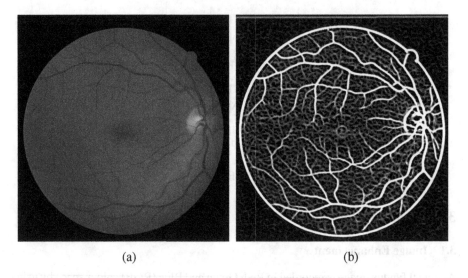

(a)   (b)

Fig. 3. Image enhancement: (a) grayscale image (b) matched filter image

3.2 Fuzzy Clustering

In recent years, fuzzy logic has been increasingly used in various applications, such as, medical imaging and instrumentation, industrial management and data systems, control and process engineering techniques, decision support systems and medical cameras. Medical image segmentation is one of important aspects to process medical imaging data. Fuzzy clustering approach is an efficient method to separate an object of interest in digital image. Since, fuzzy logic has been introduced in the form of logic values to determine the true and false statements, which are the variables between '1' and '0', respectively. In this context, the pixel values of matched filter image in Fig. 3(a) ranging from 0–255 was mapped to the Boolean logic, in which pixel values of image ranged from 0–1. Fuzzy c-median clustering was purposed to segment blood vessels in rental image and a simple algorithm was defined as follows [2, 3, 14]:

Step 1: Let consider matched filter image as a 2 dimensional matrix P

$$P = \begin{bmatrix} p_{0,0} & \cdots & p_{0,N-1} \\ \vdots & \ddots & \vdots \\ p_{M-1,0} & \cdots & p_{M-1,N-1} \end{bmatrix}$$

Step 2: Convert it into 1 dimensional array $P = \{p_1, p_2, p_3, \dots, p_L\}$. L is size of array that equal to $M \times N$, and the pixel values p were sorted orderly, connected each M row (line array) continuously to L.

Step 3: Calculate a centroid of P by considering median value and histogram of P and a fuzzy partition of C groups.

Step 4: Let consider to cluster P into c groups. Such that each $p_k \in R^T$, $k = 1, 2, 3, \dots, N$ is a feature vector of T consisting of objects represented by p_k. A fuzzy c-partition of given dataset is considered as matrix $U = [u_{ij}]$, $i = 1, 2, 3, \dots, C$ and $j = 1, 2, 3, \dots, N$, and $u_{ij} \in [0, 1]$

Step 5: Remap data of each c-partition with its pixel values according to pixel positions in 2-dimensional image.

For instance, the retinal blood vessels in Fig. 3(b) are segmented into 2 groups with calculated centroids being 0.2029 and 0.5654, respectively. The images of each fuzzy classified group are shown in Fig. 4.

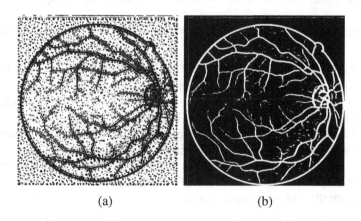

(a)   (b)

Fig. 4. Fuzzy clustering: (a) classified image group a (b) classified image group b

3.3   Searching Vascular Bifurcation

In our preliminary work the vascular bifurcations were estimated including the blood vessel branches with an input and 2 outputs, and vascular crossover with an input and 3 outputs as shown in Fig. 5.

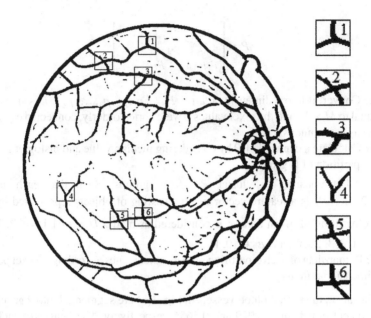

Fig. 5. Bifurcations in retinal image are linked to the boxed images in right column: bifurcations and vascular crossover display no. 1, 3 and 4, and no 2, 5, and 6, respectively.

A simplified algorithm to search bifurcations in retinal images was defined as follows:

Step 1: Convert blood vessels (Fig. 4(b)) into binary image which is "0" and "1" are referred to black and white pixels, respectively.
Step 2: Cleaning unconnected and unwanted pixels and then apply morphological thinning approach to erode thickness of blood vessels.
Step 3: Create a 3 × 3 mask of 9 pixels to scan over each point of pixels in binary image together with collecting cumulative sum in all passing pixel points.
Step 4: Search to scan all points of pixels that have had a total cumulative number of 3–4 and then plot the blue circles to identify bifurcation points.
Step 5: Print the total cumulative numbers.

For example, a segmented blood vessel in Fig. 4(b) was processed automatic detection of bifurcations and the results are shown in Fig. 6.

4   Results

Bifurcation locations in retinal images were performed for computation. The proposed algorithm was coded and written with scripting in Matlab, on a Microsoft Windows 7 32-bit machine, 4 GB of RAM with Intel i5 processor 2.50 GHz CPU. Time computation of all images was estimated to be around 90 s.

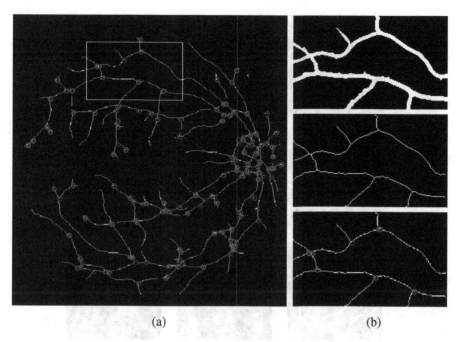

Fig. 6. Identifying bifurcations: (a) blue circles plotted over vascular bifurcations (b) zooming top white rectangle area, images are in column: blood vessels (top), morphological thinning (middle), 7 bifurcations were automatically identified (bottom) and hand calculation was 7 bifurcations. (Color figure online)

4.1 Bifurcations in Cropped Retinal Images

The segmented blood vessels in fundus images within the DRIVE database [5] were randomly cropped and used to test with proposed algorithm. The sequential images showing the pathways of algorithm and its performing results are shown in Fig. 7.

4.2 Bifurcations in Retinal Images

The digital colour retinal images were randomly chosen within the STARE database [13]. The purposed algorithm was used to locate bifurcations and the pathways of its process and results are shown in Fig. 8.

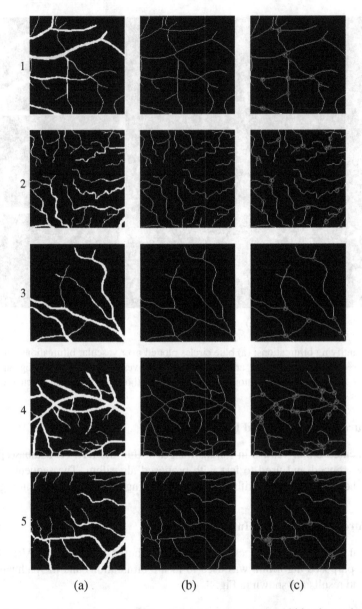

Fig. 7. Automatically locating bifurcations in cropped the segmented blood vessels in fundus images: (a) hand labelled blood vessels, (b) morphological thinning and (c) printed bifurcations.

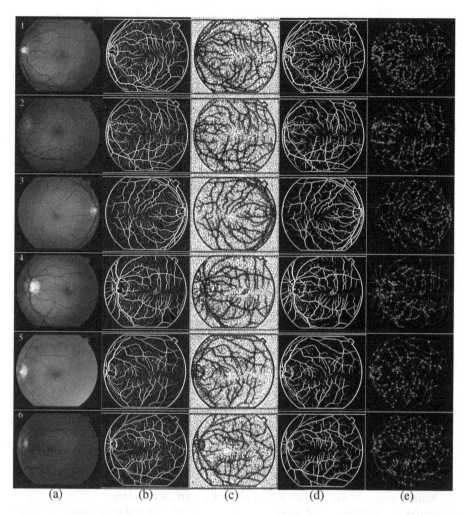

Fig. 8. Automatically locating bifurcations in digital colour fundus images: (a) grayscale, (b) matched filter, (c) fuzzy segmentation group one and (d) group two, and (e) printed bifurcations.

5   Conclusion

In this paper, we presented a new algorithm to locate vascular bifurcations in the fundus images. The purposed algorithm has shown enhancement in accuracy and reliability of automatically locating vascular bifurcations. This work was a preliminary study estimating vascular crossover and bifurcation junctions as a bifurcation location. In all cases, we found that there are two bifurcation junctions which were located at vascular crossover, for example, as shown in Fig. 5 (in boxed no 2) linked with Fig. 6(b). Our results have shown a good agreement with hand calculation and our findings deserve further research in retina diagnosis and comprehensive identification of main features in the fundus images.

Acknowledgements. The authors gratefully acknowledge the Higher Education Innovation Fund (HEIF), Liverpool Hope University, for funding and financial support.

References

1. Chaichana, T., Yoowattana, S., Sun, Z., Tangjitkusolmun, S., Sookpotharom, S., Sangworasil, M.: Edge detection of the optic disc in retinal images based on identification of a round shape. In: 2008 International Symposium on Communications and Information Technologies, ISCIT 2008, pp. 670–674. IEEE Press, Lao PDR (2008)
2. Sookpotharom, S., Chaichana, T., Pintavirooj, C., Sangworasil, M.: Automatic segmentation of blood vessels in retinal image based on fuzzy k-median clustering. In: 2007 IEEE International Conference on Integration Technology, ICIT 2007, pp. 584–588. IEEE Press, Shenzhen (2007)
3. Chaichana, T., Wiriyasuttiwong, W., Reepolmaha, S., Pintavirooj, C., Sangworasil, M.: Automatic segmentation of blood vessels in retinal image based on fuzzy k-median clustering. In: 2007 IEEE Fourth International Conference on Fuzzy Systems and Knowledge Discovery, FSKD 2007, pp. 144–148. IEEE Press, Haikou (2007)
4. Welikala, R.A., Fraz, M.M., Foster, P.J., Whincup, P.H., Rudnicka, A.R., Owen, C.G., Strachan, D.P., Barman, S.A.: Automated retinal image quality assessment on the UK Biobank dataset for epidemiological studies. Comput. Biol. Med. **71**, 67–76 (2016)
5. Digital Retinal Images for Vessel Extraction. www.isi.uu.nl/Research/Databases/DRIVE
6. Fundus Images of Rangsit University (FIRSU) database. www.rsu.ac.th
7. Fatepuria, H., Singhania, G., Shah, N., Singh, R., Roy, N.D.: Detection of crossover and bifurcation points on a retinal fundus image by analyzing neighbourhood connectivity of non-vascular regions around a junction point. IJRET **3**, 479–484 (2014)
8. Calvo, D., Ortega, M., Penedo, M.G., Rouco, J.: Automatic detection and characterisation of retinal vessel tree bifurcations and crossovers in eye fundus images. Comput. Methods Programs Biomed. **103**, 28–38 (2011)
9. Retinal Images for Authentication Database. www.varpa.es/varia.html
10. Azzopardi, G., Petkov, N.: Automatic detection of vascular bifurcations in segmented retinal images using trainable COSFIRE filters. Pattern Recogn. Lett. **34**, 922–933 (2013)
11. Paplinski, A.: Directional filtering in edge detection. IEEE Trans. Image Process. **7**, 611–615 (1998)
12. Waheed, Z., Akram, M.U., Waheed, A., Khan, M.A., Shaukat, A., Ishaq, M.: Person identification using vascular and non-vascular retinal features. Comput. Electr. Eng. **53**, 1–13 (2016)
13. STructured Analysis of the Retina Project. http://cecas.clemson.edu/~ahoover/stare
14. Sookpotharom, S., Chaichana, T., Pintavirooj, C., Sangworasil, M.: Segmentation of magnetic resonance images using discrete curve evolution and fuzzy clustering. In: 2007 IEEE International Conference on Integration Technology, ICIT 2007, pp. 697–700. IEEE Press, Shenzhen (2007)

Feasibility of Lingo Software for Bi-Level Programming Problems (BLPPs): A Study

Kailash Lachhwani[1(✉)], Abhishek Dwivedi[3], and Deepam Goyal[2]

[1] Department of Applied Science, National Institute of Technical Teachers Training
and Research, Chandigarh 160 019, India
kailashclachhwani@yahoo.com
[2] Department of Mechanical Engineering, National Institute of Technical Teachers Training
and Research, Chandigarh 160 019, India
bkdeepamgoyal@outlook.com
[3] Department of Mathematics, Desh Bhagat University, Mandi Gobindgarh 147 301,
Punjab, India
abhishek1.hiet@gmail.com

Abstract. Lingo© is a basic programming tool for solving linear programming problems (LPPs), non-linear programming problems (NLPPs) and other related programming problems. But this software has certain limitations like it can not be employed on complex hierarchical programming problems like bi-level programming problems (BLPPs), multi-level programming problems (MLPPs) etc. In this context researchers namely Kuo and Huang [1] have presented a method based on particle swarm optimization (PSO) algorithm for solving bi-level programming problems and performance of developed method have been analysed with the results of other methods like fuzzy neural networks (ANN), genetic algorithm (GA), lingo (software) on same numerical examples. In this article, the feasibility of Lingo© software for solving bi-level linear programming problems (BLPPs) has been studied and proved that this software cannot be used for solving BLPPs. The aim of this paper is to identify the results and performance analysis given by Kuo and Huang with Lingo are incorrect and Lingo© is not feasible for solving bi-level linear programming problems (BL-LPPs).

Keywords: Bi-level programming problems · Lingo · Particle swarm optimization

1 Introduction and Results

Lingo© (developed by LINDO systems Inc.) is a basic software to solve linear, non-linear and integer optimization models. Kuo and Huang [1] used lingo for obtaining the best solution of bi-level programming problems (numerical problems (1), (2), (3) and (4) of [1]) (see Appendix A) for comparative analysis. But due to technical limitations of this package, lingo does not deal directly to bi-level programming problems (BLPPs) and does not provide any solution to BLPPs. Even, if we try to solve these BLPPs with

© Springer Nature Singapore Pte Ltd. 2017
K. Deep et al. (eds.), *Proceedings of Sixth International Conference on Soft Computing for Problem Solving*, Advances in Intelligent Systems and Computing 547,
DOI 10.1007/978-981-10-3325-4_34

lingo, we have most common error (as shown in Fig. 1 while solving problem (4) of [1]) is:

68: Multiple objective functions in model.

Only one is allowed, please.

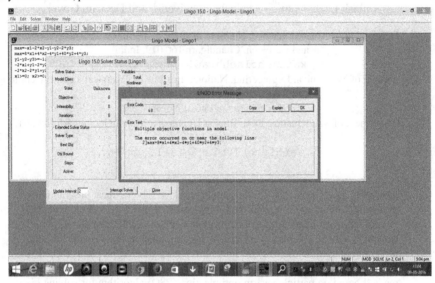

Fig. 1. Use of Lingo (trial version) for BLPP (Problem (4) of [1])

Further, if we solve individual upper level's problem and lower level's problem without considering the bi-level hierarchical structure (interactive decision making units) with lingo software, then only we can find different individual optimal solution to each level's problem. Again, if we assume that the hierarchical structure is not considered by Kuo and Huang [1] in obtaining individual best solutions of numerical problems, still the results for all these problems (problems (1), (2), (3) and (4)) given in [1] are different from the exact individual optimal solution to the problems by lingo. These all are unclear and creating confusions over the significance of suggested comparative analysis and conclusions. In order to avoid such pitfalls, it is necessary to know at least how these results are obtained with the help of lingo.

Appendix A: Problem (1), (2), (3) and (4) and Their Solution by Lingo as Suggested by Kuo and Huang [1]

S.No.	Problem Statement	Best Solution by Lingo Software		
Problem 1	Max $f_1 = -2x_1 + 11x_2$ where x_2 solves Max $f_2 = -x_1 - 3x_2$ s.t. $x_1 - 2x_2 \leq 4$ $2x_1 - x_2 \leq 24$ $3x_1 + 4x_2 \leq 96$ $x_1 + 7x_2 \leq 126$ $-4x_1 + 5x_2 \leq 65$ $x_1 + 4x_2 \geq 8$ $x_1, x_2 \geq 0$	Parameter	Lingo	
		x_1	17.4545	
		x_2	10.90909	
		f_1	85.0909	
		f_1 error rate	N/A	
		f_2	-50.18182	
		f_2 error rate	N/A	
Problem 2	Max x_2 where x_2 solves Max $-x_2$ s.t. $-x_1 - 2x_2 \leq 10$ $x_1 - 2x_2 \leq 6$ $2x_1 - x_2 \leq 21$ $x_1 + 2x_2 \leq 38$ $-x_1 + 2x_2 \leq 18$ $x_1, x_2 \geq 0$	Parameter	Lingo	
		x_1	16	
		x_2	11	
		f_1	11	
		f_1 error rate	N/A	
		f_2	-11	
		f_2 error rate	N/A	
Problem 3	Max $x_1 + 3x_2$ where y solves Max $-x_2$ s.t. $-x_1 + x_2 \leq 3$ $x_1 + 2x_2 \leq 12$ $4x_1 - x_2 \leq 12$ $x_1, x_2 \geq 0$	Parameter	Lingo	
		x_1	4	
		x_2	4	
		f_1	16	
		f_1 error rate	N/A	
		f_2	- 4	
		f_2 error rate	N/A	
Problem 4	Max $8x_1 + 4x_2 - 4y_1 + 40y_2 + 4y_3$ where y solves Max $-x_2 - 2x_2 - y_1 - y_2 - 2y_3$ s.t. $y_1 - y_2 - y_3 \geq -1$ $-2x_1 + y_1 - 2y_2 + 0.5y_3 \geq -1$ $-2x_2 - 2y_1 + y_2 + 0.5y_3 \geq -1$ $x_1, x_2, y_1, y_2, y_3 \geq 0$	Parameter	Lingo	
		x_1	0	
		x_2	0.9	
		y_1	0	
		y_2	0.6	
		y_3	0.4	
		f_1	26	
		f_1 error rate	N/A	
		f_2	-3.2	
		f_2 error rate	N/A	
		f_1 error rate	N/A	

Reference

1. Kuo, R.J., Huang, C.C.: Application of particle swarm optimization algorithm for solving bi-level linear programming problem. Comput. Math Appl. **58**, 678–685 (2009)

Time Series Analysis and Prediction of Electricity Consumption of Health Care Institution Using ARIMA Model

Harveen Kaur[1(✉)] and Sachin Ahuja[2]

[1] School of Computer Sciences, Chitkara University, Rajpura, India
harveen.kaur@chitkara.edu.in
[2] Chitkara University Research and Innovation Network, Rajpura, India
sachin.ahuja@chitkara.edu.in

Abstract. The purpose of this research is to find a best fitting model to predict the electricity consumption in a health care institution and to find the most suitable forecasting period in terms of monthly, bimonthly, or quarterly time series. The time series data used in this study has been collected from a health care institution Apollo Hospital, Ludhiana for the time period of April 2005 to February 2016. The analysis of the time series data and prediction of electricity consumption have been performed using ARIMA (Autoregressive Integrated Moving Average) model. The most suitable candidate model for the three time series is selected by considering the lowest value of two relative quality measures i.e. AIC (Akaike Information Criterion) and SBC (Schwarz Bayesian Criterion). The appropriate forecasting period is selected by considering the lowest value of RMSE (Root Mean Square Error) and MPE (Mean Percentage Error). After building the final model a two-year prediction of electricity consumption of the health care institution is performed.

Keywords: Time series · Akaike Information Criterion · Schwarz Bayesian Criterion · Autoregressive integrated moving average model · SAS University Edition

1 Introduction

It is critical for electricity traders [1] to predict electricity consumption for balancing the electricity purchase and sales portfolios. Electricity consumption is a significant commodity to promote society's economic development and raise people's standard of living [2]. Difficulties associated with electricity storage necessitate accurate prediction of its consumption. This further emphasizes the use of apt and correct approaches for electricity consumption prediction. Amongst the methods for performing the prediction, ARIMA is a statistical method for analyzing and building the forecasting model which best represents a time series by modeling the correlations in the Data.

© Springer Nature Singapore Pte Ltd. 2017
K. Deep et al. (eds.), *Proceedings of Sixth International Conference on Soft Computing for Problem Solving*, Advances in Intelligent Systems and Computing 547,
DOI 10.1007/978-981-10-3325-4_35

1.1   Autocorrelation Functions

- Autocorrelation: Autocorrelation, also called a lagged correlation or serial correlation is the way in which the observations in time series are related to each other and is calculated as the simple correlation between current observation y_t and the observation from p periods before the current one i.e. y_{t-p} for a given interval.
- Partial Autocorrelation: Partial autocorrelation is the correlation calculated after removing the linear relationship between the two consecutive observations. Basically, it is the correlation between y_t and y_{t-p} when the effect of y at other time lags 1, 2, 3, ..., p−1 is removed.
- Both autocorrelations and partial autocorrelations are computed for lags in sequence for the given time series. The autocorrelation at lag 1 is between Y_{t-1} and Y_t, the autocorrelation and partial autocorrelation at lag 2 is between Y_{t-2} and Y_t and so on for n lags.
- Autocorrelation function (ACF) and Partial autocorrelation function (PACF): The identification of the basic model is done by observing the patterns of ACFs and PACFs (Autocorrelations versus lags). ACFs and PACFs act as a tool for finding the values of the orders of autoregressive and moving average components i.e. p and q of ARIMA(p, d, q) respectively.

1.2   Autoregressive Integrated Moving Average Model

The notation for ARIMA model is ARIMA(p, d, q) where p, q, d denotes the order of autoregression, the order of moving average and the degree of differencing respectively. The final prediction depends on these parameters. Hence, the prediction equation for ARIMA (p, d, q) process is denoted by:

$$\Delta y_t = \beta_0 + \beta_1 \Delta y_{t-1} + \varepsilon_t + \lambda_1 \varepsilon_{t-1} \tag{1}$$

Here β_0, β_1 and λ_1 are the parameters of the model and ε_t is the white noise error term (the unidentifiable form of data which doesn't exhibit any patterns).

2   Literature Review

Currently, ARIMA model is avidly used for building a forecasting model for the time series of the electricity consumption. Katara et al. [3] used this method to forecast electricity demand in Tamale, Ghana using data from the Northern Electricity Department Tamale during 1990 to 2013. The model provides a seven-year forecast of the electricity demand in the city. Kandananond [4] implemented three forecasting techniques— autoregressive integrated moving average (ARIMA), multiple linear regression (MLR) and artificial neural network (ANN) to built forecasting model of the electricity demand in Thailand. Yasmeen and Sharif [5] studied, (the least out of sample forecast performance) i.e. the minimum forecast standard deviation value and Mean Absolute Percentage Error (MAPE) value of the four competing time series models. On the basis of these accuracy measures, the most suitable model to predict electricity consumption in Pakistan was built. Erdogdu [6] used co-integration analysis and implemented ARIMA modeling to estimate electricity demand and prediction respectively. Bianco et al. [7] evolved a long-term forecasting time series model by developing different regression models using historical electricity consumption, gross domestic product per capita (GDP per capita), gross domestic product (GDP) and population for the time period 1970 to 2007.

3   Methodology

The objective of this research is to develop a set of programs using SAS University Edition [8, 9] to find the best fitted model out of monthly, bimonthly and quarterly time series data [10–12] to predict the Electricity Consumption using ARIMA [13–16] model by considering the lowest value of Root Mean Square Error and Mean Percentage error for all three models. The selected forecasting model should have the lowest prediction error.

This work is purely based on ARIMA Model. The time series variable being used is the electricity consumption of the health care institution.

3.1   Data Preprocessing

Monthly data over a long period of time from April 2005 to February 2016 has been used for this research. SAS University Edition is used for building the model. Initially, data was not in a proper format and consisted of some missing values and so wasn't ready for building the forecasting model.

The data is monthly, but we have to find the suitable time period to predict the future values among three time periods i.e. monthly, bimonthly and quarterly time series.

The following pre-processing steps has been performed:

3.1.1   Converting the Data in Proper Format
By default, SAS displays the date values in a numeric format which is difficult to recognize.

Table 1 represents the sample of the raw data displaying the date values in numeric format, when it is imported to SAS. Hence, a proper date format is assigned to the variable. Table 2 represents a sample of the final data ready to be analyzed on the SAS.

3.1.2   Filling the Missing Information
In a particular observation if no value exists for a particular variable, then a missing value is said to occur. Missing data are very common and can have a huge effect on the data analysis. If not handled properly, these can lead to unstable or inaccurate results. There are various methods to fill the missing data with mean, median or previous value. Table 3 represents the sample of the current data set with some missing data. In this paper, previous values have been used to fill the current value by assuming that current value will be similar to the previous value.

Table 1. Sample of raw data displaying the date values in numeric format

Date	Electricity consumption
42465	137420
42495	240170
42526	287080
42556	295480
42587	304430
42618	285850

Table 2. Sample of final data ready to be analyzed

Date	Electricity consumption
Apr-05	137420
May-05	240170
Jun-05	287080
Jul-05	295480
Aug-05	304430
Sep-05	285850

Table 3. Sample of data with missing values

Date	Electricity consumption
Jan-06	170600
Feb-06	
Mar-06	204860
Apr-06	283940
May-06	
Jun-06	366000

3.2   Seasonal Adjustment of Data

For the seasonal adjustment of the data, a time series consists of three components: trend-cycle, combined seasonal effects, and irregular component. Here, Fig. 1 shows the seasonally adjusted data, while Figs. 2 and 3 depict the trend-cycle and irregular data in the time series respectively. The main purpose of the seasonal adjustment of the data is to identify the seasonal effects and remove them from the time series data.

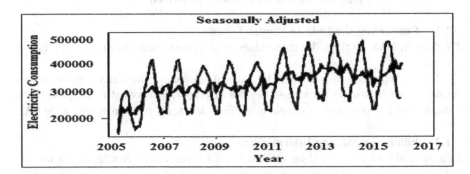

Fig. 1. Seasonal adjusted data

The time series is then called seasonally adjusted series and constitutes the trend-cycle component and irregular component.

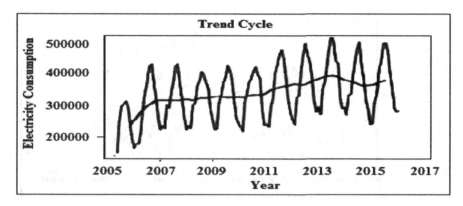

Fig. 2. Trend cycle of data

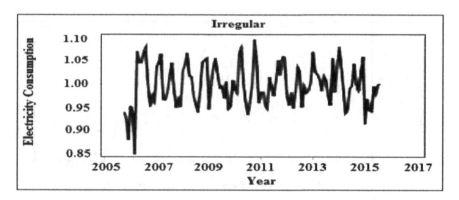

Fig. 3. Irregular form of data

3.3 Steps for Building the Model

Following are the steps for building the forecasting model using ARIMA:

3.3.1 Stationarity Checking

The Electricity Consumption of the health care institution is modeled variable. The time series is monthly time series from April 2005 to February 2016. The Fig. 4 represents the property of stationarity of the monthly data i.e. stationary or nonstationary times series. After the visual inspection of the graphs displayed in the figure, we can see that the time series is nonstationary and needs to be transformed into stationary time series; which is the first condition for applying ARIMA model. Here dB is the stationary time series. This is achieved by taking the first difference of the time series variable in the data. The blue line represents the nonstationary time series and redline represents the stationary time series. Data is prepared into two more formats i.e. bimonthly and quarterly time series.

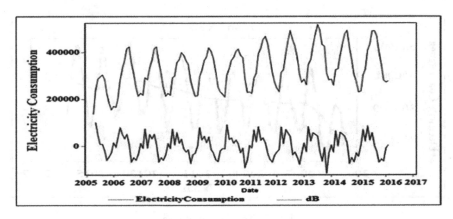

Fig. 4. Stationary vs nonstationary monthly time series.

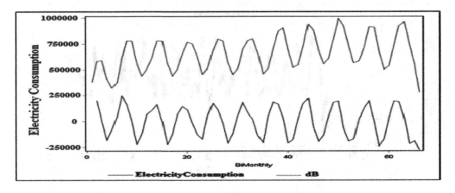

Fig. 5. Stationary vs nonstationary bimonthly time series

The Fig. 5 given below represents the property of stationarity of the bimonthly data while Fig. 6 represents the property of stationarity of the quarterly data.

3.3.2 Model Identification and Model Estimation

By examining the Autocorrelation Function (ACF) plot and Partial Autocorrelation Function (PACF) plot of the stationary time series, the basic ARIMA(p, d, q) model can be identified, but final models can be estimated by conditional least squares estimation method. In this estimation method if there is unit root in any of the AR (autoregressive) or MA (Moving Average) terms i.e. sum of parameter estimates is 1 or close to 1 then the current model is stabilized by adding or removing the number of AR or MA terms so that the unit roots can be removed. By considering the lowest value of AIC and SBC of the candidate models the most suitable model for the three time series can be decided. PACF and ACF plot of stationary monthly time series have been shown in Fig. 7. By examining the PACF plot the value of "p" - the order of autoregressive component – can be estimated. As PACF function plot drops to zero after first two lags, the value of "p" is 2. By examining the ACF plot we can estimate

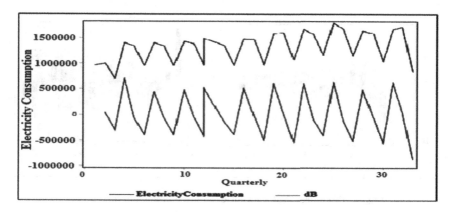

Fig. 6. Stationary vs nonstationary quarterly time series

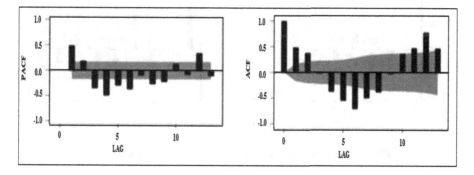

Fig. 7. PACF and ACF plot for differenced monthly time series.

the value of "q" - the order of moving average component. ACF function plot drops to zero after first three lags and hence, the value of "q" is 3. Since, the time series is differenced once, hence the value of "d" is 1. Basic model for the monthly time series of electricity consumption is ARIMA (2, 1, 3).

Similarly, referring to Fig. 8 the values of "p" and "q" can be estimated for differenced bimonthly time series data as 1 and 2 respectively. As illustrated in Fig. 9 "p" and "q" values for differenced quarterly time series data can be ascertained as 2 and 1 respectively.

The basic models are estimated by examining the ACF and PACF plots for all the three models. However, for bringing stability in the models, the existence of the unit roots is checked by conditional least squares estimation method.

If any unit root is found, then the current model is stabilized by removing or adding the number of AR or MA terms. The estimates for the parameters of all three models are calculated using the conditional least squares estimation method.

Tables 4, 5 and 6 display the parameter estimates for monthly, bimonthly and quarterly series respectively. It is worthwhile to note that in SAS, the estimate for moving average is always taken with opposite sign as displayed in table. Hence, after

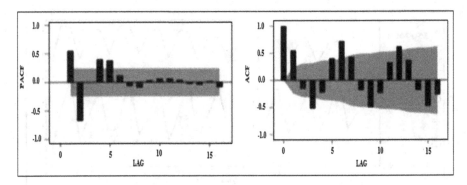

Fig. 8. PACF and ACF plot for differenced bimonthly time series

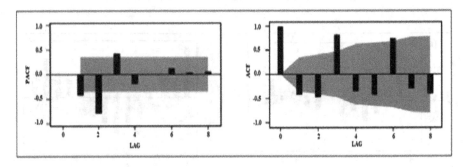

Fig. 9. PACF and ACF plot for differenced quarterly time series

Table 4. Conditional least squares estimation for monthly time series.

Parameter	Estimate	Standard error	t value	Approx Pr > |t|	Lag
MU	1588.1	898.69711	1.77	0.0797	0
MA1,1	1.99693	0.08688	22.98	<0.0001	1
MA1,2	−1.42805	0.15598	−9.16	<0.0001	2
MA1,3	0.32658	0.08748	3.73	0.0003	3
AR1,1	1.72513	0.01148	151.09	<0.0001	1
AR1,2	−1	0.01144	−87.42	<0.0001	2

Table 5. Conditional least squares estimation for bimonthly time series.

Parameter	Estimate	Standard error	t value	Approx Pr > |t|	Lag
MU	3638.5	2935.1	1.24	0.2199	0
MA1,1	0.68181	0.12167	5.6	<0.0001	1
AR1,1	1.00597	0.06042	16.65	<0.0001	1
AR1,2	−0.95325	0.05881	−1621	<0.0001	2

Table 6. Conditional least squares estimation for quarterly time series.

Parameter	Estimate	Standard error	t value	Approx Pr > |t|	Lag
MU	14328.6	10062	1.42	0.1655	0
MA1,1	−0.17119	0.21144	−0.81	0.4231	1
AR1,1	−0.99001	0.07057	−14.03	<0.0001	1
AR1,2	−1	0.06747	−14.82	<0.0001	2

Table 7. Possible models for monthly series.

Model	AIC	SBC
ARIMA(1,1,2)	3106.59	3118.057
ARIMA(2,1,2)	3098.66	3112.995
ARIMA(2,1,3)	3014.03	3031.238

Table 8. Possible models for bimonthly series.

Model	AIC	SBC
ARIMA(1,1,1)	1696.89	1703.413
ARIMA(1,1,2)	1695.576	1704.274
ARIMA(2,1,1)	1630.904	1639.602

Table 9. Possible models for quarterly series.

Model	AIC	SBC
ARIMA(1,1,0)	917.7574	920.6889
ARIMA(1,1,1)	922.5976	926.9948
ARIMA(2,1,1)	853.8797	859.7425

analysis of time series and performing the conditional least squares estimation methods, some possible models for the three series has been found. The possible candidate models for monthly, bimonthly and quarterly time series are depicted below in Tables 7, 8, and 9, respectively:

The values of the relative quality measures AIC and SBC are also mentioned. The best model is selected by considering the lowest value of AIC and SBC for all the time series. Hence the final model selected for monthly series is ARIMA(2, 1, 3)

$$\Delta y_t = 1588.1 + 1.72513\Delta y_{t-1} - \Delta y_{t-2} + \varepsilon_t - 1.99693\varepsilon_{t-1} + 1.42805\varepsilon_{t-2} - 0.32658\varepsilon_{t-3} \tag{2}$$

The final model selected for bimonthly series is ARIMA(2, 1, 1)

$$\Delta y_t = 3638.5 + 1.00597\Delta y_{t-1} - 0.95325\Delta y_{t-2} + \varepsilon_t - 0.68181\varepsilon_{t-1} \tag{3}$$

The final model selected for quarterly series is ARIMA(2, 1, 1)

$$\Delta y_t = 14328.6 - 0.99001\Delta y_{t-1} - \Delta y_{t-2} + \varepsilon_t + 0.17119\varepsilon_{t-1} \tag{4}$$

4   Results and Discussion

The actual electricity consumption and that forecasted by ARIMA method are compared from April 2005 to February 2016. An ARIMA based forecasting of electricity consumption has been performed and applied to the practical power system of the health care institution. ARIMA based method is more reliable and better than the traditional forecasting methods.

The data has been analyzed for all the three series. Table 10 shows the calculated RMSE (Root Mean Square Error) and MPE (Mean Percentage Error) of the best ARIMA models selected for monthly, bimonthly and quarterly time series

Table 10. RMSE and MPE for monthly, bimonthly and quarterly time series.

ARIMA	RMSE	MPE
Monthly	40021.25	0.22
Bimonthly	63891	0.39
Quarterly	132976.98	1.22

As depicted in Table 10, the most suitable period for forecasting is the monthly time series model with the lowest value of RMSE and MPE. On the basis of these forecasting accuracy measures of RMSE and MPE, the monthly forecasting model will be used to perform the final prediction of the electricity consumption of the health care institution.

Figure 10 shows the comparison of the actual and forecasted observations for the time period of April 2005 to February 2016, while, Fig. 11 illustrates the prediction of electricity consumption for 2 years ahead. Hence, the resultant prediction of electricity

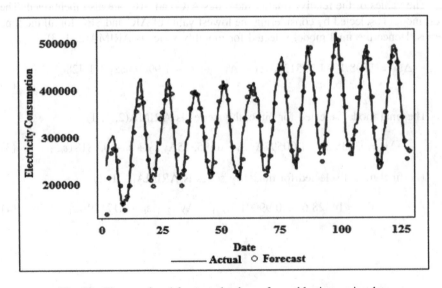

Fig. 10. The actual and forecasted values of monthly time series data

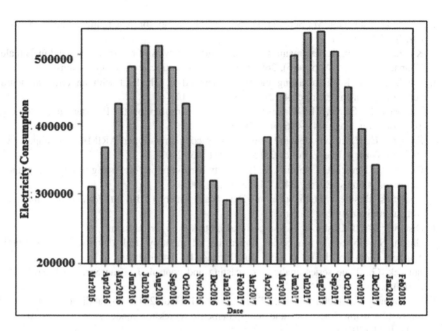

Fig. 11. Prediction of electricity consumption for two years ahead

consumption of the health care institution has been performed for the time period of March 2016 to February 2018.

5 Conclusion

The ARIMA based forecasting of electricity consumption has been performed and applied to the practical power system of the health care institution. From the results it can be concluded that this methodology is very efficient, and is more accurate and reliable than any other single forecasting methods which bring out imprecise predictions. Electricity being a crucial requirement of society needs to be predicted in advance as it is difficult to store. This forecasting will benefit the health care institution in many ways. It will help the management to anticipate the increase in capacity of the power system to cater to the futuristic needs, in addition to allocating the budget for planning and updating of their power system.

References

1. Sauhats, A., Varfolomejeva, R., Lmkevics, O., Petrecenko, R., Kunickis, M., Balodis, M.: Analysis and prediction of electricity consumption using smart meter data. In: IEEE 5th International Conference on Power Engineering, Energy and Electrical Drives (POWER-ENG) (2015)

2. Ahuja, D., Tatsutani, M., Schaffer, D.: Sustainable energy for developing countries. S.A.P.I. EN. S. (2.1) (2009)
3. Katara, S., Faisal, A., Engmann, G.: A time series analysis of electricity demand in Tamale, Ghana. Int. J. Stat. Appl. **4**(6), 269–275 (2014). Scientific & Academic Publishing
4. Kandananond, K.: Forecasting electricity demand in Thailand with an artificial neural network approach. Energies **4**(12), 1246–1257 (2011)
5. Yasmeen, F., Sharif, M.: Forecasting electricity consumption for Pakistan. Int. J. Emerg. Technol. Adv. Eng. **4**(4), 496–503 (2015)
6. Erdogdu, E.: Electricity demand analysis using cointegration and ARIMA modeling: a case study of Turkey. Energ. Policy **35**(2), 1129–1146 (2007)
7. Bianco, V., Manca, O., Nardini, S.: Electricity consumption forecasting in Italy using linear regression models. Energy **34**(9), 1413–1421 (2009)
8. Analytics, Business Intelligence and Data Management (n.d.). http://www.sas.com/en_us/software/university-edition.html.
9. Discover SAS® Customer Resources (n.d.). https://support.sas.com
10. Ghosh, S.: Univariate time-series forecasting of monthly peak demand of electricity in Northern India. Int. J. Indian Cult. Bus. Manag. **1**(4), 466 (2008)
11. Janacek, G.: Time series analysis forecasting and control. J. Time Ser. Anal. **31**(4), 303 (2010)
12. Dordonnat, V., Pichavant, A., Pierrot, A.: GEFCom2014 probabilistic electric load forecasting using time series and semi-parametric regression models. Int. J. Forecast. **2**(3), 1005–1011 (2016)
13. Camara, A., Feixing, W., Xiuqin, L.: Energy consumption forecasting using seasonal ARIMA with artificial neural networks models. Int. J. Bus. Manag. **11**(5), 231 (2016)
14. Razak, A.F., Shitan, M., Hashim, A.H., Abidin, I.: Load forecasting using time series models. Jurnal Kejuruteraan **21**(1), 53–62 (2009)
15. Nie, H., Liu, G., Liu, X., Wang, Y.: Hybrid of ARIMA and SVMs for short-term load forecasting. Energ. Procedia **16**, 1455–1460 (2012)
16. Lee, C.-M., Ko, C.-N.: Short-term load forecasting using lifting scheme and ARIMA models. Expert Syst. Appl. **38**(5), 5902–5911 (2011)

Recommendation System with Sentiment Analysis as Feedback Component

R. Jayashree and Deepa Kulkarni[(✉)]

Department of CSE, P.E.S University, Bangalore, India
dipa497@gmail.com, jayashree@pes.edu

Abstract. In today's world Artificial intelligence (AI) is known for deploying human like intelligence in to computers, so that they behave like humans. One of specialization areas of AI is expert systems. This area focuses on programming machines to take real life decisions. System with its intelligence helps users by suggesting them with variety of choices and making it easier for people to take best decisions while purchasing items. This work is intended to develop and deploy a Hotel Recommendation System. The work makes use of Collaborative user and item filtering techniques in combination with sentiment classification for generating recommendations. To improve the recommendations results, sentiment classification results are used as the feedback. There is also performance comparison between two different classifiers "Naïve Bayesian" (NB) and "K-Nearest Neighbor" (K-NN) with respect to their ability to recommend. This hybrid technique helps us in the case where an item has no ratings but has only textual reviews. Since this technique draws conclusion based on reviews along with the ratings, recommendation results are more accurate compared to recommendation systems based solely on filtering techniques.

Keywords: User and item based collaborative filtering · Sentiment analysis · Naïve Bayesian · K Nearest Neighbor classifier

1 Introduction

With introduction of World Wide Web lots of information is being contributed and shared through the internet. As the growth of information is exponentially increasing machines are used to handle this information. Right from the time when the computers or machines were invented, their ability to do the different types tasks has been growing at an exponential rate. It has been noticed that there is a growing trend of people purchasing products on the internet via e-commerce websites. This kind of trend has created the pre requisite for developing the systems that can help people by assisting them while they are purchasing goods online. This has led to the need for Intelligent Recommendation Systems. Artificial intelligence is responsible for creating the computers or machines as intelligent as human beings. Natural Language Processing is a methodology provided by Artificial intelligence using which we can communicate or interact with intelligent systems in the any of the

© Springer Nature Singapore Pte Ltd. 2017
K. Deep et al. (eds.), *Proceedings of Sixth International Conference on Soft Computing for Problem Solving*, Advances in Intelligent Systems and Computing 547,
DOI 10.1007/978-981-10-3325-4_36

languages spoken by the humans like English. In this work we are using a hybrid approach for developing a recommendation system.

2   Background and Related Work

This section gives a thorough understanding about the variety of techniques involved in designing and deploying the recommendation frameworks as well as techniques in involved in sentiment analysis.

Bucurab [1] has performed extensive work regarding sentiment analysis. The work intends to recognize and study the sentiments contained in the surveys which are put forth by the tourists on the websites. This work discusses performance of unsupervised sentiment analysis on hotel reviews. Accuracy of the technique shows to be around 74% which is good for unsupervised method.

Ghorpade, et al. [2] research focuses "Natural Language Processing" (NLP) and Bayesian classification, the study is on improving the extraction of sentiments from reviews or opinions that is, to minimise loss of information while extracting sentiments. The study here has developed some ontology modules to easily refine the attributes depending on the user requirements using "Jolly and Pleasant Exercise" (JAPE) mathematical techniques. Study reveals that improved word dictionary for a certain domain ontology produces well trained training sets which in turn helps in easy classification of reviews into positive and negative that is done using machine algorithm.

Suresh et al. [3] from their research work has stated that many text mining techniques like sentiment analysis and question answering system, the query matching score is affected by the presence of noise. To overcome this problem study in this area generates the bag of words using "context", this may result in generating less in number but the highly probable members which helps in easy look up and process of spelling correction of query and is made efficient.

Kumar et al. [4] proposed a system which is improved version of movie recommendation by combining collaborative filtering and sentiment analysis. The study explains in detail about the types of recommendations, how to train the classifier and how the sentiment analysis can be done using Naïve Bayesian algorithm in movie review context. The results of the study say that using combined filtering and sentiment analysis leads to better accuracy.

As per the study by Cane Wing-ki Leung et al. [5], an attempt is made transform user's opinions which are expressed in natural language in textual form in to ratings that is understood by collaborative filtering. Here all the reviews are first pre processed. Then reviews are classified as positive and negative. Next the polarity or orientation of reviews are transformed in to ratings scale. Thus experiment provides a new approach for mapping sentiment results to filtering results.

As per study of Bagchi [6] performance and quality aspects of recommendation using different types of similarity measures are analyzed using Apache Mahout. The work is based on recommendations collaborative filtering. And to calculate similarity different techniques like Euclidean distance, cosine similarity, Pearson correlation, Tanimoto and Log Likelihood similarity measures are used and the performance of recommendation system with each of these is measured. The work confirms that performance of system is best with use of Euclidean distance similarity measure.

As per work done by Subramaniyaswamy V et al. [7] purposes a system that assists user to look out for tourist locations that they might likes to visit a place from available user contributed photos of that place available on photo sharing websites. This work suggests an algorithm ADA-Boost to classify data and Bayesian Learning model for predicting desired location to a user based on his/her preferences. The Work includes collecting geo tagged photos from social media and identify destinations. Next group tourists based on age, gender and travel season. Trips generation or route planning is done using attributes of people and by building recommendation model is built

3 Recommendation System – An Overview

Basically, a recommendation system does the analysis of the set of items, towards which users may be attracted and finally suggest the most suitable items to users. Hence recommendation systems help users to take better decisions in situations where they are supposed to choose from huge variety of options.

Recommending hotels to a tourist given a particular tourist location is the one of best applications of the recommendation systems. Recommendation systems can be used in different domains apart from tourism like recommending video clips for a particular user in You-tube based on user's past video preferences. This type of recommendation is done by collaborative filtering approaches. To do the recommendation we need to identify user's attributes or features as well as item's attributes or features of. Using these attributes, recommendation systems will analyze which set of users are attracted to which set of items. This type of analysis is done by content based filtering systems and these systems do not rely on users past preferences. In this paper we make use of collaborative filtering techniques.

This includes

- User Based Collaborative Filtering
- Item Based Collaborative Filtering.

Hybrid Recommendation System
We can create a hybrid recommendation system by integrating collaborative filtering systems with Sentiment Analysis. Training and testing these hybrid systems using the concepts of machine learning. Sentiment analysis is process of identifying and analysing opinions expressed by the people in a certain piece of text. Here we use sentiments expressed by tourists in their reviews about a particular hotel in order to recommend that hotel. We then combine these results with collaborative results to refine list of final recommended items.

4 Proposed Work

This work focuses on enhancing the already existing recommendation systems. The goal is to perform sentiment analysis on the hotel reviews using Naive Bayesian and K Nearest Neighbour classification algorithms. The results of classification are used as feedback for recommending the hotels using collaborative filtering technique.

The proposed approach is divided in to three modules. The Fig. 1 below gives understating of System Architecture and Fig. 2 below provides clear knowledge of all steps performed

1. **Parser Module**

   This module does the primary task of reading the reviews file that has been loaded. After parsing the file it does the tokenization of review text which is followed by POS tagging. Purpose of tagging the review text is to choose the opinion word and form a "bag of words" model.

2. **Classification Module**

   This module consists of creating training dataset. Next step here is to build the classifier. Once classifier is built next step is to train classifier. Then finally test the classifier and determine the accuracy of the classifier. This helps us to know how well classifier can assign label of "positive" or "negative" to the review.

3. **Recommendation Module**

   This module computes the user based similarity and item based similarity using the rating provided. Hence finds the similar users and items. And finally predicts or recommendations the items for new users based on computations.

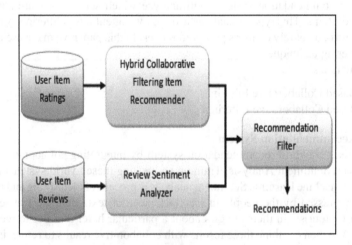

Fig. 1. System Architecture

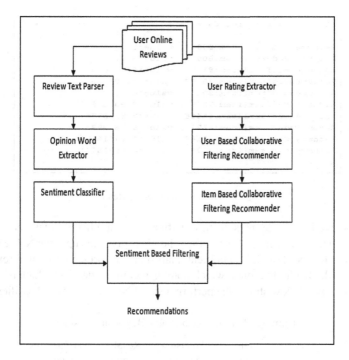

Fig. 2. Detailed Design

Steps Involved

1. Parsing the review dataset file
2. Creating bag of words or opinion word extraction
3. Creating training dataset
4. Build Classifier
5. Train and Test Classifier using textual reviews
6. Generating user based recommendations using ratings
7. Generating item based recommendations using ratings
8. Integrating classification and filtering results for recommendation

5 Results

For analyzing the performance of the hybrid recommender we consider the TripAdvisor dataset. Here first we use the ratings of hotels as input to traditional recommendation system. By performing user and item based filtering we get list of hotels that are recommended. Results are as in the figure below

```
Recommendations for User ID 2

RecommendedItem[item:80879, value:5.0]
RecommendedItem[item:80925, value:5.0]
RecommendedItem[item:81170, value:4.0308137]
RecommendedItem[item:73787, value:4.005326]
RecommendedItem[item:80836, value:4.0]
RecommendedItem[item:73757, value:3.9973497]
RecommendedItem[item:81466, value:3.9959776]
RecommendedItem[item:81251, value:3.7673535]
RecommendedItem[item:81165, value:3.6471615]
RecommendedItem[item:80930, value:3.6271572]
BUILD SUCCESSFUL (total time: 5 minutes 7 seconds)
```

Fig. 3. Traditional Filtering Results

In the above Fig. 3, numbers displayed after "item" are list of hotels recommended for USER 2. The review files of list of these hotels is given as input the Naïve Bayesian classifier and K-NN classifier. Sentiment analysis is performed on textual reviews for each hotel in the list to determine which hotel is recommended and which is not.

Table 1 shown below shows the performance of Naïve Bayesian Classifier.

Table 1. Performance of Naïve Bayesian Classifier

	TP rate	FP rate	Precision	Recall	F-measure	Class
	0.788	0.219	0.814	0.788	0.801	Positive
	0.781	0.212	0.751	0.781	0.766	Negative
Weighted average	0.785	0.216	0.786	0.785	0.785	

Table 2 shown below shows the performance of K-Nearest Neighbor Classifier.

Table 2. Performance of K-Nearest Neighbor Classifier

	TP rate	FP rate	Precision	Recall	F-measure	Class
	0.977	0.483	0.711	0.977	0.823	Positive
	0.517	0.023	0.948	0.517	0.669	Negative
Weighted average	0.77	0.276	0.818	0.770	0.754	

Comparative Study of Classifiers

Looking at the numbers in the table above we can see that the K-NN classifier has good precision value as compared with Naive Bayesian (NB). But when we take look at time taken by both classifiers NB Classifier performs very well compared to K-NN.

Table 3 shown below shows the performance comparison of K-NN and Naïve Bayesian classifier.

Table 3. Performance comparison of K-NN and Naïve Bayesian classifier

Parameters	Naive Bayesian	K-Nearest Neighbor
TP rate	0.723	0.842
FP rate	0.327	0.460
Precision	0.769	**0.846**
Recall	0.723	0.842
F-measure	0.738	0.819
Time taken	**12.84** s	184.072 s

Considering the fact that we have taken dataset having 15 hotels where each hotel file has around 100–150 reviews which turns out to be data of 1500 to 2000 reviews where in a single review may be of sentence or paragraph. Thus if we want better accuracy it is advisable to use K-NN classifier by compromising time taken.

If the concern is related processing speed, then NB classifier is recommended, also NB gives better tradeoffs between time taken and accuracy since it has precision is not totally unacceptable when compared to K-NN classifier.

Table 4. Results of sample dataset

Hotel name	Overall rating	No. of positive reviews	No. of negative reviews	Ratio N/P	Recommended
Barcelona Catalonia	4	95	23	0.242	Yes
Affinia50 New York City	3.5	471	157	0.333	Yes
San Juan Hotel	3	27	14	0.51	No
Casa Fuster Hotel	**3.5**	**62**	**11**	**0.177**	**No**
Country Inn Suites New Orleans	**3.5**	**129**	**51**	**0.39**	**Yes**
Cremorne Point Manor	3.5	28	12	0.42	No
Rydges World Square	3.5	41	33	0.80	No
Simpsons of Potts Point Hotel	5	89	37	0.41	No
Suite Hotel Berlin	4	86	16	0.18	Yes
The Soho Hotel London	5	116	26	0.22	Yes

Accuracy of Recommendation
The details recorded in the above Table 4, and values are obtained by keeping the user specified Negative to Positive (NP) ratio to 30%. This NP ratio is threshold value for determining whether or not to recommend the hotel. In the above data the recommender has recommended the hotels with quite good accuracy. System has incorrectly recommended the hotel highlighted with boldface. This recommendation is obtained by using filtering with trained Naïve Baysiean classifier, and with rating threshold value set to 3.

6 Conclusion

Recommender Frameworks seems to have taken an extremely predominant place in the web world of the people. We have considered a hotel review dataset which contains reviews and different ratings for different aspects of hotel although we take only overall rating. The major objective of the paper is deploying a "Hybrid Recommender Framework" that performs thorough refining of products to be recommended for clients. This proposed model works in combination with "sentiment classification" or "opinion mining" and "collaborative filtering". The recommendation of products that have not been rated earlier is main issue that is resolved. Sentiment classification results are used as feedback to improvise the results of "user-item" filtering. Ultimately top best 'n' results are recommended.

Major difficulty is to find overall sentiments that lie within reviews, that are written in more general form like reviews whose first few lines show positive side and last few lines depict negative side of the product. This kind of reviews make it hard classify them. Also we cannot put them under "neutral" class as they do express sentiments. Be that as it may, this strategy without a doubt exhibits better proposals than utilizing synergistic separating alone and consequently is a superior recommender framework outline. Any recommendation framework that gathers user's reviews can be adjusted to utilize hybrid proposal approach.

References

1. Bucurab, C.: Using opinion mining techniques in tourism. Procedia Econ. Finance **23**, 1666–1673 (2015). 2nd Global Conference on Business, Economics, Management and Tourism 2013
2. Ghorpade, T., Ragha, L.: Featured based sentiment classification for hotel reviews using NLP and Bayesian classification. In: International Conference on Communication, Information & Computing Technology (ICCICT 2012), pp. 1–5 (2012)
3. Ashokjadhav, S., Nageshbhattu, S., Subramanyam, R.B.V., Suresh, P.: Context dependent bag of words generation. In: International Conference on Advances in Computing, Communications and Informatics (ICACCI 2013), pp. 1526–1531 (2013)
4. Singh, V.K., Mukherjee, M., Mehta, G.K.: Combining collaborative filtering and sentiment classification for improved movie recommendations. In: Sombattheera, C., Agarwal, A., Udgata, S.K., Lavangnananda, K. (eds.) MIWAI 2011. LNCS, vol. 7080, pp. 38–50. Springer, Heidelberg (2011). doi:10.1007/978-3-642-25725-4_4

5. Leung, C.W., Chan, S.C., Chung, F.: Integrating collaborative filtering and sentiment analysis: a rating inference approach. In: ECAI 2006 Workshop on Recommender Systems, Riva del Garda, Italy, pp. 62–66 (2006)
6. Bagchi, S.: Performance and quality assessment of similarity measures in collaborative filtering using Mahout. In: 2nd International Symposium on Big Data and Cloud Computing, ISBCC 2015 (2015)
7. Subramaniyaswamy, V., Vijayakumar, V., Logesh, R., Indragandhi, V.: Intelligent travel recommendation system by mining attributes from community contributed photos. Procedia Comput. Sci. **50**, 447–455 (2015). 2nd International Symposium on Big Data and Cloud Computing ISBCC 2015

Natural Language Processing Based Question Answering Using Vector Space Model

R. Jayashree and N. Niveditha[✉]

Department of CSE, P.E.S University, Bangalore, India
niveditha.n.atnidhi@gmail.com, jayashree@pes.edu

Abstract. Natural Language Processing (NLP) is a technique used to build computational models which deals with the interaction between computers and human languages. Question answering is expected to give the precise results for the query instead of a group of links or references which might contain an answer. The information in the web is basically growing and users are finding difficult to look for the answers through the search engines. In this research, a new approach is used to build the question answering system which uses vector space model by using unstructured data. In this proposed work, Keywords are generated by calculating the tf-idf score for each keyword and they are indexed to every file and query. The query vectors and Document vectors are compared and similarity values are generated using term frequency. Highest ranked documents are generated as per the similarity values and NER tagging is done to produce candidate answers from which best answer is chosen.

Keywords: Natural Language Processing · Information retrieval · Information extraction

1 Introduction

Natural Language Processing is a technique used to build computational models which deals with the interaction between computers and human languages. The idea of Natural language Processing is to make computers to have the knowledge about the human understanding or input of natural language. It is mainly concerned with how humans can communicate using Natural Language. The challenge of Natural Language Processing is that computers needs humans to speak to them in a precise and highly structured Programming language.

Question answering has recently gained popularity from retrieval of information and natural language processing. As internet users are increasing day by day, the requirement for the QA system is going too forward. Question answering system is expected to give answers directly to the question given in the Natural language instead of providing a series of results. Usually a computer program builds the answers for the queries by querying the structured database of knowledge. Question answering system is a technique of deriving or extracting answers from knowledge base to the queries given by the users in Natural language. Generally, search engines gives a ranked list of answers

© Springer Nature Singapore Pte Ltd. 2017
K. Deep et al. (eds.), *Proceedings of Sixth International Conference on Soft Computing for Problem Solving*, Advances in Intelligent Systems and Computing 547, DOI 10.1007/978-981-10-3325-4_37

where user needs to select the answer from those results which is a difficult thing but Question Answering gives you the exact results for the user question. A set of Wikipedia pages is an example of document collection of natural language used for Question answering systems.

The QA systems are broadly classified as mainly two domains: 1. Closed domain, 2. Open domain. Closed domain Question answering system is meant to a specific domain. Only limited types of questions are accepted under closed domain. Answers are retrieved for the questions from only one specific domain. Open domain Question answering system is meant to multi domain. There is no limitation in the types of questions and it answers to the question asked from any domain. Much information is needed here to retrieve the results.

A basic Question answering system contains a set of modules. In most of the basic question answering system there are three basic set of modules for the functionality and they are described as:

1. Question Analysis module. This module goal is to process the input query, understanding the input query and identify the type of the question. Also keyword extraction is done from the query as tokens or terms or keywords. This module performs the query classification. Some analytical operations are done for the representation of the query. Question is analyses using some NLP techniques. Open NLP, Stanford Core NLP Parser are some of the tools used for this Processing. This module is also called as Question Processing module or Question classification module.
2. Document processing module. This module takes the tokens or the keywords identified in the question processing module and use some of the popular search engine to perform document processing which retrieves the information related to the identified keywords. This module recognizes a set of related documents and extracts a set of paragraphs as per the focus of the question.
3. Answer Extraction module. This module tries to extract the answer from the series of related relevant results that are obtained from document processing module. It mainly deals with Answer filtering, Answer ranking, and Answer generation. Once the answers are identified, they are extracted and ranked. Ranking is done based on the similarity. The top most answer will be taken as the best answer for the given question.

2   Background and Related Work

Question Answering systems are the easiest way to gain the answers by asking the questions. The greater part of the QA frameworks give answers to Factoid and List type questions and some give answers to complicated, thinking based and descriptive type questions. BASEBALL and LUNAR [2] are the Question answering systems which came early and were confined to domain specific systems and they worked over databases. BASEBALL QA system took the domain US baseball league and LUNAR Question answering framework took the domain about geographical analysis of rocks and answered questions based on that specific domain. TREC question answering system provided the way to Open domain Question answering system. A machine learning methodology can create a better classifier naturally which is more adaptable than a manual one since it can be effectively adjusted to another area. Through the use of machine learning approaches, we got to know about the technique of Support vector machines that they perform well when compared to other classifiers [2].

There is another type of question answering system called Rule based QA system [1] which provides answers for different type of questions like What, Who, Where, When, and Why. Who type questions generally expect the answer to be a person, Where type questions generally expect the answer to be the Location, When type of questions generally expect the answer to be the Date, What type of questions will expect the answer to be one word, or one sentence or descriptive answers and finally the Why type of questions is answered based on reasoning [1].

Tiansi Dong, Ingo Glockner, et al. [3] have proposed an QA framework goes for consequently finding succinct responses to subjective inquiries expressed in normal dialect. One such question answering system is Log answer. It is a web based system which was mainly proposed for German language. It extracted answers for arbitrary queries by using the techniques of Artificial Intelligence. The client enters a query into the interface, and LogAnswer then shows a rundown of answers. By translating the German Wikipedia into MultiNet representation, knowledge bases are created. Through this knowledge bases, the answers are extracted to the given queries. Here machine learning approach technique was used to solve the problem of Wrong answer avoidance by placing the meta classifier to the results of term based classifiers. ML based filtering technique was used to filter the answers by finding useful and not useful answers. Keeping in mind the end goal to extend the utilization of LogAnswer and to accomplish a superior true assessment, it was decided to adjust the framework to QA Forums so that users can get answers properly. Also in future, the author has told that he will include the opencyc ontology to the knowledge base [3].

Another Question answering system uses heuristic approach [2] which was proposed by Varsha Bhoir, et al. works by considering a specific domain of tourism. Here the author has considered some key points like query, source for retrieving answer, response, Processing of answer, and representation of answer. The input query can be any natural language query, like list type, factoid type, or yes or no type. The source could be structured database, or text of documents, or semi-structured database, or even web. Next is the type of answer needed by the user whether list of answers, or descriptive, or the one word answers. Here in this paper, parsed web pages are collected from the web using crawler from the tourism websites. Then Pre-processing is done on these web pages. Pre-processing mainly comes with tokenization, stop-word removal and stemming. Also Pre-processing is done to the user query. Tokenization means it divides the sentences into words and it is taken as tokens. Stop-word removal removes the frequently occurring words like a, an, the, etc. Stemming takes the lemma of the word which means the base form of the word. Finally, answer is extracted using Procedure Programming and Natural Language Processing techniques. The answers are assessed using Maximum Reciprocal Ratio. The goal of the system was to increase the accuracy and retrieving the exact results. The author concluded by saying that a better classifier must be used in the future in order to provide the results for different type of questions. Also, he mentioned to use SVM to find the type of questions as future enhancement.

A Naïve Bayesian [4] classifier is based on Bayes theorem and it uses conditional independence. It expects that the presence or absence of a specific element of a class is random to the presence or absence of some other element, given the class variable. The classifier is guided by the group of labelled training samples.

3 Methodology

There are mainly three modules in our Proposed system.

1. Document VSM generation module and Query VSM generation module
2. Similarity measure module
3. Document ranking and answer generation module

1. **Document VSM generation module and Query VSM generation module**
   Documents are read from the repository. Keywords are generated for all the documents by calculating the tf-idf value for each keyword. These keywords are indexed in every file with their keyword id values. If a particular keyword is present in a file, then that particular keyword's id is indexed to that file. This is the way to generate the Document vectors. Keywords are indexed to the query. If a particular keyword is present in the query then that particular keyword's id is assigned to the query. This is the way to generate the Query vectors.

2. **Similarity measure module**
   Having document vectors and query vectors, similarity measure is done between them based on the term frequency.

3. **Document ranking and answer generation module**
   Based on the similarity values from the previous module, documents are ranked with the highest similarity values. Top ranked documents are performed with NER tagging in order to generate the candidate answers. Having candidate list of answers, the most frequently occurring answer is selected to be the best answer.

Figure 1 shows the overall architecture of the proposed system. Keywords are generated from the documents and all the document files are indexed with unique id's of keywords. Query is also indexed with id's of keywords. Then similarity is measured between Query vectors and Document vectors. Documents are ranked and finally best answer is chosen from the candidate answers.

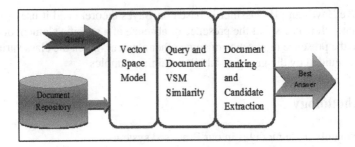

Fig. 1. Proposed System Architecture

The above Fig. 2 shows the detailed workflow of the Proposed system. Through the following steps, the best answer is generated for the given input.

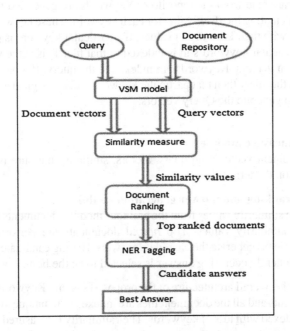

Fig. 2. Work Procedure of overall system

4   Results

The results are obtained by taking the dataset of some scientific inventions. The Question answering system gives the results for the questions within the domain. When a user gives an input query, number of steps will take place and final best answer is chosen.

For example:

1. Who invented the Electric Bulb?

Answer: Edison
2. Who invented the theory of relativity?
   Answer: Einstein.
3. Where is Karnataka?
   Anwer: Please enter the question related to domain.

The results obtained from our proposed work are shown through screenshots.

1. Document VSM generation

The above Table 1 describes the VSM generation for Documents. All keywords are indexed in every particular file with their Keyword id numbers and term frequency of each word is displayed along with that.

2. Query VSM generation for the given query

Table 1. Document VSM generation

Document VSM generation
Folder name : File name: keyword id : word frequency,..........
A:6.txt : 26:1, 35:2, 55:1, 57:3, 59:1, 78:2, 79:1, 80:1, 82:5,......
P:1.txt : 45:1, 47:1, 49:2, 55:3, 59:1, 61:1, 62:1, 103:1, 105:3,....
C:3.txt : 75:5, 82: 2, 123:5, 125:2, 130:1, 133:1, 135:2, 146:2,.....

The above Tables 2a and 2b shows the input given in the field and generation of query VSM.

3. Document ranking and similarity values generation

The above Table 3 shows the ranking of documents and generation of similarity values.

4. Answer generation

The above Table 4 shows the best answer generation for the given query.

Table 2a. Query VSM generation

Query VSM generation
Query: Who invented Electric Bulb?
Query VSM: 48:1, 784:1

Table 2b. Query VSM generation

Query VSM generation
Query: Who invented theory of relativity?
Query VSM: 26:1, 104:1

Table 3. Similarity value generation

Documents	Similarity values
Document: A:50.txt	Similarity: 48.0
Document: A:06.txt	Similarity: 28.0
Document: A:15.txt	Similarity: 8.0
Document: A:05.txt	Similarity: 6.0
Document: B:14.txt	Similarity: 5.0
Document: B:18.txt	Similarity: 3.0
Document: B:05.txt	Similarity: 2.0
Document: A:31.txt	Similarity: 1.0
Document: B:10.txt	Similarity: 1.0
Document: B:17.txt	Similarity: 1.0
Document: B:11.txt	Similarity: 1.0
Document: A:2.txt	Similarity: 1.0
Document: A:7.txt	Similarity: 0.0
Document: B:20.txt	Similarity: 0.0
Document: B:16.txt	Similarity: 0.0
Document: B:19.txt	Similarity: 0.0

Table 4. Best answer generation

Candidate answers
Edison
Francis robbins
Samuel
Fleming
Alexander
Andrew
Howard florey
Best Answer : Edison

5   Conclusion

Question answering system is naturally used to query the things and gain the knowledge which is a direct approach for the users to acquire the information to any data regardless of structure and format. The QA system in our project perform Vector space model technique on user query and document repository and generate similarity values by the use of document and query vectors. And finally best answer is formulated through document ranking and NER tagging. Currently this Question answering system answers for the simple questions and it is closed domain. It identifies the answer type and chooses the best answer from the candidate answers. In future, the QA system can be extended to answer complex questions. It can be extended by using Open domain. Also, since it is unstructured data, it is required to use some sophisticated techniques and approaches to extract the answers in future. Also it can be extended by asking the multiple questions at a time and get summarized results for all the questions.

References

1. Linda, Jovita, Hartawan, A., Suhartono, D.: Using vector space model in question answering. Procedia Comput. Sci. **59**, 305–311 (2015)
2. Bhoir, V., Potey, M.A.: Question answering system: a heuristic approach. In: 2014 Fifth International Conference on Applications of Digital Information and Web Technologies (ICADIWT 2014), pp. 165–170. IEEE Conference Publications (2014). doi:10.1109/ICADIWT.2014.6814704
3. Dong, T., Glockner, I., Furbach, U., Pelzer, B.: A natural language question answering system as a participant in human Q&A portals. In: Proceedings of the Twenty-Second International Joint Conference on Artificial Intelligence, pp. 2430–2435 (2011)
4. Yadav, R., Mishra, M.: Question classification using Naïve Bayes machine learning approach. Int. J. Eng. Innovative Technol. (IJEIT) **2**(8) (2013)

Effects of Delay and Drug on HIV Infection

Saroj Kumar Sahani$^{(\boxtimes)}$

South Asian University, Akbar Bhawan, Chanakyapuri, New Delhi 110021, India
sarojkumar@sau.ac.in
http://www.sau.int

Abstract. This article discusses delayed model of HIV infection with combination therapy consisting of RTI and PI drug. The delay included in this article two kinds of delays viz. immune response delay and intracellular delay. A well known growth law so called logistic growth is assumed for uninfected and healthy T cell. Local properties of the infection free equilibrium point is discussed in terms of R_0, the basic reproduction number. The existence of Hopf bifurcation with respect to delayed parameter is verified using geometric switching conditions numerically because of delay dependent parameters in the model. Extensive numerical simulations have been carried out on the model to ascertain the effects of drug on viral dynamic and disease progression.

Keywords: Delay · Bifurcation · HIV · Stability

1 Introduction

The HIV unlike other virus is very specific virus which infects human and has many puzzling features as compared to other viral infection. one of them is that there it take approximately 10 years to fully developed and regarded as AIDS. The virus so called *human immunodeficiency virus*, as the name implies, it destroy the immune system. The immune system in human body is mostly due to presence of a kind of white blood cells called CD4$^+$T cell [49]. The count of this cell classifies the different stages of the infection. In a healthy person, the usual count of CD4$^+$T is around $1000\,\text{mm}^{-3}$. When this count fall around $200\,\text{mm}^{-3}$, the individual is classified as having AIDS [51] which is the last phase of infection. The other intermediate phases are primary infection stage and followed by chronic infection phase. They are also called acute symptom phase and latency phase respectively.

The literatures on the study of HIV infection through mathematical models are many. A very excellent account of work done by some of the authors in this direction is presented by Perelson et al. [51] as a review article. Many studies are confined to the dynamics of the disease progression [3,5,12,27,31,

S.K. Sahani—Author is grateful to South Asian University for providing financial assistance to present this paper in the conference.

© Springer Nature Singapore Pte Ltd. 2017
K. Deep et al. (eds.), *Proceedings of Sixth International Conference on Soft Computing for Problem Solving*, Advances in Intelligent Systems and Computing 547,
DOI 10.1007/978-981-10-3325-4_38

44, 45, 49, 50, 56, 57, 59] but other deals with the effects of drug on the viral loads [23–25, 29, 60, 62]. The mathematical model and their studies are very popular and efficient way of handling complex process such as HIV infection and has been able to produce some important aspects of long time dynamics of the infection [1–3, 5, 8, 12, 27, 31, 37, 40, 44–46, 50–52, 57, 59, 61, 62].

Many authors have included the drug therapy into account to explain their importance on the overall life of an individual [11, 13, 17–19, 21, 30, 31, 33, 41, 60]. These studies have helped the microbiologist to come up with the combination drug therapy for cure of HIV infection. All of these drug actually acts on some key aspects of the infection process thereby hindering the growth of HIV in the body.

Considering the delayed effects in any disease model are very challenging and its outcome are well appreciated. Many authors now a days includes these delay effects in mathematical model [4, 7, 14, 15, 17, 26, 28, 32, 34, 39, 43, 53, 58, 63]. Many of these includes the effect of drug and delay on the HIV infection [47, 54, 64]. These models have made some important contribution in better understanding of life cycle of the HIV [3, 14, 15, 39, 43, 55, 56].

Out of all delays that usually occurs in HIV infection, intracellular delay is most important delay, and in past, many authors have taken into account [6, 17, 22, 35, 36, 38, 42, 56]. Other delays effects such as immunological delays that can also be taken into consideration to give a better representation of the infection process. If both delays are taken into consideration, it become more suitable to model such interaction between the virus and helper T cell but at the same time it becomes more changeling to analyse such model [48, 55].

This article deals with simultaneous effects of two kinds of delay viz. immune response and intracellular delay. This particular model also consider the effect of RTI and PI drug so that the model is more appropriate to explain the emergence of disease and also the elimination of virus from the body.

2   The Proposed Model

It is well known fact that immune response play a prominent role in fighting against disease and fight against viruses. The Immune system mainly composed of two types of cells called B cells and T cells. Immunity due to T cells are called cellular immunity whereas the immunity because of B cells are called humoral immunity. In many types of viral infection, humoral immunity is found to be superior than cellular immunity [16] in fighting some infection. Therefore, this article incorporate the effects of B cell on HIV lifecycle. So, to form the model, assume that u_1, u_2, u_3 and u_4 be the densities of uninfected T cells, infected T cells, HIV and B cells respectively in the body. The growth of healthy CD4+ T cells in the body have been modelled by logistic proliferation rate with α as a intrinsic growth rate of T cell. Body also generate the T cell by internal mechanism and for simplicity it is assumed to be at a constant Q. So, in absence of virus, the total growth rate is given by $Q + \alpha u_1 \left(1 - \frac{u_1}{K}\right)$ where K represents the maximum density of healthy T cell at which further generation of T cell stops. Since T cells are also dies in the process, the parameter δ_1 is assumed to account for the per capita death rate of T cell. If a person gets infected with HIV, the T cell interact with HIV and this rate of interaction is supposed to be as per well known *law of mass action* with β being rate of infection. The over all rate law of uninfected cell is therefore

$$\frac{du_1}{dt} = Q - \delta_1 u_1 + \alpha u_1 \left(1 - \frac{u_1}{K}\right) - \beta u_1 u_3$$

In this article, the effect of PI and RTI drug effect is also incorporated. Assume that RTI drug efficacy is denoted by ϵ_r, then after the interaction of HIV with T cell, the effective rate of conversion of healthy T cell to infected T cell is given by $\beta(1 - \epsilon_r)u_1(t)u_3(t)$. Thus the net healthy T cell proliferation rate can be assumed to be:

$$\frac{du_1}{dt} = Q - \delta_1 u_1 + \alpha u_1 \left(1 - \frac{u_1}{K}\right) - \beta(1 - \epsilon_r)u_1 u_3$$

The term $\beta(1 - \epsilon_r)u_1(t)u_3(t)$ also account for the growth of infected T cell per unit time. But this term is not instantaneously added to rate equation of infected T cell but instead with a lag of delay τ_1 units of time which account for the intracellular delay. Once infected, the infected T cell is assumed to follow exponential distribution and during time span of τ_1 time, the probability of surviving of infected T cell is $e^{-\delta_2\tau_1}$, where δ_2 is the clearance rate of infected cell. Therefore, number of infected cell at time t is just the summation of all infected cell which has been infected by HIV and hence alive during $[t - \tau_1, t]$ time interval [9] i.e.

$$u_2(t) = \int_0^{\tau_1} \beta e^{-\delta_2\tau_1} u_1\left(t - T\right) u_3\left(t - T\right) dT$$

The above equation is equivalent to the following equation which can be easily obtained by changing the variable using linear transformation $t - T = \phi$

$$u_2(t) = \int_{t-\tau_1}^{t} \beta e^{-\delta_2(t-\phi)} u_1\left(\phi\right) u_3\left(\phi\right) d\phi$$

The differential form of the above growth equation can now be obtained using differential under integral sign which gives

$$\frac{du_2}{dt} = \beta_1(1 - \epsilon_r)u_1(t)u_3(t) - \beta_1(1 - \epsilon_r)e^{-\delta_2\tau_1}u_1(t - \tau_1)u_3(t - \tau_1) - \delta_2 u_2(t)$$

The term $\beta_1(1 - \epsilon_r)e^{-\delta_2\tau_1}u_1(t - \tau_1)u_3(t - \tau_1)$ in above equation represents the lysis rate of infected T cell which will be releasing newly formed copies of virus. Therefore, let the death of one infected helper T cell produce on an average N number of virus which are capable of infecting other helper T cells. A constant per capita death rate δ_3 of HIV is also assumed. The immune response in this model is assumed to be mainly due to humoral cell and therefore HIV can also be removed from the body due B cell and rate of removal is assumed to be proportional to densities of HIV particle and B cell. This term is commonly known as viral loss due to immunity. So the equation of growth of HIV is given by

$$\frac{du_3}{dt} = N\beta e^{-\delta_2 \tau_1} u_1(t-\tau_1)u_3(t-\tau_1) - \delta_3 u_3(t) - \beta_2 u_3(t)u_4(t).$$

Lastly, the effector cell or immune response is assumed to get activated with delay τ_2 and is proportional to infected T cell density and virus density. At the same time, the effector cell concentration also decrease due to its natural death rate δ_4 and due to fighting with virus. It is assumed that β_4 represents the death rate of B cell per HIV cell and per B cell. Hence, the proposed model for HIV infection dynamics is

$$\frac{du_1}{dt} = s - \delta_1 u_1(t) + \alpha u_1(t)\left(1 - \frac{u_1(t)}{K}\right) - \beta_1(1-\epsilon_r)u_1(t)u_3(t)$$

$$\frac{du_2}{dt} = \beta_1(1-\epsilon_r)u_1(t)u_3(t) - \beta_1(1-\epsilon_r)e^{-\delta_2\tau_1}u_1(t-\tau_1)u_3(t-\tau_1) - \delta_2 u_2(t)$$
$$\tag{1}$$
$$\frac{du_3}{dt} = N\beta_1(1-\epsilon_p)(1-\epsilon_r)u_1(t-\tau_1)u_3(t-\tau_1)e^{-\delta_2\tau_1} - \delta_3 u_3(t) - \beta_2 u_3(t)u_4(t)$$

$$\frac{du_4}{dt} = \beta_3 u_2(t-\tau_2)u_3(t-\tau_2) - \delta_4 u_4(t) - \beta_4 u_3(t)u_4(t)$$

where $N \geq 1$, $\tau = \max\{\tau_1, \tau_2\}$, $\tau \in (0, \infty)$, α, K, β_1, β_2, β_3, β_4, ϵ_r, ϵ_p, s, δ_1, δ_2, δ_3 and δ_4 are nonnegative numbers with initial functions

$$u_1(\theta) = \phi_1(\theta)$$
$$u_3(\theta) = \phi_3(\theta)$$
$$u_4(\theta) = \phi_4(\theta) \tag{2}$$
$$u_2(0) = \int_{-\tau}^{0} \beta e^{\delta_2\phi} u_1(\phi)\, u_3(\phi)\, d\phi$$

where $\phi_i(\theta) \in C([-\tau, 0])$, $\theta \in [-\tau, 0]$, $\tau = \max\{\tau_1, \tau_2\}$ is such that $\phi_i(\theta) \geq 0$, $\phi_i(0) > 0$ for $i = 1, 3, 4$.

3   Preliminary Results

The following Lemmas state the positivity and boundedness of the solution of the system (1).

Lemma 1. *All solutions of the system (1) which starts in positive domain* $R_+^4 = \{(u_1, u_2, u_3, u_4) : u_1 > 0, u_2 > 0, u_3 > 0, u_4 > 0\}$ *remain positive for all time t.*

Proof. The first equation of system (1) gives

$$\frac{du_1}{dt} \geq u_1(t)\left(-\delta_1 + \alpha\left(1 - \frac{u_1(t)}{K}\right) - \beta_1(1-\epsilon_r)u_3(t)\right)$$

which on simple integration gives

$$u_1(t) \geq u_1(0)\exp\left\{\int_0^t \left(-\delta_1 + \alpha\left(1 - \frac{u_1(t)}{K}\right) - \beta_1(1 - \epsilon_r)u_3(t)\right)\right\}.$$

Now, consider the third equation of system (1) in interval $[0, \tau_1]$ with initial condition (2) which gives

$$\frac{du_3}{dt} = N\beta_1(1 - \epsilon_p)(1 - \epsilon_r)\phi_1(t - \tau_1)\phi_3(t - \tau_1)e^{-\delta_2\tau_1} - \delta_3 u_3(t) - \beta_2 u_3(t)u_4(t)$$

$$\geq -u_3(t)(\delta_3 + \beta_2 u_4(t)),$$

and by direct integration gives $u_3(t) \geq u_3(0)\exp\left\{-\int_0^t (\delta_3 + \beta_2 u_4(t))\right\}$ if integral is finite. This process can again be repeated to prove the positivity of u_3 in interval $[\tau_1, 2\tau_1]$ and hence in any finite interval $[0, t]$.

Again, to prove positivity of u_4, consider the interval $[0, \tau_2]$ with initial condition (2) and fourth equation system (1), then

$$\frac{du_4}{dt} = \beta_3 u_2(t - \tau_2)u_3(t - \tau_2) - \delta_4 u_4(t) - \beta_4 u_3(t)u_4(t)$$

$$= \beta_3 \phi_2(t - \tau_2)\phi_3(t - \tau_2) - \delta_4 u_4(t) - \beta_4 u_3(t)u_4(t)$$

$$\geq -u_4(t)(\delta_4 + \beta_4 u_3(t)),$$

which again on integration gives $u_4(t) \geq u_4(0)\exp\left\{-\int_0^t (\delta_4 + \beta_4 u_3(t))\right\}$, proving the positivity of u_4 in interval $[0, \tau_2]$ as long as integral is finite. Repeating the same way as in case of u_3, the positivity of u_4 can be proved for any interval $[0, t]$. Now, finally the second equation in (1) can be written as

$$u_2(t) = \int_{t-\tau_1}^t \beta e^{-\delta_2(t-\phi)} u_1(\phi) u_3(\phi) \, d\phi.$$

The non-negativity of u_1 and u_3 with above equation for u_2 implies that $u_2(t) \geq 0$ for all time t. This finally proves the Theorem.

Theorem 1. *If $\eta = K(\alpha - \delta_1) + \sqrt{K^2 (\alpha - \delta_1)^2 + 4\alpha K s}$, then in the domain*

$$\Delta = \left\{(u_1, u_2, u_3, u_4) : \|u_1\| \leq \frac{\eta}{2\alpha}, \|u_1 + u_2\| \leq \frac{2s + \eta}{2\delta'}, \right.$$

$$\left. \left\|u_1 + u_2 + \frac{u_3}{N(1 - \epsilon_p)}\right\| \leq \frac{2s + \eta}{2\delta''}, \|u_4\| \leq \frac{\beta \tilde{u}_2 \tilde{u}_3}{\delta_4}\right\},$$

all solutions of the system (1) that initiate in positive orthant, remain ultimately bounded.

Proof. If $\eta = K(\alpha - \delta_1) + \sqrt{K^2 (\alpha - \delta_1)^2 + 4\alpha K s}$, then it can be easily proved that $\eta > 0$ always for all values of the parameters. Then, using first equation of system (1),

$$\frac{du_1}{dt} = s - \delta_1 u_1 + \alpha u_1 \left(1 - \frac{u_1}{K}\right) - \beta u_1 u_3$$

$$\leq s - \delta_1 u_1 + \alpha u_1 \left(1 - \frac{u_1}{K}\right)$$

which yields,

$$\lim_{t \to +\infty} \sup \ u_1(t) \leq \frac{\eta}{2\alpha}. \tag{3}$$

Now on adding first two equations,

$$\frac{d}{dt}(u_1 + u_2) = s - \delta_1 u_1 + \alpha u_1 \left(1 - \frac{u_1}{K}\right) - \beta e^{-\delta_2 \tau} u_1(t - \tau) u_3(t - \tau)$$

$$\leq s + \frac{\eta}{2} - \delta_1 u_1 - \delta_2 u_2$$

$$\leq Q + \frac{\eta}{2} - \delta'(u_1 + u_2),$$

where $\delta' = \min\{\delta_1, \delta_2\}$ and therefore,

$$\lim_{t \to +\infty} \sup \ (u_1(t) + u_2(t)) \leq \frac{s + \eta/2}{\delta'}. \tag{4}$$

Similarly, using $u_1 + u_2 + \frac{u_3}{N(1-\epsilon_p)}$ and first three equations of (1)

$$\frac{d}{dt}\left(u_1 + u_2 + \frac{u_3}{N(1 - \epsilon_p)}\right) = s + \alpha u_1 \left(1 - \frac{u_1}{K}\right) - \delta_1 u_1 - \delta_2 u_2$$

$$- \frac{\delta_3}{N(1 - \epsilon_p)} u_3 - \frac{\beta}{N(1 - \epsilon_p)} u_1 u_3$$

$$\leq s + \frac{\eta}{2} - \delta_1 u_1 - \delta_2 u_2 - \frac{\delta_3}{N(1 - \epsilon_p)} u_3$$

$$\leq s + \frac{\eta}{2} - \delta'' \left(u_1 + u_2 + \frac{u_3}{N(1 - \epsilon_p)}\right)$$

where $\delta'' = \min\{\delta_1, \delta_2, \delta_3\}$, and this again implies

$$\lim_{t \to +\infty} \sup \ \left(u_1(t) + u_2(t) + \frac{u_3(t)}{N(1 - \epsilon_p)}\right) \leq \frac{s + \eta/2}{\delta''}. \tag{5}$$

Finally, Eqs. (3)–(5) prove that there exit positive constant $\tilde{u}_2$ and $\tilde{u}_3$ such that $u_2 \leq \tilde{u}_2$ and $u_3 \leq \tilde{u}_3$ and using this in last equation of system (1) gives

$$\frac{du_4}{dt} = \beta_3 u_2(t - \tau_2) u_3(t - \tau_2) - \delta_4 u_4(t) - \beta_4 u_3(t) u_4(t)$$

$$\leq \beta_3 \tilde{u}_2 \tilde{u}_3 - \delta_4 u_4.$$

This gives

$$\lim_{t \to +\infty} \sup \ u_4(t) \leq \frac{\beta_3 \tilde{u}_2 \tilde{u}_3}{\delta_4}. \tag{6}$$

Hence, Eqs. (3)–(6) prove that all the solutions of the system in the domain Δ, is bounded.

4   Local Stability Behaviour of Solutions

Now denote

$$b_1 = \beta_1 \left(1 - e_r\right),$$
$$b_2 = \beta_1 \left(1 - e_r\right) e^{-\delta_2 \tau_1},$$
$$b_3 = \beta_1 \left(1 - e_r\right) \left(1 - e_p\right) e^{-\delta_2 \tau_1},$$

and as stated above, the quantity $\eta = K(\alpha - \delta_1) + \sqrt{K^2 \left(\alpha - \delta_1\right)^2 + 4\alpha K s}$ remain positive for all parametric values and therefore, the system (1) permits two types of critical points:

1. an unique infection free critical point $E_0 \left(\bar{u}_1, 0, 0, 0\right)$, with $\bar{u}_1 = \eta/2\alpha$,
2. and at least one endemic equilibrium point $E_1 \left(\tilde{u}_1, \tilde{u}_2, \tilde{u}_3, \tilde{u}_4\right)$ where $\tilde{u}_1$ is one of the positive root of biquadratic

$$c_0 \tilde{u}_1^4 + c_1 \tilde{u}_1^3 + c_2 \tilde{u}_1^2 + c_3 \tilde{u}_1 + c_4 = 0$$

$$\tilde{u}_3 = \frac{ks + \alpha k \tilde{u}_1 - \delta_1 k \tilde{u}_1 - \alpha \tilde{u}_1^2}{\beta_1 (1 - \epsilon_r) k \tilde{u}_1}$$

$$\tilde{u}_2 = \frac{\beta_1 (1 - \epsilon_r) \left(1 - e^{-\delta_2 \tau_1}\right) \tilde{u}_1 \tilde{u}_3}{\delta_2}$$

$$\tilde{u}_4 = \frac{\beta_3 \tilde{u}_2 \tilde{u}_3}{\delta_4 + \beta_4 \tilde{u}_3}.$$

The coefficients c_0, c_1, c_2, c_3 and c_4 are given by

$$c_0 = (b_1 - b_2)\, \alpha^2 \beta_2 \beta_3$$
$$c_1 = K\alpha \left(2 \left(b_2 - b_1\right) \beta_2 \beta_3 \left(\alpha - \delta_1\right) + Nb_1 b_3 \beta_4 \delta_2\right)$$
$$c_2 = -K \left(b_2 \beta_2 \beta_3 \left(-2s\alpha + K \left(\alpha - \delta_1\right)^2\right) + b_1 \left(2s\alpha \beta_2 \beta_3 \right.\right.$$
$$\left.\left. + K \left(\alpha - \delta_1\right) \left(\beta_2 \beta_3 \left(-\alpha + \delta_1\right) + Nb_3 \beta_4 \delta_2\right) + \alpha \beta_4 \delta_2 \delta_3\right) + Nb_1^2 b_3 K \delta_2 \delta_4\right)$$
$$c_3 = K^2 \left(2b_2 s \beta_2 \beta_3 \left(-\alpha + \delta_1\right) + b_1 s \left(2\beta_2 \beta_3 \left(\alpha - \delta_1\right) - Nb_3 \beta_4 \delta_2\right)\right.$$
$$\left. + b_1 \beta_4 \left(\alpha - \delta_1\right) \delta_2 \delta_3 + b_1^2 \delta_2 \delta_3 \delta_4\right)$$
$$c_4 = K^2 s \left(\left(b_1 - b_2\right) s \beta_2 \beta_3 + b_1 \beta_4 \delta_2 \delta_3\right).$$

4.1   Local Stability Behaviour of $E_0 \equiv (\bar{u}, 0, 0, 0)$

The Jacobian Matrix at E_0 is

$$\begin{pmatrix} \frac{s}{\bar{u}} + \frac{\bar{u}\alpha}{K} + \lambda & 0 & b_1 \bar{u} & 0 \\ 0 & \delta_2 + \lambda & e^{-\lambda \tau_1} b_2 \bar{u} - b_1 \bar{u} & 0 \\ 0 & 0 & -e^{-\lambda \tau_1} Nb_3 \bar{u} + \delta_3 + \lambda & 0 \\ 0 & 0 & 0 & \delta_4 + \lambda \end{pmatrix}$$

and corresponding characteristic equation has obvious negative roots $-\left(\frac{s}{\bar{u}} + \frac{\bar{u}\alpha}{K}\right)$, $-\delta_2$ and $-\delta_4$ while other infinite roots are given by the transcendental equation

$$\lambda - \frac{(1 - \epsilon_p)(1 - \epsilon_r)N\eta\beta_1 e^{-\delta_2\tau_1}}{2\alpha} e^{-\lambda\tau_1} + \delta_3 = 0. \tag{7}$$

The above equation is transcendental equation and therefore have infinite number of roots in $\mathbb{C}$. Define the basic reproduction number(R_0) [1] as

$$R_0 \overset{\text{def}}{=} \frac{(1 - \epsilon_p)(1 - \epsilon_r)N\eta\beta_1 e^{-\lambda\tau_1}}{2\alpha\delta_3}. \tag{8}$$

The following theorem provides the conditions for the critical point E_0 to be locally stable.

Theorem 2. *The disease free critical point $(\bar{u}_1, 0, 0, 0)$ is always locally asymptotically stable for all $\tau_1 > 0$ if and only if $R_0 < 1$*

Proof. Suppose that the E_0 is stable at $\tau_1 = 0$, and therefore putting $\tau_1 = 0$ in Eq. (7),

$$\delta_3 - \beta_1(1 - \epsilon_p)(1 - \epsilon_r)N\bar{u}_1 + \lambda = 0$$

which gives $\lambda = \beta_1(1 - \epsilon_p)(1 - \epsilon_r)N\bar{u}_1 - \delta_3 < 0$ if and only if

$$\delta_3 > \beta_1(1 - \epsilon_p)(1 - \epsilon_r)N\bar{u}_1.$$

Now, to ascertain whether there could be any root of Eq. (7) with positive real part, continuously increase the τ_1 so that some of the root now cross the imaginary axis i.e. there exist at least one pair of purely imaginary roots $\lambda = iw$ of the equation, then

$$\begin{aligned}\delta_3 - N\bar{u}_1 b_3 \cos(w\tau_1) &= 0 \\ w + N\bar{u}_1 b_3 \sin(w\tau_1) &= 0.\end{aligned} \tag{9}$$

Now, elimination of $w\tau_1$ from Eq. (9) gives

$$w^2 = N^2\bar{u}_1^2 b_3^2 - \delta_3^2.$$

If it further assume that Eq. (7) should not possess any root with positive real part for any value of parameter, it is necessary and sufficient that w should be always negative i.e. $\delta_3 > N\bar{u}_1 b_3$ or $\delta_3 > \frac{(1-\epsilon_p)(1-\epsilon_r)N\eta\beta_1 e^{-\lambda\tau_1}}{2\alpha}$ or $R_0 < 1$.

4.2 Local Stability Behaviour of $E_1 \equiv (\tilde{u}_1, \tilde{u}_2, \tilde{u}_3, \tilde{u}_4)$ and Bifurcation

The Characteristic Equation is given by

$$P_4(\lambda) + e^{-\lambda\tau_1} e^{-\delta_2\tau_1} P_3(\lambda) + e^{-\lambda\tau_2} P_2(\lambda) + e^{-\lambda(\tau_1+\tau_2)} e^{-\delta_2\tau_1} P_1(\lambda) = 0, \tag{10}$$

where the polynomials are given by

$$P_4(\lambda) = \lambda^4 + a_{13}\lambda^3 + a_{12}\lambda^2 + a_{11}\lambda + a_{10}$$
$$P_3(\lambda) = a_{23}\lambda^3 + a_{22}\lambda^2 + a_{21}\lambda + a_{20}$$
$$P_2(\lambda) = a_{32}\lambda^2 + a_{31}\lambda + a_{30}$$
$$P_1(\lambda) = a_{41}\lambda + a_{40}.$$

It is clear that the coefficients of above characteristic equation explicitly contains the delay parameter τ_1 and this poses a challenge to analytically study the local behaviour of the solution. There is no method available in literature by which above characteristic equation can be studied for roots having negative real part.

The following section is therefore devoted to numerically explore the existence of Hopf bifurcation [10, 20, 32] by examining the presence of roots of characteristic Eq. (10) having negative real parts.

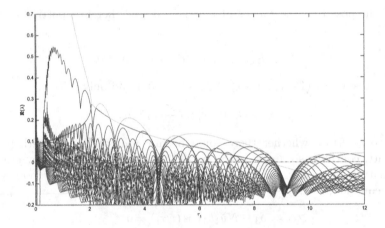

Fig. 1. Variation of Re(λ) with respect to τ_1

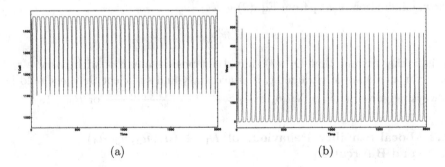

(a)   (b)

Fig. 2. Time series plot of u_1 and u_3 when $\tau_1 = 5$

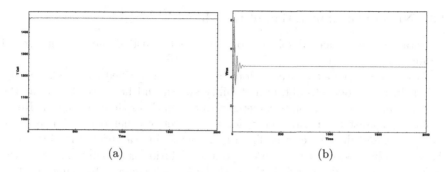

Fig. 3. Time series plot of u_1 and u_3 when $\tau_1 = 9.5$

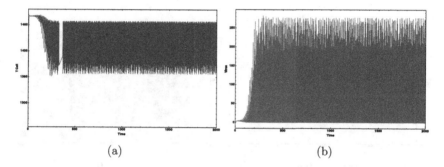

Fig. 4. Time series plot of u_1 and u_3 when $\tau_1 = 11$

Table 1. Model parameters

Parameters	Descriptions	Value
Q	T cell production rate	$16.0\,\mathrm{mm}^{-3}\,\mathrm{day}^{-1}$
α	T cell proliferation rate	$0.95\,\mathrm{day}^{-1}$
K	Maximum T cell concentration level	$1500\,\mathrm{mm}^{-3}$
β_1	Rate of infection of T cell	$0.0011\,\mathrm{mm}^3\mathrm{day}^{-1}$
β_2	Virus death rate due to immune response	$0.8046\,\mathrm{mm}^3\mathrm{day}^{-1}$
β_3	Production rate of effector cells	$0.2129\,\mathrm{mm}^3\mathrm{day}^{-1}$
β_4	Effector cell death due to Virus	$0.2129\,\mathrm{mm}^3\mathrm{day}^{-1}$
δ_1	Healthy T cell death rate	$0.03\,\mathrm{day}^{-1}$
δ_2	Infected T cell death rate	$0.2\,\mathrm{day}^{-1}$
δ_3	Viral death rate	$1.0\,\mathrm{day}^{-1}$
δ_4	Effector cell death rate	$3.0\,\mathrm{day}^{-1}$
N	Virion produced per infected T cell	$60\ \mathrm{particle\ cell}^{-1}$

4.3   Numerical Exploration of the Model

To obtain some numerical solution of the model, consider the following set of parameters.

In first case, the existence of Hopf bifurcation is verified by simulating the model. In this model, to obtain a stability switch and hence to determine the critical values of bifurcation parameter τ_1 at which this switch occurs, the $Re(\lambda)$ is plotted against the parameter τ_1 for a fixed set of other parameter as given the Table 1 and drug efficacies. The Fig. 1 shows this variation and this shows that a stability switch occurs, here a periodic solution is initially present in the model. But as the τ_1 is varied, the periodic branch changes to stable solution and upon further increase in τ_1, periodic solution reappears. This clearly shows

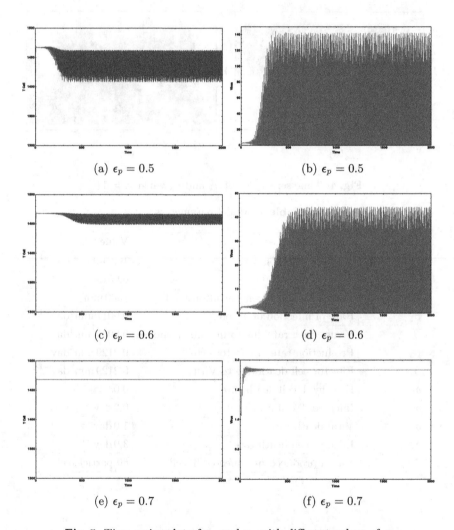

(a) $\epsilon_p = 0.5$ (b) $\epsilon_p = 0.5$

(c) $\epsilon_p = 0.6$ (d) $\epsilon_p = 0.6$

(e) $\epsilon_p = 0.7$ (f) $\epsilon_p = 0.7$

Fig. 5. Time series plot of u_1 and u_3 with different values of ϵ_p

that the model exhibit Hopf bifurcation with respect to parameter τ_1 as evident from time series plots Figs. 2, 3 and 4.

The next two set of simulations are performed to justify the importance of drug therapy in HIV infection. The Fig. 5 shows that as the efficacy of PI drug increases, there is oscillation in the level of T cell and virus decrease and ultimately dies out. At the same time, the virus level also decreases and hence individual has more chance of surviving with combination drug therapy. Similarly Fig. 6 shows the effect of RTI drug efficacy on HIV infection. In this case too, the similar trend as in case of PT drug can be seen. Moreover, these simulation show that a combination drug is more useful that a single drug in HIV infection.

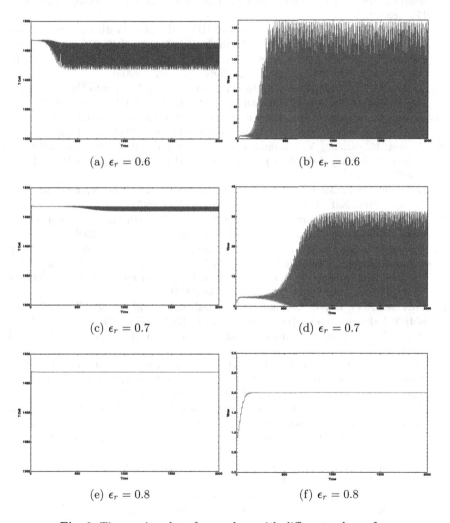

(a) $\epsilon_r = 0.6$ (b) $\epsilon_r = 0.6$

(c) $\epsilon_r = 0.7$ (d) $\epsilon_r = 0.7$

(e) $\epsilon_r = 0.8$ (f) $\epsilon_r = 0.8$

Fig. 6. Time series plot of u_1 and u_3 with different values of ϵ_r

References

1. Allen, L.J., Brauer, F., Van den Driessche, P., Wu, J.: Mathematical Epidemiology. Springer, Heidelberg (2008)
2. Andersen, R.M., May, R.M.: Epidemiological parameters of HIV transmission. Nature **333**(6173), 514–519 (1988)
3. Anderson, R.M.: Mathematical and statistical studies of the epidemiology of HIV. AIDS **3**(6), 333–346 (1989)
4. Bachar, M., Dorfmayr, A.: HIV treatment models with time delay. C.R. Biol. **327**(11), 983–994 (2004)
5. Bailey, J.J., Fletcher, J.E., Chuck, E.T., Shrager, R.I.: A kinetic model of CD4+ lymphocytes with the human immunodeficiency virus (HIV). BioSystems **26**(3), 177–183 (1992)
6. Bairagi, N., Adak, D.: Global analysis of HIV-1 dynamics with hill type infection rate and intracellular delay. Appl. Math. Model. **38**(21), 5047–5066 (2014)
7. Banks, H., Bortz, D.: A parameter sensitivity methodology in the context of HIV delay equation models. J. Math. Biol. **50**(6), 607–625 (2005)
8. Banks, H., Bortz, D., Holte, S.: Incorporation of variability into the modeling of viral delays in HIV infection dynamics. Math. Biosci. **183**(1), 63–91 (2003)
9. Beretta, E., Kuang, Y.: Modeling and analysis of a marine bacteriophage infection with latency period. Nonlinear Anal. Real World Appl. **2**(1), 35–74 (2001)
10. Beretta, E., Kuang, Y.: Geometric stability switch criteria in delay differential systems with delay-dependent parameters. SIAM J. Math. Anal. **33**(31), 144–1165 (2002)
11. Bonhoeffer, S., May, R.M., Shaw, G.M., Nowak, M.A.: Virus dynamics and drug therapy. Proc. Nat. Acad. Sci. **94**(13), 6971–6976 (1997)
12. Brauer, F., Castillo-Chavez, C., Castillo-Chavez, C.: Mathematical Models in Population Biology and Epidemiology, vol. 1. Springer, New York (2001)
13. Chiyaka, C., Garira, W., Dube, S.: Modelling immune response and drug therapy in human malaria infection. Comput. Math. Methods Med. **9**(2), 143–163 (2008)
14. Culshaw, R.V., Ruan, S.: A delay-differential equation model of HIV infection of CD4$^+$ T-cells. Math. Biosci. **165**(1), 27–39 (2000)
15. Culshaw, R.V., Ruan, S., Webb, G.: A mathematical model of cell-to-cell spread of HIV-1 that includes a time delay. J. Math. Biol. **46**(5), 425–444 (2003)
16. Deans, J.A., Cohen, S.: Immunology of malaria. Annu. Rev. Microbiol. **37**(1), 25–50 (1983)
17. Dixit, N.M., Markowitz, M., Ho, D.D., Perelson, A.S.: Estimates of intracellular delay and average drug efficacy from viral load data of HIV-infected individuals under antiretroviral therapy. Antivir. Ther. **9**, 237–246 (2004)
18. Granich, R.M., Gilks, C.F., Dye, C., De Cock, K.M., Williams, B.G.: Universal voluntary HIV testing with immediate antiretroviral therapy as a strategy for elimination of HIV transmission: a mathematical model. Lancet **373**(9657), 48–57 (2009)
19. Grossman, Z., Polis, M., Feinberg, M.B., Grossman, Z., Levi, I., Jankelevich, S., Yarchoan, R., Boon, J., de Wolf, F., Lange, J.M., et al.: Ongoing HIV dissemination during haart. Nat. Med. **5**(10), 1099–1104 (1999)
20. Hale, J.K.: Functional Differential Equations. Springer, New York (1971)

21. Haynes, B.F., Gilbert, P.B., McElrath, M.J., Zolla-Pazner, S., Tomaras, G.D., Alam, S.M., Evans, D.T., Montefiori, D.C., Karnasuta, C., Sutthent, R., et al.: Immune-correlates analysis of an HIV-1 vaccine efficacy trial. N. Engl. J. Med. **366**(14), 1275–1286 (2012)

22. Herz, A., Bonhoeffer, S., Anderson, R.M., May, R.M., Nowak, M.A.: Viral dynamics in vivo: limitations on estimates of intracellular delay and virus decay. Proc. Nat. Acad. Sci. **93**(14), 7247–7251 (1996)

23. Hethcote, H.W.: The mathematics of infectious diseases. SIAM Rev. **42**(4), 599–653 (2000)

24. Ho, D.D.: Toward HIV eradication or remission: the tasks ahead. Science **280**(5371), 1866–1867 (1998)

25. Ho, D.D., Neumann, A.U., Perelson, A.S., Chen, W., Leonard, J.M., Markowitz, M., et al.: Rapid turnover of plasma virions and CD4 lymphocytes in HIV-1 infection. Nature **373**(6510), 123–126 (1995)

26. Holder, B.P., Beauchemin, C.A.: Exploring the effect of biological delays in kinetic models of influenza within a host or cell culture. BMC Public Health **11**(Suppl 1), S10 (2011)

27. Hraba, T., Doležal, J., čelikovský, S.: Model-based analysis of CD4+ lymphocyte dynamics in HIV infected individuals. Immunobiology **181**(1), 108–118 (1990)

28. Jiang, X., Zhou, X., Shi, X., Song, X.: Analysis of stability and hopf bifurcation for a delay-differential equation model of HIV infection of CD4+ T-cells. Chaos, Solitons Fractals **38**(2), 447–460 (2008)

29. Kermack, W.O., McKendrick, A.G.: A contribution to the mathematical theory of epidemics. Proc. Roy. Soc. Lond. A: Math. Phys. Eng. Sci. **115**, 700–721 (1927). The Royal Society

30. Kirschner, D.: Using mathematics to understand HIV immune dynamics. AMS Not. **43**(2), 191–202 (1996)

31. Kirschner, D.E., Webb, G.F.: A mathematical model of combined drug therapy of HIV infection. Comput. Math. Methods Med. **1**(1), 25–34 (1997)

32. Kuang, Y.: Delay Differential Equations: With Applications in Population Dynamics. Academic Press, Boston (1993)

33. Law, M.G., Prestage, G., Grulich, A., Van de Ven, P., Kippax, S.: Modelling the effect of combination antiretroviral treatments on HIV incidence. AIDS **15**(10), 1287–1294 (2001)

34. Li, D., Ma, W.: Asymptotic properties of a HIV-1 infection model with time delay. J. Math. Anal. Appl. **335**(1), 683–691 (2007)

35. Li, M.Y., Shu, H.: Impact of intracellular delays and target-cell dynamics on in vivo viral infections. SIAM J. Appl. Math. **70**(7), 2434–2448 (2010)

36. Liu, S., Wang, L.: Global stability of an HIV-1 model with distributed intracellular delays and a combination therapy. Math. Biosci. Eng. **7**(3), 675–685 (2010)

37. Merrill, S.J.: Modeling the interaction of HIV with cells of the immune system. In: Castillo-Chavez, C. (ed.) Mathematical and Statistical Approaches to AIDS Epidemiology. Lecture Notes in Biomathematics, vol. 83, pp. 371–385. Springer, Heidelberg (1989)

38. Mittler, J.E., Markowitz, M., Ho, D.D., Perelson, A.S.: Improved estimates for HIV-1 clearance rate and intracellular delay. AIDS **13**(11), 1415 (1999)

39. Mittler, J.E., Sulzer, B., Neumann, A.U., Perelson, A.S.: Influence of delayed viral production on viral dynamics in HIV-1 infected patients. Math. Biosci. **152**(2), 143–163 (1998)

40. Murray, J.M., Emery, S., Kelleher, A.D., Law, M., Chen, J., Hazuda, D.J., Nguyen, B.Y.T., Teppler, H., Cooper, D.A.: Antiretroviral therapy with the integrase inhibitor raltegravir alters decay kinetics of HIV, significantly reducing the second phase. AIDS **21**(17), 2315–2321 (2007)

41. Nelson, P.W., Mittler, J.E., Perelson, A.S.: Effect of drug efficacy and the eclipse phase of the viral life cycle on estimates of HIV viral dynamic parameters. JAIDS J. Acquir. Immune Defic. Syndr. **26**(5), 405–412 (2001)

42. Nelson, P.W., Murray, J.D., Perelson, A.S.: A model of HIV-1 pathogenesis that includes an intracellular delay. Math. Biosci. **163**(2), 201–215 (2000)

43. Nelson, P.W., Perelson, A.S.: Mathematical analysis of delay differential equation models of HIV-1 infection. Math. Biosci. **179**(1), 73–94 (2002)

44. Nowak, M., May, R.M.: Virus Dynamics: Mathematical Principles of Immunology and Virology. Oxford University Press, Oxford (2000)

45. Nowak, M.A., Bangham, C.R.: Population dynamics of immune responses to persistent viruses. Science **272**(5258), 74–79 (1996)

46. Nowak, M.A., May, R.M.: Virus Dynamics (2000)

47. Ouifki, R., Witten, G.: Stability analysis of a model for HIV infection with RTI and three intracellular delays. BioSystems **95**(1), 1–6 (2009)

48. Pawelek, K.A., Liu, S., Pahlevani, F., Rong, L.: A model of HIV-1 infection with two time delays: mathematical analysis and comparison with patient data. Math. Biosci. **235**(1), 98–109 (2012)

49. Perelson, A.S.: Modelling viral and immune system dynamics. Nat. Rev. Immunol. **2**(1), 28–36 (2002)

50. Perelson, A.S., Kirschner, D.E., De Boer, R.: Dynamics of HIV infection of CD4$^+$ T cells. Math. Biosci. **114**(1), 81–125 (1993)

51. Perelson, A.S., Nelson, P.W.: Mathematical analysis of HIV-1 dynamics in vivo. SIAM Rev. **41**(1), 3–44 (1999)

52. Perelson, A.S., Neumann, A.U., Markowitz, M., Leonard, J.M., Ho, D.D.: HIV-1 dynamics in vivo: virion clearance rate, infected cell life-span, and viral generation time. Science **271**(5255), 1582–1586 (1996)

53. Pitchaimani, M., Monica, C.: Global stability analysis of HIV-1 infection model with three time delays. J. Appl. Math. Comput. **48**, 1–27 (2014)

54. Pitchaimani, M., Monica, C., Divya, M.: Stability analysis for HIV infection delay model with protease inhibitor. Biosystems **114**(2), 118–124 (2013)

55. Sahani, S.K.: Effects of intracellular delay and immune response delay in HIV model. Neural Parallel Sci. Comput. **23**, 357–366 (2015)

56. Sahani, S.K.: A delayed model for HIV infection incorporating intracellular delay. Int. J. Appl. Comput. Math., 1–20 (2016). DOI:10.1007/s40819-016-0190-7

57. Smith, H.L., De Leenheer, P.: Virus dynamics: a global analysis. SIAM J. Appl. Math. **63**(4), 1313–1327 (2003)

58. Sun, Z., Xu, W., Yang, X., Fang, T.: Effects of time delays on bifurcation and chaos in a non-autonomous system with multiple time delays. Chaos, Solitons Fractals **31**(1), 39–53 (2007)

59. Wang, L., Li, M.Y.: Mathematical analysis of the global dynamics of a model for HIV infection of CD4$^+$ T cells. Math. Biosci. **200**(1), 44–57 (2006)

60. Wein, L.M., Zenios, S.A., Nowak, M.A.: Dynamic multidrug therapies for HIV: a control theoretic approach. J. Theoret. Biol. **185**(1), 15–29 (1997)

61. Wodarz, D., Lloyd, A.L.: Immune responses and the emergence of drug-resistant virus strains in vivo. Proc. R. Soc. Lond.-B **271**(1544), 1101–1110 (2004)

62. Wodarz, D., Nowak, M.A.: Mathematical models of HIV pathogenesis and treatment. BioEssays **24**(12), 1178–1187 (2002)
63. Xiang, H., Feng, L.X., Huo, H.F.: Stability of the virus dynamics model with beddington-deangelis functional response and delays. Appl. Math. Model. **37**(7), 5414–5423 (2013)
64. Zhu, H., Zou, X.: Dynamics of a HIV-1 infection model with cell-mediated immune response and intracellular delay. Discrete Contin. Dyn. Syst. Ser. B **12**(2), 511–524 (2009)

A Second Order Non-uniform Mesh Discretization for the Numerical Treatment of Singular Two-Point Boundary Value Problems with Integral Forcing Function

Navnit Jha[✉]

Faculty of Mathematics and Computer Science, South Asian University,
Chanakyapuri 110 021, New Delhi, India
navnitjha@sau.ac.in

Abstract. In the present work, we examine the three-point numerical scheme for the non-linear second order ordinary differential equations having integral form of forcing function. The approximations of solution values are obtained by means of finite difference scheme based on a special type of non-uniform meshes. The derivatives as well as integrals are approximated with simple second order accuracy both on uniform meshes and non-uniform meshes. A brief convergence analysis based on irreducible and monotone behaviour of Jacobian matrix to the numerical scheme is provided. The scheme is then tested on linear and non-linear examples that justify the order and accuracy of the new method.

Keywords: Finite difference scheme · Non-uniform mesh · Trapezoidal rule · Irreducible and monotone matrix · Maximum absolute error · Convergence order

1 Introduction

Many differential equations do not possess closed form of solutions and may not be solved by symbolic computations. For practical purpose, an appropriate numerical approximation to the solution values is often sufficient. In the past years, there has been tremendous development on high order accurate numerical algorithm for the solution of ordinary and partial differential equations. The various approaches are finite element, spline, wavelets and Galerkin method. Among many available solution techniques, finite difference method is easier to program and have easy error analysis based on finite Taylor's series expansion. The detailed stability and convergence for the difference schemes developed on non-uniform meshes for the second order boundary value problems have been developed by [1]. The two-point boundary value problems (BVPs) having integral form of forcing functions using cubic spline on non-uniform meshes and differential transform method has been described in the past by [2, 3]. The high order finite difference methods have more functional evaluations and restricted to the application of mildly non-linear differential equations [4]. Thus, a second order accurate finite difference method is equally important because, it can be directly applied to fully non-linear problems and easy implementation of algorithms. For a uniform mesh with spacing h the divided difference operators yields truncation error of $O(h^2)$,

© Springer Nature Singapore Pte Ltd. 2017
K. Deep et al. (eds.), *Proceedings of Sixth International Conference on Soft Computing for Problem Solving*, Advances in Intelligent Systems and Computing 547,
DOI 10.1007/978-981-10-3325-4_39

but for non-uniform mesh with maximum interval h, the truncation error may or may not be $O(h^2)$. There is an added advantage of non-uniform meshes compared with uniform meshes. The truncation error of the scheme depends upon derivatives of the dependent variable as well as mesh spacing. Therefore, the uniform distribution of the discretization error is difficult to achieve if the meshes are uniformly taken. It is more suitable to choose larger mesh spacing where function behaviour is smooth and smaller mesh spacing where behavior of function and their derivatives changes sharply. Thus, it is possible to distribute the error uniformly on the domain of integration and have more accurate solution for a given number of mesh-points [5]. In the present work, we shall show that for a class of mesh function the solution error remains $O(h^2)$ and we call such enhancement of truncation error as supra-convergence and classifies in the range of optimal accurate method [6, 7].

Now, we consider the non-linear two point boundary value problems with forcing function in integral form

$$U^{(2)}(x) = F(x, U(x), U^{(1)}(x)) + \int_0^1 K(x, r)dr, 0 < x, r < 1, \tag{1.1}$$

along with Dirichlet boundary values

$$U(0) = A, U(1) = B, A, B \text{ are finite constants} \tag{1.2}$$

Assuming $K(x, r)$ is a real valued function and

$$I(x) = \int_0^1 K(x, r)dr, G(x, U(x), U^{(1)}(x)) = F(x, U(x), U^{(1)}(x)) + I(x). \tag{1.3}$$

Therefore, the Eq. (1.1) can be re-written as

$$U^{(2)}(x) = G(x, U(x), U^{(1)}(x)), 0 < x < 1. \tag{1.4}$$

The existence and uniqueness of solutions values to (1.4) subject to Dirichlet boundary values can be assured by assuming continuity of $G(x, U(x), V(x))$, $\partial G/\partial U$, $\partial G/\partial V$ and $\partial G/\partial U > 0$ along with bounded $\partial G/\partial U$ (see [8]).

2 Sundqvist's Non-uniform Mesh

Consider the non-uniform partition of the interval $I = [0, 1] = \{x_l : l = 0(1)L + 1\}$, where $x_0 = 0, x_l = x_{l-1} + h_l, l = 1(1)L + 1$ and the subsequent mesh spacing are determined by the recursive relation $h_{l+1} = h_l(1 + \alpha h_l), l = 1(1)L$ and $\alpha = \bar{\alpha}/h_1$. Here, $\alpha \in (-1, 1)$ is a mesh parameter in particular if $\alpha = 0$, the mesh distribution is uniform (see [7, 9–11]). Since $\sum_{l=1(1)L+1} h_l = 1$, thus for a given value of $\bar{\alpha}$, it is easy to determine the first mesh step size h_1. For example, $h_1 = 1/(1 + 3\bar{\alpha} + 2\bar{\alpha}^2 + \bar{\alpha}^3)$ if $L = 3$. Let us denote $H = [h_1, h_2, \ldots, h_{N+1}]^T$. To proceed further, we need to prove some results pertaining to the non-uniform meshes ($\alpha \neq 0$).

Lemma 2.1 The mesh-step size is inversely proportional to the number of mesh-points and the sequence $\{h_l\}_{l=1}^{L+1}$ is monotonic.

Proof: Consider the uniform partition of reference domain $Q = [0,1] = \{q_l = lh, l = 0(1)L+1\}, h = 1/(L+1)$. Following [12], we define a one-one onto map $g : Q \to I$ such that $g(q_l) = x_l, l = 0(1)L+1$ and the Jacobian $J(q) = \frac{dg(q)}{dq}$ is bounded above and below by some positive constants as $0 < m \leq J(q) \leq M < \infty \, \forall q \in Q$.

Now, $J(q) > 0 \Rightarrow \frac{dg(q)}{dq} > 0 \Rightarrow \frac{g(q_{l+1}) - g(q_l)}{q_{l+1} - q_l} > 0 \, \forall l \Rightarrow \frac{x_{l+1} - x_l}{(l+1)h - lh} > 0, \forall l$

and hence $h_{l+1} > 0 \, \forall l$.

Also, $J(q) \leq M \Rightarrow \frac{dg(q)}{dq} \leq M \Rightarrow \frac{g(q_{l+1}) - g(q_l)}{q_{l+1} - q_l} \leq M \, \forall l \Rightarrow \frac{x_{l+1} - x_l}{(l+1)h - lh} \leq M$

$$\Rightarrow h_{l+1} \leq Mh, \forall l \Rightarrow \text{Max}_l |h_{l+1}| \leq Mh \Rightarrow \|\boldsymbol{H}\|_\infty \leq Mh = \frac{M}{L+1} \leq \frac{M}{L} \qquad (2.1)$$

$$\Rightarrow \|\boldsymbol{H}\|_\infty = O(h) = O\left(\frac{1}{L}\right) \text{ and hence, } \|\boldsymbol{H}\|_\infty \to 0 \text{ as } L \to \infty. \qquad (2.2)$$

Thus, the maximum step size tends to zero in the limiting case as $L \to \infty$. Further, the increasing mesh step-size

$$\Leftrightarrow h_l < h_{l+1} \, \forall l = 1(1)L \Leftrightarrow h_l < h_l(1 + \alpha h_l)$$
$$\Leftrightarrow 1 < 1 + \alpha h_l, \because h_l > 0 \, \forall l \Leftrightarrow \alpha > 0$$

In a similar manner, it is easy to establish that mesh step-size is decreasing when $\alpha < 0$.

3   Finite Difference Approximations

Let the exact solution value of $U(x)$ at the mesh-point $x = x_l, l = 1(1)L$, be denoted by U_l and the approximate solution value is denoted by u_l. By the help of linear combination of three solution values $U_{l-1} = U(x_l - h_l)$, $U_l = U(x_l)$ and $U_{l+1} = U(x_l + h_l + \alpha h_l^2)$ defined at one central mesh-point x_l and two neighbouring points $x_{l\pm1}$, we construct the following operator

$$\mathcal{F}U_l = [U_{l+1} - (2\alpha h_l + 1)U_{l-1} + 2\alpha h_l U_l]/[2 + 3\alpha h_l], \qquad (3.1)$$

$$\mathcal{S}U_l = 2[U_{l+1} - (2 + \alpha h_l)U_l + (1 + \alpha h_l)U_{l-1}]/[(1 + \alpha h_l)(2 + \alpha h_l)]. \qquad (3.2)$$

Then, by the help of Taylor's expansion it is easy to see that

$$\mathcal{F}U_l = h_l U^{(1)}(x) + h_l^3 U^{(3)}(x)/6 + O(h_l^4), \qquad (3.3)$$

$$SU_l = h_l^2 U^{(2)}(x) + O(h_l^4). \tag{3.4}$$

Thus, the three-point approximations

$$\bar{U}_l^{(1)} = h_l^{-1} \mathcal{F} U_l \text{ and } \bar{U}_l^{(2)} = h_l^{-2} SU_l \tag{3.5}$$

provides the second order accuracies to the first and second-order derivatives respectively on a non-uniform meshes. In particular, if $\alpha = 0, \mathcal{F}U_l = (2\mu\delta)U_l$ and $SU_l = \delta^2 U_l$, where μ is averaging operator and δ is central difference operator.

Therefore, the discrete equation corresponding to the second order differential equation

$$U^{(2)}(x) = F(x, U(x), U^{(1)}(x)) \tag{3.6}$$

in terms of compact operators is given by

$$h_l^{-2} SU_l = F(x_l, U_l, h_l^{-1} \mathcal{F} U_l) + T_l, l = 1(1)L. \tag{3.7}$$

where the truncation error T_l to the Eq. (3.7) is calculated as $T_l = O(h_l^2)$. As a result, the discretization scheme (3.7) is second order accurate for all α (may not be zero).

4 Numerical Integration

In this section, we describe the formula for approximating the integral

$$I(x) = \int_{r=0}^{1} K(x, r) dr \tag{4.1}$$

on non-uniform meshes, and as a particular case, the formula reduces to Trapezoidal rule of integration. Here, the domain of integration $[0, 1]$ are partitioned with the mesh-points $r_i = x_i, i = 0(1)L+1$, as discussed in Sect. 2 and we shall use the notation $K_{l,i} = K(x_l, r_i)$.

For a given mesh-point $x = x_l$, we shall approximate the integral (4.1) by a linear combination of the values of $K(x_l, r)$ as follows

$$I(x_l) = \int_{r=0}^{1} K(x_l, r) dr = \sum_{i=1}^{L+1} \int_{r=r_{i-1}}^{r_i} K(x_l, r) dr. \tag{4.2}$$

Let

$$\int_{r=r_{i-1}}^{r_i} K(x_l, r) dr = aK_{l,i-1} + bK_{l,i} \tag{4.3}$$

Then, for $K(x_l, r) = 1$ and r, one obtain

$$a+b = r_i - r_{i-1}, ar_{i-1} + br_i = \frac{1}{2}(r_i^2 - r_{i-1}^2) \tag{4.4}$$

Solving (4.4) for the unknown a and b, and by the help of (4.3), we find

$$\int_{r=r_{i-1}}^{r_i} K(x_l, r)dr = \frac{1}{2}h_i(1 + \alpha h_i)(K_{l,i-1} + K_{l,i}). \tag{4.5}$$

Using (4.5) and (4.2), it is easy to determine the integral formula on the interval $[0, 1]$ as

$$I(x_l) = \frac{1}{2}\sum_{i=1}^{L+1}[h_i(1 + \alpha h_i)(K_{l,i-1} + K_{l,i})]. \tag{4.6}$$

The error in the integral formula (4.5) obtained on the interval $[r_{i-1}, r_i]$, is given by

$$E_i = \left|-\frac{\theta h_i^3}{12}\right| + O(h_i^4), \theta = \left(\frac{\partial^2 K}{\partial r^2}\right)_{(x_l, \bar{r})}, r_{i-1} < \bar{r} < r_i \tag{4.7}$$

Thus, by the help of inequality (2.1), the composite error in the integral (4.6) on the interval $[0, 1]$ is given by

$$\begin{aligned}E = \sum_{i=1}^{L+1} E_i &\le \sum_{i=1}^{L+1} \theta h_i^3/12 \le \sum_{i=1}^{L+1} \bar{M}h^3 \\ &= \bar{M}(L+1)h^3 = \bar{M}h^2 = O(h^2), \bar{M} = \theta M^3/12\end{aligned} \tag{4.8}$$

Thus, a second-order accurate formula for the numerical integration is derived on a non-uniform mesh spacing, $\alpha \neq 0$. Moreover, when $\alpha = 0$, the integral formula (4.6) is Trapezoidal rule of integration.

Now we are available with the second-order accurate formula for the derivatives and integral both on a uniform meshes as well as on non-uniform meshes. Therefore, by the help of Formulas (3.7) and (4.6), we can write the difference scheme for differential-integral Eq. (1.1) as follows

$$h_l^{-2}SU_l = F(x_l, U_l, h_l^{-1}\mathcal{F}U_l)) + I(x_l) + \bar{T}_l, l = 1(1)L, \tag{4.9}$$

and the error is given by $\bar{T}_l = T_l + E = O(h_l^2) + O(h^2) \approx O(h^2)$.

The method (4.9) is obtained using minimum number of mesh-points necessary to discretize the highest order derivative appearing in the Eq. (1.1) and thus, it is a compact scheme of second order accuracy irrespective of the mesh distribution being uniform or non-uniform. The method requires no special attention for singular problems. For the practical implementation, we neglect the error term $\bar{T}_l$ and the resulting difference equations have tri-diagonal Jacobian and thus, easily computed by means of Thomas algorithm or any iterative method [13].

5 Convergence Analysis

In this section, we shall prove that the proposed numerical scheme (4.9) gives us second order accurate solution values for (1.1)–(1.2). At the mesh point $x = x_l$, the differential-integral Eq. (1.1) can be written as

$$U^{(2)}(x_l) = F(x_l, U(x_l), U^{(1)}(x_l)) + I(x_l), l = 1(1)L. \tag{5.1}$$

The numerical scheme (4.9) can be written as

$$\phi_l + \hat{T}_l = 0, l = 1(1)L, \tag{5.2}$$

where

$$
\begin{aligned}
\phi_l = {}& -2(2 + \alpha h_l)^{-1} U_{l-1} + 2(1 + \alpha h_l)^{-1} U_l - 2(1 + \alpha h_l)^{-1}(2 + \alpha h_l)^{-1} U_{l+1} \\
& + h_l^2 F(x_l, U_l, h_l^{-1}(2 + 3\alpha h_l)^{-1}(U_{l+1} - (2\alpha h_l + 1)U_{l-1} + 2\alpha h_l U_l)) \\
& + (1/2)h_l^2 \sum_{i=1}^{L+1} h_i(1 + \alpha h_i)(K_{l,i-1} + K_{l,i})
\end{aligned} \tag{5.3}
$$

and

$$\hat{T}_l = O(h_l^2 h^2) \approx O(h^4). \tag{5.4}$$

System of difference Eq. (5.3) can be expressed in matrix-vector form as

$$\phi(U) + T = 0, \phi = \begin{bmatrix} \phi_1 \\ \vdots \\ \phi_L \end{bmatrix}, U = \begin{bmatrix} U_1 \\ \vdots \\ U_L \end{bmatrix}, T = \begin{bmatrix} \hat{T}_1 \\ \vdots \\ \hat{T}_L \end{bmatrix} \tag{5.5}$$

The approximate solution vector $u = [u_1, \ldots, u_L]^{\mathrm{T}}$ must satisfy

$$\phi(u) = 0. \tag{5.6}$$

From the Eqs. (5.5) and (5.6), one obtains

$$\phi(u) - \phi(U) = T. \tag{5.7}$$

Let $\epsilon_l = u_l - U_l, l = 1(1)L$ be the discretization error and $\varepsilon = u - U$ be the discretization error vector.
Let

$$\bar{g}_l = F\left(x_l, U_l, \bar{U}_l'\right) + I(x_l) \approx \bar{G}_l. \tag{5.8}$$

and

$$P_l = \bar{g}_l - \bar{G}_l. \tag{5.9}$$

Then, by the application of Mean value theorem and neglecting higher order terms of discretization error, we get

$$P_l = a_l \bar{\epsilon}_l^{(1)} + b_l \epsilon_l \,, a_l = \left(\frac{\partial g}{\partial U^{(1)}}\right)_{x=x_l}, \quad b_l = \left(\frac{\partial g}{\partial U}\right)_{x=x_l}, \tag{5.10}$$

where $\bar{\epsilon}_l^{(1)}$ is obtained by replacing U to ϵ in $\bar{U}_l^{(1)}$ discussed in Eq. (3.5). In view of the relation (5.10), we obtain

$$[\phi(u) - \phi(U)] = R\epsilon, \tag{5.11}$$

where $R = (R_L, R_M, R_U)$ is a tri-diagonal matrix of order L and

$$R_L = -1 + \frac{h_l}{2}(\alpha - a_l) + O(h_l^2), \tag{5.12}$$

$$R_M = 2 - 2\alpha h_l + O(h_l^2), \tag{5.13}$$

$$R_U = -1 + \frac{h_l}{2}(3\alpha + a_l) + O(h_l^2). \tag{5.14}$$

The Eqs. (5.7) and (5.11), yields the matrix equation

$$R\epsilon = T. \tag{5.15}$$

Let $\bar{h} = \text{Max}_{l=1(1)L+1}|h_l|$, then for sufficiently small value of $\bar{h}$,

$$R_L = -1 \neq 0, R_M = 2 \neq 0, R_U = -1 \neq 0. \tag{5.16}$$

Therefore, the graph of matrix R is strongly connected and thus, irreducible [14, 15].
Let $a = \text{Min}_{l=1(1)L}|a_l|$, $b = \text{Min}_{l=1(1)L}|b_l|$ and δ_l be the sum of lth row elements of the matrix R. Then,

$$\delta_1 \geq 1 + \frac{h_l}{2}(a - \alpha) + O(h_l^2), \tag{5.17}$$

$$\delta_l \geq b h_l^2 + O(h_l^3), l = 2(1)L - 1, \tag{5.18}$$

$$\delta_L \geq 1 - \frac{h_l}{2}(3\alpha + a) + O(h_l^2). \tag{5.19}$$

As $h_l \to 0^+$, the following inequalities holds

$$\delta_1 \geq 1, \delta_l \geq 0, l = 2(1)L - 1, \text{provided } b \geq 0, \delta_L \geq 1 \qquad (5.20)$$

and hence, the matrix R is monotone and hence R^{-1} exists with positive entries [16].

Let $R_{i,j}^{-1}$ be the $(i,j)^{th}$ element of the matrix R^{-1} and define mate matrix-norm $\|R^{-1}\|_\infty = \max_{1 \leq i \leq L} \sum_{j=1}^{L} |R_{i,j}^{-1}|$. Then the relation $\sum_{l=1}^{L} R_{i,l}\delta_l = 1, i = 1(1)L$ yields the following bounds on the elements of matrix R^{-1}:

$$R_{i1}^{-1} \leq \delta_1^{-1} \leq 1 + \frac{h_l}{2}(\alpha - a) + O(h_l^2), \qquad (5.21)$$

$$R_{ij}^{-1} \leq \delta_j^{-1} \leq \frac{1}{h_j^2 b}, j = 2(1)L - 1 \text{ provided } b \neq 0, \qquad (5.22)$$

$$R_{iL}^{-1} \leq \delta_L^{-1} \leq 1 + \frac{h_l}{2}(\alpha + a) + O(h_l^2). \qquad (5.23)$$

As a result from the Eqs. (5.15) and (5.21)–(5.23), the bounds of the error is given by

$$\|\varepsilon\|_\infty \leq \|R^{-1}\|_\infty \cdot \|\bar{T}\|_\infty \leq \frac{h_l^2}{b} + O(h_l^4) \approx O(h^2). \qquad (5.24)$$

Therefore, as $h_l \to 0^+$, $\|\varepsilon_\infty\| \to 0$.

Summarizing, we have proved second order accuracy of the proposed numerical scheme on a non-uniform meshes and the errors in approximate and exact solution values approaches to zero for the sufficiently small values of the mesh step-size. The condition $b \geq 0$ in Eq. (5.20) and $b \neq 0$ in Eq. (5.22) jointly holds if $b > 0$ and b corresponds to $\partial g / \partial U$. Hence, $\partial g / \partial U$ must be positive for the numerical solution to converge.

6 Computational Results

By the help of proposed scheme, we have examined linear and non-linear problems whose analytical solutions are known to us. The maximum absolute errors and root mean squared errors are respectively obtained by the formula $\|\epsilon\|_\infty^L = \max_{l=1(1)L} |U_l - u_l|$, $\|\epsilon\|_2^L = \left(\sum_{l=1}^{L} |U_l - u_l|^2 / L \right)^{1/2}$ and computational order of convergence are determined by $\Theta_\infty = log_2 \left(\|\epsilon\|_\infty^L / \|\epsilon\|_\infty^{2L+1} \right)$, $\Theta_2 = log_2 \left(\|\epsilon\|_2^L / \|\epsilon\|_2^{2L+1} \right)$. The difference schemes are obtained by Maple for algebraic computations and numerical values are obtained with C programming using long double precisions.

Example 6.1 [2, 17]. Consider the linear singular equation

$$U^{(2)}(x) = -\frac{\tau}{x} U^{(1)}(x) + \frac{\tau}{x^2} U(x) + 12x^2 + \left[4x^2(\tau + 4x^4) - \frac{\tau}{x^2}\right]e^{x^4} + \int_{r=0}^{1} K(r,x)dr = 0,$$

where $K(r,x) = 48r^3x^6e^{x^4r^4}$ and the analytical solution is given by $U(x) = e^{x^4}$. For the numerical computations, we have taken $\tau = 0, 1$ and 2 at various mesh spacing and the respective results are shown in Tables 1, 2 and 3.

Example 6.2 [18]. Consider the non-linear singular equation appearing in physiological phenomenon

$$U^{(2)}(x) + \left(1 + \frac{\tau}{x}\right)U^{(1)}(x) - \frac{25x^8}{x^5 + 4}e^{U(x)} + \int_0^1 K(r,x)dr = 0, 0 < x, r < 1$$

Table 1. Accuracies and computational order of convergence at $\tau = 0$ in example 1.

$L+1$	$\bar{\alpha}$	$\|\varepsilon\|_\infty$	$\|\varepsilon\|_2$	Θ_∞	Θ_2
4	0	1.67e-01	1.30e-01	–	–
8	0	4.84e-02	3.58e-02	1.79	1.86
16	0	1.26e-02	9.04e-03	1.94	1.99
32	0	3.19e-03	2.25e-03	1.98	2.01
4	-0.1500	4.01e-02	3.19e-02	–	–
8	-0.0200	9.48e-03	6.10e-03	2.08	2.39
16	-0.0035	2.17e-03	1.16e-03	2.13	2.39
32	-0.0006	2.88e-04	1.86e-04	2.91	2.64

Table 2. Accuracies and computational order of convergence at $\tau = 1$ in example 1.

$L+1$	$\bar{\alpha}$	$\|\varepsilon\|_\infty$	$\|\varepsilon\|_2$	Θ_∞	Θ_2
4	0	1.99e-01	1.59e-01	–	–
8	0	5.83e-02	4.38e-02	1.77	1.86
16	0	1.52e-02	1.11e-02	1.94	1.98
32	0	3.86e-03	2.75e-03	1.98	2.01
4	-0.1800	2.15e-02	1.59e-02	–	–
8	-0.0300	6.16e-03	3.81e-03	1.81	2.06
16	-0.0052	1.47e-03	7.11e-04	2.07	2.42
32	-0.0008	2.67e-04	1.12e-04	2.46	2.66

Table 3. Accuracies and computational order of convergence at $\tau = 2$ in example 1.

$L+1$	$\bar{\alpha}$	$\|\varepsilon\|_\infty$	$\|\varepsilon\|_2$	Θ_∞	Θ_2
4	0	2.20e-01	1.76e-01	–	–
8	0	6.52e-02	4.92e-02	1.75	1.84
16	0	1.71e-02	1.25e-02	1.93	1.98
32	0	4.35e-03	3.11e-03	1.98	2.01
4	-0.206	1.35e-02	9.03e-03	–	–
8	-0.042	3.57e-03	2.20e-03	1.92	2.03
16	-0.007	7.85e-04	4.77e-04	2.19	2.21
32	-0.001	2.81e-04	1.12e-04	1.48	2.09

Table 4. Accuracies and computational order of convergence at $\tau = 0$ in example 2.

$L+1$	$\bar{\alpha}$	$\|\varepsilon\|_\infty$	$\|\varepsilon\|_2$	Θ_∞	Θ_2
4	0	1.40e-02	1.01e-02	–	–
8	0	5.58e-03	4.11e-03	1.33	1.30
16	0	2.05e-03	1.47e-03	1.44	1.48
32	0	7.28e-04	5.20e-04	1.49	1.50
4	0.1600	5.92e-03	3.68e-03	–	–
8	0.0500	1.22e-03	8.74e-04	2.28	2.07
16	0.0059	2.50e-04	1.38e-04	2.28	2.66
32	0.0011	3.79e-05	2.48e-05	2.72	2.48

Table 5. Accuracies and computational order of convergence at $\tau = 1$ in example 2.

$L+1$	$\bar{\alpha}$	$\|\varepsilon\|_\infty$	$\|\varepsilon\|_2$	Θ_∞	Θ_2
4	0	1.17e-02	8.21e-03	–	–
8	0	5.54e-03	4.51e-03	1.08	0.86
16	0	2.25e-03	1.85e-03	1.30	1.29
32	0	8.55e-04	7.10e-04	1.39	1.38
4	0.0700	8.46e-03	5.30e-03	–	–
8	0.0500	1.59e-03	1.17e-03	2.41	2.18
16	0.0059	3.97e-04	2.31e-04	2.01	2.34
32	0.0011	5.73e-05	3.32e-05	2.79	2.80

Table 6. Accuracies and computational order of convergence at $\tau = 2$ in example 2.

$L+1$	$\bar{\alpha}$	$\|\varepsilon\|_\infty$	$\|\varepsilon\|_2$	Θ_∞	Θ_2
4	0	9.66e-03	6.16e-03	–	–
8	0	5.11e-03	4.41e-03	0.92	0.48
16	0	2.17e-03	1.93e-03	1.24	1.19
32	0	8.47e-04	7.61e-04	1.36	1.34
4	0.0300	8.44e-03	5.13e-03	–	–
8	0.0300	2.37e-03	1.55e-03	1.83	1.73
16	0.0059	5.50e-04	3.52e-04	2.10	2.14
32	0.0011	1.25e-04	8.13e-05	2.14	2.11

where $K(r,x) = 40x^3 r(24xr^4 + 11\tau + 44)\sin^{-1}(r)/[\pi(x^5 + 4)]$ and analytical solution is given by $U(x) = 1/\log(x^5 + 4)$. For the computations, we have taken $\tau = 0, 1$ and 2 at various mesh spacing and respective results are shown in Tables 4, 5 and 6.

The numerical experiments in both the above examples show that non-uniform meshes improve the order and accuracies of the solution values with almost same wall-clock time compared with uniform meshes finite difference approximations.

7 Conclusion

A simple numerical scheme for the solution of non-linear second order differential equations having integral form of forcing functions are developed. It is straight forward to apply such a supra-convergent compact scheme for a wide class of non-linear problems $\phi(x, U(x), U^{(1)}(x), U^{(2)}(x), \int_0^1 K(x,r)dr) = 0, 0 < x, r < 1,$ along with Dirichlet or Neumann or mixed boundary values. The proposed descritization of derivatives can be regarded as generalization of averaging and central-differencing operators while the numerical integration is that of Trapezoidal rule. It is possible to extend such mesh-network for higher order scheme, for example fourth order method for mildly linear two- and three-dimensions elliptic equations.

Acknowledgments. Science and Engineering Research Board, Department of Science and Technology, Government of India (No. SR/FTP/MS-020/2011) has supported the present research work.

References

1. Manteuffel, T.A., White, A.B.: The numerical solution of second-order boundary value problems on nonuniform meshes. Math. Comput. **47**(176), 511–535 (1986)
2. Mohanty, R.K., Jain, M.K., Dhall, D.: A cubic spline approximation and application of TAGE iterative method for the solution of two point boundary value problems with forcing function in integral form. Appl. Math. Model. **35**, 3036–3047 (2011)
3. Aruna, K., Kanth, A.S.V.R.: Solution of singular two point boundary value problems with forcing function in integral form via differential transform method. World J. Modell. Simul. **9**(4), 262–268 (2013)
4. Bieniasz, L.K.: A set of compact finite-difference approximations to first and second derivatives, related to the extended Numerov method of Chawla on non-uniform grids. Computing **81**, 77–89 (2007)
5. Ferziger, J.H., Peric, M.: Computational Methods for Fluid Dynamics. Springer, Berlin (2002)
6. Kreiss, H.O., Manteuffel, T.A.: Supra-convergent schemes on irregular grids. Math. Comput. **47**(176), 537–554 (1986)
7. Saul'yev, V. K.: Integration of Equations of Parabolic Type by the Method of Nets, vol. 54. Elsevier (2014)
8. Keller, H.B.: Numerical Methods for Two-Point Boundary-Value Problems. Blaisdell, Waltham (1968)
9. Sundqvist, H., Veronis, G.: A simple finite-difference grid with non-constant intervals. Tellus A. **22**(1), 26–31 (1970)
10. Britz, D.: Digital Simulation in Electrochemistry. Springer, Berlin (2005)
11. Samarskii, A.A., Matus, P.P., Vabishchevich, P.N.: Difference schemes with operator factors. In: Difference Schemes with Operator Factors. Mathematics and its Applications, vol. 546, pp. 55–78. Springer, Dordrecht (2002)
12. Khakimzyanov, G., Dutykh, D.: On supra convergence phenomenon for second order centered finite differences on non-uniform grids. *arXiv preprint* arXiv:1511.02770 (2015)
13. Hageman, L.A., Young, D.M.: Applied Iterative Methods. Dover Publication, New York (2004)
14. Young, D.M.: Iterative Solution of Large Linear Systems. Elsevier, Amsterdam (2014)

15. Varga, R.S.: Matrix Iterative Analysis. Springer Series in Computational Mathematics. Springer, Berlin (2000)
16. Henrici, P.: Discrete Variable Methods in Ordinary Differential Equations. Wiley, New York (1962)
17. Rashidinia, J., Mohammadi, R., Jalilian, R.: The numerical solution of nonlinear singular boundary value problems arising in physiology. Appl. Math. Comput. **185**, 360–367 (2007)
18. Jha, N., Mohanty, R.K.: TAGE iterative algorithm and nonpolynomial spline basis for the solution of nonlinear singular second order ordinary differential equations. Appl. Math. Comput. **218**, 3289–3296 (2011)

Author Index

© Springer Nature Singapore Pte Ltd. 2017
K. Deep et al. (eds.), *Proceedings of Sixth International Conference on Soft Computing for Problem Solving*, Advances in Intelligent Systems and Computing 547, DOI 10.1007/978-981-10-3325-4

Printed in the United States
By Bookmasters